Pulse Chemistry and Technology

Pulse Chemistry and Technology

Brijesh K. Tiwari
Manchester Food Research Centre, Manchester Metropolitan University, UK
E-mail: B.Tiwari@mmu.ac.uk

Narpinder Singh
Department of Food Science, Guru Nanak Dev University, Amritsar, India

RSC Publishing

ISBN: 978-1-84973-331-1

A catalogue record for this book is available from the British Library

Published by The Royal Society of Chemistry,
Thomas Graham House, Science Park, Milton Road,
Cambridge CB4 0WF, UK

Registered Charity Number 207890

For further information see our web site at www.rsc.org

Printed and bound in Great Britain by CPI Group (UK) Ltd, Croydon, CR0 4YY, UK

Preface

Pulse grains are grown in various countries under a diverse range of agro-climatic conditions. The terms "legumes" and "pulses" are used interchangeably because all pulses are considered legumes but not all legumes are considered pulses. According to the Food and Agriculture Organization, "Pulses play an important role in the nutritional food security of a large number of peoples and represent a major source of protein in many developing countries". Worldwide protein malnutrition has continued to become a global issue and highlights the increase need of protein to feed the people live in developing countries. Pulses are important source of proteins to a large segment of the world's population living in developing countries where consumption of animal protein is limited. In recent years, consumer demand for pulses and pulse fractions is on rise for healthier and nutritious food. Pulses are considered as health foods because of their many health benefits including low glycaemic index, cardiovascular and renal benefits of pulses. Although cereal chemistry, processing and technology have several dedicated textbooks, no dedicated text on pulse processing is currently available for food science and technology graduates. Pulse Chemistry and Technology provides a comprehensive description of pulse grain structure, chemical composition, properties, post-handling, storage, processing, quality evaluation and various pulse based food products. This textbook also describes how pulse grain properties are related to the processing and utilization of pulses and pulse fractions. This textbook also outlines the recent advance knowledge about chemical constituents and processing of pulse, which has advanced extensively in the last two decades. This book is the first textbook, which lays the foundation to address the gap in the literature and provide a dedicated and in depth reference for pulse chemistry, processing and utilisation bringing together essential information on the science and technology of pulses. Grain chemists and processors have made considerable progress over past few decades, which are scattered in the scientific literature or available as a standalone chapter in any standard postharvest technology textbook. The Pulse Chemistry and Technology textbook is structured in 14 chapters. The first chapter addresses the importance, production and consumption trend of

Pulse Chemistry and Technology
Brijesh K. Tiwari and Narpinder Singh

Published by the Royal Society of Chemistry, www.rsc.org

pulses, followed by chapters describing grain structure and major and minor constituents of pulses. Dedicated chapters on pulse starch, proteins and their function properties are also included to provide insight into, and highlight the importance of pulse chemical constituents. Physical, thermal, aerodynamic properties of pulses are covered in a chapter. The second part of the book discusses postharvest handling including drying, storage, processing technologies and utilisation of pulses. The range of pulse based products, pulse grain quality evaluation criteria and techniques are also included as part of the comprehensive coverage of this textbook. Each chapter is appropriately cross-referenced to ensure that the reader is able to easily locate information if required further analysis.

Pulse Chemistry and Technology is designed especially for students, academics, food product developers working in the area of pulse processing. This textbook is essential for the understanding and application of pulses to satisfy the needs of society for sustainable food processing, quality and security. Agriculture, food science and technology programmes at various universities worldwide have a component of pulse grain processing. Practically no attempt has been made to develop a comprehensive dedicated textbook of pulse processing that deals with grain chemistry and processing. The authors are proud to provide a dedicated textbook for the food science and technology graduates to understand and integrate the scientific disciplines relevant to pulse grains.

Brijesh K. Tiwari
Narpinder Singh

Contents

Pulse Chemistry and Technology
Brijesh K. Tiwari and Narpinder Singh

Published by the Royal Society of Chemistry, www.rsc.org

Chapter 11 Storage of Pulses

Chapter 12 Processing of Pulses

Chapter 13 Pulse Products and Utilisation

Chapter 14 Pulse Grain Quality Criteria

CHAPTER 1
Introduction

1.1. IMPORTANCE OF PULSES

The word "pulse" is derived from the Greek word "poltos", which means porridge. Pulses are important crops, serving as an important source of nutrition for billions of people around the world. Pulses encompass those species of plants that belong to the Fabaceae (Leguminoseae) family and are consumed by human beings or domestic animals, commonly in the form of dry matter seeds, *i.e.* as grain legumes. A few oil-bearing seeds such as groundnut (*Arachis hypogaea*) and soybean (*Glycine max*) are also categorised as legume crops. However, these are grown primarily for processing into edible oil. Grain pulses are important foodstuffs in tropical and subtropical countries, where they are second in importance to cereals as a source of protein (20–25% protein by weight). Pulse grains are an excellent source of protein, carbohydrates, dietary fibre, vitamins, minerals and phytochemicals. Much of the world's population relies on pulses as staple food, particularly in combination with cereals. Cereals, being deficient in lysine, are commonly consumed along with pulses, thus completing the dietary protein intake.

Pulses are the edible seeds of leguminous plants, which have the ability to fix atmospheric nitrogen via root nodules. Most of the economically important legumes belong to the subfamily Papilionoideae (Faboideae), which has about 700 genera and 18 000 species. although not all species are consumed by humans. The legumes used by humans are commonly called food legumes or grain legumes or pulses. A classification of pulses is given in Figure 1.1.

Pulses are defined by the Food and Agricultural Organization (FAO) of the United Nations as annual leguminous crops yielding from one to twelve grains or seeds of variable size, shape and colour within a pod. The terms "legumes" and "pulses" are used interchangeably because all pulses are considered legumes (although not all legumes are considered pulses). The term "pulse", as used by the FAO, is reserved for crops harvested solely for use as dry grains.

Pulse Chemistry and Technology
Brijesh K. Tiwari and Narpinder Singh

Published by the Royal Society of Chemistry, www.rsc.org

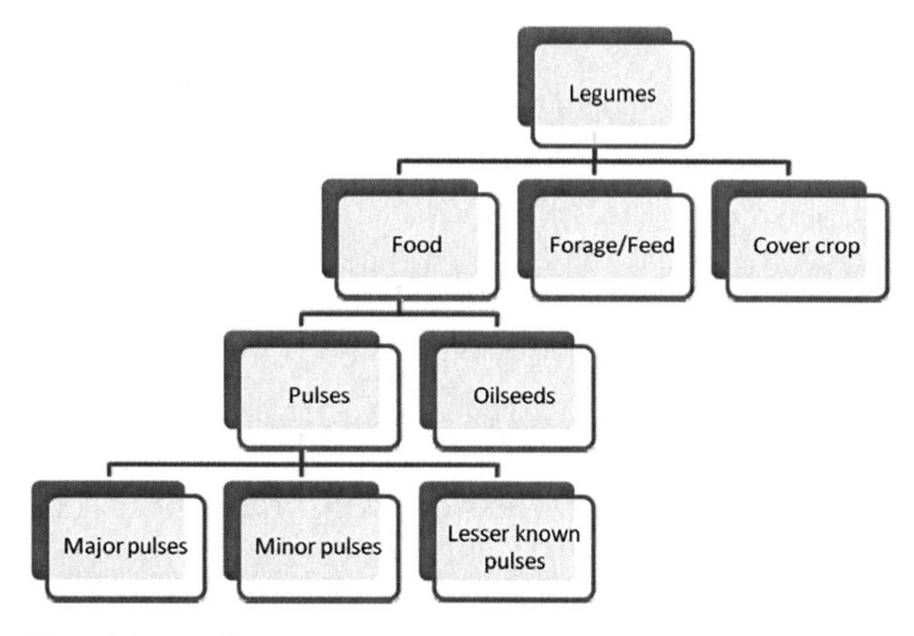

Figure 1.1 Classification of pulses

This term therefore excludes green beans and green peas, which are considered as vegetable crops. The crops which are mainly grown for oil extraction (oilseeds like soybeans and peanuts), and those which are used exclusively for sowing (clovers, alfalfa) are also excluded from the term pulses. FAO recognises 11 primary pulses: (1) dry beans; (2) dry broad beans; (3) dry peas; (4) chickpea; (5) dry cowpea; (6) pigeon pea; (7) lentil; (8) bambara groundnut; (9) vetch; (10) lupins and (11) minor pulses. Some of the commonly grown and consumed pulses are show in Figure 1.2.

1.2. TRENDS IN PULSE PRODUCTION AND CONSUMPTION

1.2.1. Pulse Production

Pulse crops have a relatively small share in the total agricultural area and production when compared to cereals. Pulse production data for major pulse producing countries from 2000 to 2009 is presented in Table 1.1.

The global average yield of pulse crops during 2008 was estimated to be around 1 tonne ha^{-1}, which is significantly lower than of average yields of cereals. The global pulse production during 2009 was 61.5 million tonnes (Mt) from an area of about 70.6 million hectares (Mha), with an average yield of 871 kg ha^{-1}. Developing and developed countries contribute about 74% and 26% respectively to the global production of pulses. India, China, Brazil, Canada, Myanmar and Australia are the major pulse producing countries. India is the largest producer and consumer of pulses in the world, contributing

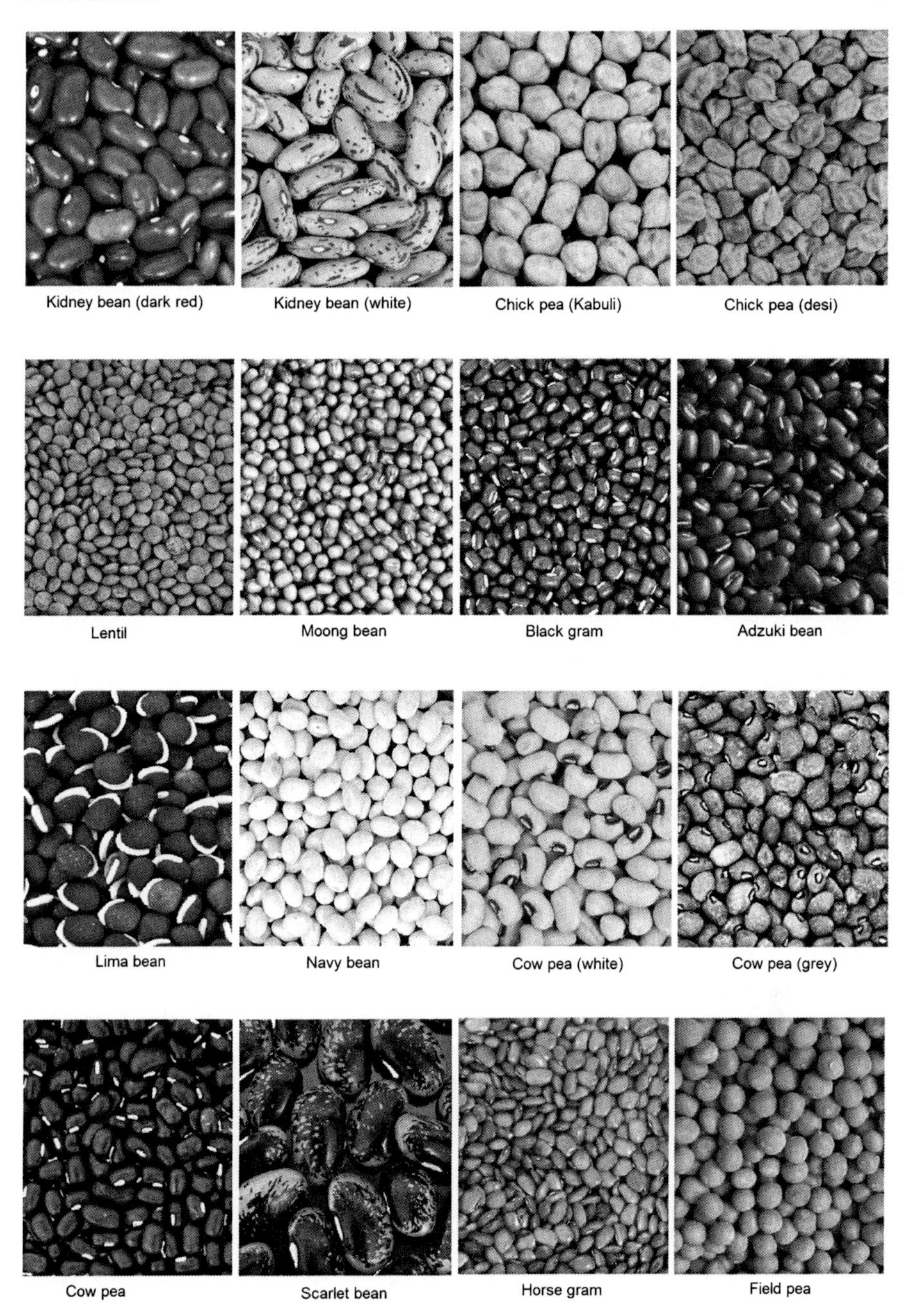

Figure 1.2 Commonly grown and consumed pulses around the world

around 25–28% of the total global pulse production. The pulse production in India is about 15 million tonnes covering an area of about 22 Mha. Canada, India, Nepal, USA, China and Ethiopia are the main lentil producing

Table 1.1 Production of important pulses during 2000–2009[a] (metric tonnes)

Pulses	*Country*	*2009*	*2008*	*2004*	*2000*
Lentils	Canada	1 510 200	1 043 200	915 800	914 100
	India	950 000	810 000	1 037 800	1 078 900
	Nepal	147 725	161 147	158 671	137 343
	USA	265 760	108 545	189 692	137 393
	Ethiopia	123 777*	94 103	55 113	59 000
	China	120 000*	150 000*	144 000*	116 000
Chickpea	India	7 060 000	5 748 600	5 717 500	5 118 100
	Australia	445 000	442 543	135 215	162 000
	Turkey	562 564	518 026	620 000	548 000
	Pakistan	740 500	474 600	611 100	564 500
	Myanmar	398 000	347 900	224 300	84 279
	Ethiopia	312 080*	286 820	162 858	164 627
Peas, dried	Canada	3 379 400	3 571 300	3 097 200	2 864 300
	China	960 000*	1 100 000*	1 060 000	102 000*
	Russia, Federation	1 348 890	1 256 830	1 242 500	815 230
	India	754 459[Im]	749 981[Im]	725 200[Im]	814 800
	USA	777 320	556 560	517 957	157 578
	France	546 846	451 416	1 680 780	1 936 500
Pigeon peas	India	2 270 000	3 075 900	2 346 400	2 694 000
	Myanmar	765 000	719 000	477 000	182 000
	Malawi	206 021[Im]	149 873	93 084	99 261
	Uganada	91 000	90 000	84 000	78 000
	Kenya	46 474	84 168	105 571	65 604
	United Republic of Tanzania	37 610[Im]	47 649[Im]	47 128[Im]	47 869
Dry beans	Brazil	3 486 760	3 461 190	2 967 010	3 038 240
	India	2 440 000	3 010 000	2 883 900	2 847 000
	Myanmar	3 000 000[f]	3 218 000	1 863 000	1 285 260
	China	1 489 135*	1 707 885*	1 758 489*	1 658 498
	USA	1 150 310	1 159 290	804 809	1 203 970
	Mexico	1 041 350	1 122 720	1 163 400	887 868
Cowpea	Nigeria	2 369 580	2 916 000	2 631 000	2 150 000
	Niger	1 550 000[f]	1 569 300	339 500	268 700
	Burkina Faso	3 250 000[f]	300 000	276 349	127 682
	Myanmar	180 000[f]	175 900[f]	129 000	77 000
	Cameroon	130 000[f]	130 101	96 735	70 000[f]

Source: [a] FAOSTAT (2011).
*, unofficial figure; Im, FAO data based on imputation methodology; f, FAO estimate.

countries, Canada being the largest producer. India, Australia, Turkey, Pakistan, Myanmar and Ethiopia are the main chickpea producing countries. India is the largest producer of chickpeas and pigeon peas. India contributes about 90% of pigeon peas, 75% of chickpeas and 37% of lentils to the area under these pulses.[1] Canada is the largest contributor to the global production of dried peas; China, Russia, India, USA and France are other major producers. Brazil, India and Myanmar are the main producers of dry beans. Brazil is the largest producer of dry beans, but does not produce much of other pulses. Turkey produces large amounts of lentils and chickpeas, but very few peas or beans. Pigeon pea is grown only in the developing world; India, Myanmar and Kenya are the main producers of this crop. Average global pulse production of 61.2 Mt during the triennium 2007–09 was recorded, representing an annual growth of about 0.7% per annum over the 55.03 Mt produced in 1997. Myanmar (11.48%), Canada (10.80%), Germany (8.27%), Sudan (8.08%), Spain (7.37%), Ethiopia (4.92%), China (4.67%) and Syria (4.12%) showed more than 4% annual pulse production growth rate. In the past, the production of pulses remained relatively low because they were less favoured by farmers due to their lower yield, greater susceptibility to diseases and less remunerative price than cereals. Worldwide, around 70% of the area harvested for pulses falls under rainfed area with low input resources, compared to only about 30% for cereals. This is mainly responsible for the low average global yields of pulse crops, which are around one-fourth the average yields of cereal crops. World average yields of pulses have increased slightly with a growth rate of 0.4% per annum over the last decade against growth of 1.5% per annum for cereals. Dry beans cover the highest proportion of total area under pulses, followed by chickpeas and cowpeas. Chickpea is the second most important pulse crop in the world, grown on 11 Mha with a total production of around 9 Mt during 2008. The producer price of pulses has also increased in both developing and developed countries at a growth rate of around 1.5% and 3% respectively during the last decade.

Large number of different pulses with diverse composition and processing quality are grown in different countries, in a wide range of agro-climate conditions. Dry beans, dry peas, chickpea, lentils and pigeon peas are the important pulses in terms of production and consumption. Dry beans contributed around 32% to global pulse production, followed by dry peas (17%), chickpea (15.9%), broad beans (7.5%), lentils (5.7%), cowpeas (6%) and pigeon pea (4.0%). Chickpeas are consumed to the largest extent in India and comprises about 45–50% of the total pulse production there. Beans accounted for the largest share of the world's pulse production followed by peas, chickpeas and lentils. Among pulses produced worldwide, dry bean, pea, chickpea, and lentil contribute about 46%, 26%, 20% and 8%, respectively.[1] Globally, South Asia is the largest producer of chickpeas (about 76%) with a share of more than 80% of area harvested. The developing countries' share in total area and production of chickpeas is 95% and 93%, respectively. The Indian subcontinent, eastern Africa and Central America/Caribbean region are the world's main pigeon pea producing regions. Worldwide, the total area under pigeon pea has increased by around 15% in the last two decades. Lentils are relatively more tolerant to drought and are grown throughout the world. Worldwide, the area under lentils has also increased, but less than that under pigeon pea. However, the production of lentils has increased more than the area under this crop. South Asia is the largest lentil growing and producing region in the world, with around 50% of growing area and 40% of production. Lentils are a crop of developed as well as developing countries. The developed countries (mostly Canada, the USA and Australia) contribute around one-third of the total area as well as one-third of total production. Figure 1.3 shows the global share of pulses in terms of acreage and production.

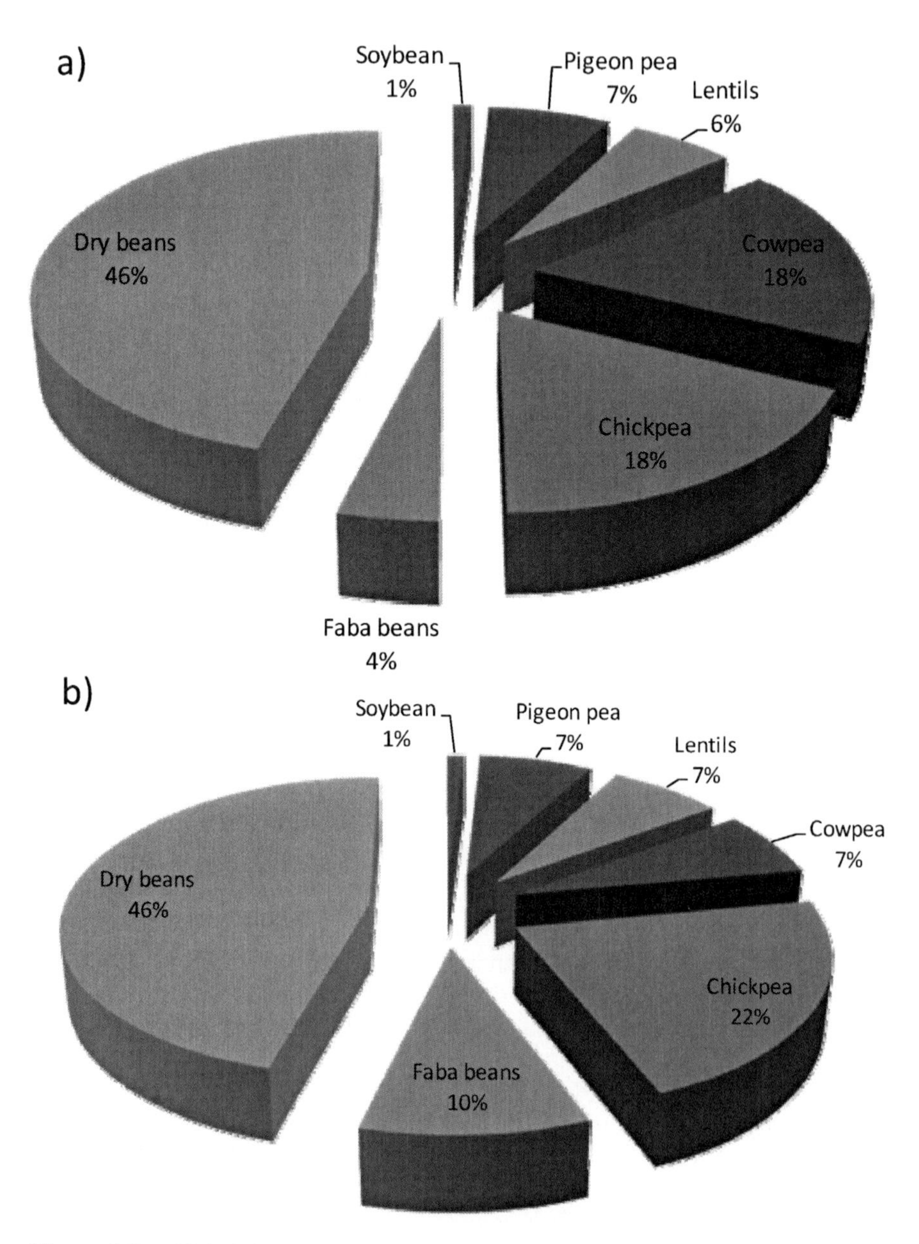

Figure 1.3 Global share of pulses in terms of (a) acreage (total world area 62 Mha) and (b) production (total world production 47 Mt)

1.2.2. Consumption of Pulses

The average *per capita* consumption of different pulses is given in Table 1.2. With a share of 25–28% in pulse production, India does not have the highest pulse consumption on a *per capita* basis. The availability of pulses in India has decreased from 60 g/day per person in 1951 to 31 g/day per person in 2008. Brazil has the largest average *per capita* consumption of pulses, followed by Ethiopia.

Many countries that do not contribute much to global pulse production, such as Ethiopia and Cameroon, have higher average *per capita* consumption than that of India and Pakistan. Brazil also has comparatively higher *per capita* pulse consumption (45.22 g/day per person). China is also one of the top pulse producing countries but has one of the lowest rates of *per capita* pulse consumption.

The changing trend in the consumption of pulses indicates that the dietary patterns have also changed around the world over the last decade and the demand for pulses has increased. The dietary pattern is influenced by many factors, such as income, prices, individual preferences, culture and environmental conditions, *etc.* The dietary energy measured in kcals *per capita* per day has been steadily increasing worldwide, as indicated by FAOSTAT data. Worldwide the contribution of cereals to *per capita* calorie intake over the last decade has decreased whereas the share of pulse crops has remained more or less constant. In developing countries, the contribution of cereals to *per capita* calorie consumption has declined. The rate of increase in pulse production in top pulse producing countries like India and China has been less than the population growth rate during the last two decades. As a result, the *per capita* consumption of pulse grains has increased in these countries.

The import of pulses by major consumer countries had little impact in stabilizing their price in last decade. The price of pulses continuously rose between 2000 and 2010, at much faster rate than that of cereals, in most developing countries. To meet the increasing global demand for pulses, pulse production must be increased in tandem with consumption demand. If the rate of increase in the

Table 1.2 *Per capita* consumption of pulses in different countries in 2000 and 2007

	Quantity (g/day per person)	
Country	*2007*	*2000*
India	35.28	30.48
Pakistan	22.14	18.65
Brazil	45.22	45.96
Canada	21.37	21.52
Myanmar	41.69	27.27
Nepal	22.89	23.02
Ethiopia	42.92	35.29
Nigeria	29.01	26.18
Cameroon	38.18	37.05

Source: FAOSTAT (2011).

area under pulse crop remains low (less than 0.5% per year), as in the last decade, then there may be acute shortage of pulses in future. This will significantly affect the nutritional status of people living in developing countries such as India, where a large sector of the population is vegetarian, relying on pulses to meet their daily protein requirement. Pulse production will have to be increased by around 20–25% by 2030 to meet the increasing global demand for pulses. Increasing the area and introducing high-yielding varieties could help in meeting the estimated future demand for pulses. It is estimated that an area of about 80–82 Mha may be needed to meet increasing demand in the next two decades. Considering the present growth rate in pulse productivity and population, the present yield of 50 kg h^{-1} needs to be increased to around 120 kg h^{-1} in the next two decades to meet the increasing global demand for pulses.

1.2.3. Trade in Pulses

The average production cost of pulse grains is higher than that of cereal grains in the major pulse producing developing countries. Additionally, pulse production costs have increased more in the last few years in most pulse producing countries and this has increased the price of pulses in the international market. Pulses are imported by different countries to meet their domestic demand which arises from the increase in population. Table 1.3 demonstrates the import and export volume of important pulses by various countries.[1] The international pulse trade has averaged around 7.4 Mt during the last decade. Table 1.3 clearly shows that peas contribute most to world pulse trade, followed by beans, lentils and chickpeas. Canada is the world's largest producer and exporter of yellow and green field peas. Canada exports food peas to South Asian countries (India, Bangladesh, Sri Lanka) as well as South and Central American countries. In Europe, peas imported from Canada used for feed purposes in the animal feed industry. Canada produces many types of beans (navy, pinto, kidney, black, cranberry, great northern) and around 70% of the bean crop is exported to the USA, UK and Italy every year. Turkey, Sri Lanka, UAE, Egypt, Algeria and Spain are the major lentil importing countries. India is ranked as the top importer of chickpeas, dried peas and dried beans. Pakistan, Spain, UAE, Bangladesh and Saudi Arabia are the other major importers of chickpeas. Brazil is the second largest importer of dried beans after India, followed by USA, Italy, UK, Japan and China. Canada exports the largest quantity of dried peas and lentils, followed by the USA. Australia is the largest exporter of chickpeas. Mexico, Turkey, Myanmar, Canada and Ethiopia are other chickpea exporting countries. These counties export less than half the amount of chickpeas exported by Australia.

The production cost of pulse grains is greater than that of cereals in both developing and developed countries. At present, world trade in pulse crops represents about 12–15% of pulse production. This indicates that about 85–88% of pulse crops are consumed in the country where they are produced and only 12–15% is either imported or exported. Both the import and the export of pulse grains have increased worldwide in the last decade. It is also worth noting

Table 1.3 Import and export of pulses by different countries during 2008 (metric tonnes)

	Import		*Export*	
Pulses	*Country*	*2008*	*Country*	*2008*
Lentils	Turkey	191 683	Canada	852 876
	Sri Lanka	102 710	USA	168 646
	UAE	82 537	Turkey	70 340
	Egypt	66 364[R]	Australia	76 026
	Algeria	65 000[F]	UAE	36 612
	Spain	48 175	China	17 965
Chickpeas	India	198 215	Australia	271 548
	Pakistan	114 682	Mexico	108 802
	Spain	54 377	Turkey	88 338
	UAE	75 249	Myanmar	82 000[R]
	Bangladesh	79 216[R]	Canada	57 879
	Saudi Arabia	47 074[R]	Ethiopia	41472
Peas, dried	India	1 215 660	Canada	1 918 310
	China	220 934	USA	499 464
	Belgium	105 578	France	278 536
	Bangladesh	113 152[R]	Australia	124 475
	UK	71 010	Ukraine	77 208
	Italy	59 528	Belgium	72 707
Dry beans	India	604 518	China	959 823
	Brazil	209 690	Myanmar	675 000[R]
	USA	166 783	USA	415 321
	Italy	109 875	Canada	293 595
	UK	148 055	Argentina	229 199
	Japan	119 113	Ethiopia	74 389*
	China	103 602	UK	61 375

Source: FAOSTAT (2011).
F, FAO estimate; R, estimated data using partner's database; U, unofficial figure.

that, in parallel, the price of pulse grains traded in the world is increasing at a much higher rate than their production.

The inability of developing countries to increase their export volume of legumes arises from several issues including post-harvest losses, particularly during storage and milling of grain legumes, and lack of support services, particularly credit and marketing. Mycotoxin contamination and pesticide residues in pulse crops are also persistent issues in both developed and developing countries. In fact, to meet the food needs of the future, it is critically important that scientific and technological advancements be accelerated and applied both in agricultural practices for pulse crops and in the food industry involved in pulse processing.

1.3. CURRENT STATUS OF PULSE PROCESSING

Pulses are generally processed before consumption. The consumption of pulses in different parts of the world varies depending on their availability, dietary pattern and the local prevailing conditions. Pulses are known to be used in a range of food preparations after suitable processing. Pulse processing has evolved from being merely a need to store pulses safely from the time and location of harvest until they reach the consumer. For example, chickpeas and beans are consumed as whole seeds as well as flour. Cooking, germination or use as composite flours for preparation of a variety of snack products (deep fried or extruded) are some of the common ways in which pulses are used for human consumption. The whole or split grains are also roasted and fried for consumption as snack products. According to the FAO, about 71% of global pulse consumption is for food, while 18% goes to feed with most of the remaining used as seed.[2] The food industry is increasingly interested in the potential to incorporate novel ingredients such as pulses into food products to improve nutritional value and health benefits. A general trend to increase the use of pulses and pulse fractions in the development of functional foods such as high fibre, gluten-free food with low glycaemic index can be evidenced in recent times. Pulse fractions are also incorporated into the different food products to improve functionality such as water and oil absorption. In general, pulses have been given lesser priority in terms of processing. Therefore, the pulse industry must advance its knowledge of the processing of pulses into ingredients and the impact of such processing on the functionality of such ingredients in food product formulations. Attention must be given to the optimisation of processing in terms of quality and functionality, in addition to other factors such as yield and energy use, in order to successfully introduce more value-added pulse processing and the incorporation of these ingredients into foods. Use of novel food processing technologies could be an alternative approach to introduce novel food products based on pulse crops in the developed countries.

1.4. CONCLUSIONS AND FUTURE TRENDS

Given the continuing population growth and the still low *per capita* consumption of pulses, the prospects for further expanding pulse processing and utilisation appear bright. Pulse production, processing and utilisation are expected to expand in the future as further economic development takes hold on a global scale and as changing lifestyles demand an increasing population to consume healthier foods. The introduction of modern technologies will invariably improve the processing and utilisation of pulses in both developed and developing countries. The potential of pulses for meeting food requirements is great. The promotion of pulse processing industries and the introduction of innovations to existing processes will require the active participation of government or policy-makers, research bodies and entrepreneurs.

REFERENCES

1. FAOSTAT. Available online: http://faostat.fao.org/default.aspx (accessed date: 26 December 2011).
2. P. Watt, in *Pulse Foods Processing, Quality and Nutraceutical Applications*, ed. B. K. Tiwari, A. Gowen and B. McKenna, Elsevier, Amsterdam, 2011.

CHAPTER 2

Pulse Grain Structure

2.1. INTRODUCTION

In general, flowering plants are broadly classified into monocotyledons (monocots) or dicotyledons (dicots). Seeds of monocots contain one cotyledon, whereas the embryo of dicots contain two cotyledons. Cereals such as wheat, rice, barley and oats are monocots and pulses are dicots; their seeds can be separated into two halves. Generally, seeds of pulses consists of three parts, *i.e.* cotyledon (endosperm), germ (embryo) and seed coat (testa). The cotyledon stores nutrients and is an important part of the grain. Upon germination, the cotyledons usually develop into the first embryonic leaves of a seedling. The germ is the "heart" of the seed kernel and is a concentrated source of several essential nutrients including vitamin E, folate (folic acid), phosphorus, thiamine, zinc, magnesium, *etc.* The seed coat is made up of several layers that enclose the cotyledon.

Grain characteristics, morphology and microstructure are important in the design of food processes and provide a basis for the study of changes induced in grains by various food processing operations. An understanding of the microstructure of cereal grains as related to hardness and processing has advanced the technology of both dry and wet milling of most cereals, including wheat, corn and rice.[1] Studies of food grain microstructure help us to understand several important grain properties and associated changes. For example, physical properties of grains, water uptake and hydration rates, kernel density, breakage susceptibility, power requirement during the milling process, and the quality of processed products can all be explained on the basis of microstructural changes. Microstructural studies of pulses and indeed of foods in general are also essential, because most of the elements that are critical in transport properties, physical and rheological behaviour, and textural and sensorial traits are below the 100 μm size range.[2] Furthermore, microstructural aspects control the nutrient bioavailability and digestibility of food products.[3,4]

Pulse Chemistry and Technology
Brijesh K. Tiwari and Narpinder Singh

Published by the Royal Society of Chemistry, www.rsc.org

Structural elements contributing to the identity and quality of pulses are plant cells and cell walls, starch granules, protein assemblies and even food polymer networks, among others. Structure–functional relationships can strongly affect the physicochemical, functional, technological, and even nutritional properties of foods obtained from pulses or pulse fractions. Among food grains the study of pulse microstructure is limited to textural defects, for example hard-to-cook phenomena, and limited studies have been conducted to establish a relationship between grain structure and milling efficiencies and other attributes.

Most of the structural features studied earlier were based on light microscopy studies. In fact, light microscopy is still a first-line choice for studying structure, considering the simplicity, low cost and minimal sample preparation required compared to electron microscopy. Phase-contrast microscopy, which converts slight differences in refractive index and cell density into easily detected variations in light intensity, is commonly used to observe the structure of cell organelles and movement of solutes in a cell. Foods are complex multicomponent systems, and microstructural elements are usually difficult to observe in their natural or transformed states.[5] Therefore, an additional sample preparation step is part of every microstructural study. The development of low vacuum scanning electron microscopy (SEM) (for dehydrating conditions) and environmental SEM (for fully hydrating conditions) has led to many food structure researchers examining samples in ways that were previously impossible.[6] A detailed discussion on the microscopy and imaging techniques employed for study of food microstructure can be found elsewhere.[7] This chapter aims at providing an in-depth understanding of the microstructure of pulse grains.

2.2. GRAIN MORPHOLOGY

In general, all pulses have a similar structure, but they differ in colour, shape, size and thickness of the seed coat. The morphology of the pulse kernel varies from species to species, as well as within a given species. Table 2.1 presents some variations in pulse grain morphology. Interspecies variation is most likely due to the variations in environmental and agronomic practices. Seed coat colour and seed size are the two main criteria that identify the numerous market classes recognized throughout the world.[8]

Unlike a cereal grain (*e.g.* wheat), which is botanically known as a "caryopsis", meaning that a single cotyledon is modified to form the scutellum that absorbs food from the endosperm, pulses (*e.g.* bean seed) exhibit more complexity in structural features (see Figure 2.1).

2.3. ANATOMICAL PARTS OF GRAIN

Pulse seeds can be best studied by considering their three main parts: seed coat, endosperm (cotyledon) and germ (embryo). Usually the cotyledon constitutes

Table 2.1. Morphological features of some pulses

Characteristics	*Seed colour*	*Seed weight*	*Seed shape*	*Seed testa texture*
Chickpeas	Brown	**Desi types**	Pea-shaped	Rough
	Yellow	Small (<150 g)	Trapezoidal	Smooth
	Dark brown	Medium (150–220 g)	Angular	Tuberculated
	Creamy white	Large (230–300 g)		
		Very large (>300 g)		
		Kabuli types		
		Small (<250 g)		
		Medium (250–300 g)		
		Large (310–400 g)		
		Very large (>400 g)		
Black gram	Green	Small (<30 g)	Globose	Rough
	Greenish-brown	Medium (30–50 g)	Oval	Smooth
	Brown	Large (>50 g)	Drum shaped	
	Black			
	Mottled			
Green gram	Yellow	Small (<30 g)	Oval	Rough
	Green	Medium (30–50 g)	Drum shaped	Smooth
	Mottled	Large (>50 g)		
	Black			
Pigeon peas	Cream	Small (<70 g)	Oval	Smooth
	Drown	Medium (70–80 g)	Globular	
	dark brown	Large (90–110 g)	Elongate	
	Grey	Very large (>110 g)		
	Purple (with uniform mottled colour pattern)			
Field peas	Green	Small (<150 g)	Spherical	Smooth
	Yellow	Medium (150–200 g)	Cylindrical Dimpled	Wrinkled
	Creamy	Large (>200 g)		
Kidney beans	White	Small (<250 g)	Circular	Smooth
	Brown	Medium (250–350 g)	Circular to elliptic	
	Red	Large (350–450 g)	Elliptic	
	Dark red	Very large (>450 g)	Kidney shaped	
	Black			
Lentils	Green	Small (<20 g)	Elliptical	Smooth
	Grey	Medium (20–25 g)	Oval	Rough
	Pink	Large (26–30 g)		
	Brown	Very large (>30 g)		
	Black			

a major part of any grain, accounting for 80–90% of the total seed. The seed coat and germ contribute about 8–16% and 1–3%, respectively. Table 2.2 shows the proportions of different components of the grain. Cowpea seed coat

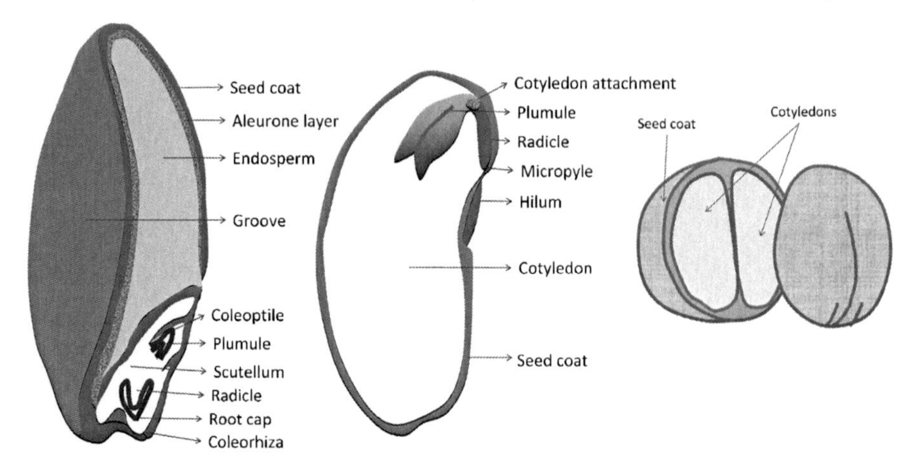

Figure 2.1 A general structure of wheat caryopsis and common bean

content varies from 9.3% to 33%,[9] with colour ranging from shades of buff, green, brown, red, purple to black. Although pulses with a range of colours are available, the predominant colours are brown and a combination of white and brown. Seed coat thickness varies from thick with a smooth surface to thin with a rough and convoluted surface.[10]

Cross-sections of a whole lentil grain examined using SEM reveal that the two halves of cotyledons are held firmly within the outer seed coat, forming an ellipsoid, as shown in Figure 2.2. Lentils have seed coat thickness between 25 and 65 μm depending on cultivar.[11] Figure 2.3 shows a higher resolution image (300 ×) of the internal structure of lentil, *i.e.* cell wall, middle lamella, inner integument and outer integument. The schematic of the pulse grain shown in Figure 2.4 provides a better insight into the cellular arrangement of the tissues in the seed coat and the cotyledon. Elongated cells can be seen in the outer and the inner integument of the seed coat that provide support to parenchyma cells in the cotyledon.[11] When the seed coat is removed from the grain, the part that

Table 2.2 Anatomical parts and proportion of some pulses

	Proportion of whole grain (%)		
Pulse	*Seed coat*	*Cotyledon*	*Germ*
Chickpeas	12.5	85.5	2.0
Pigeon peas	15.5	83.0	1.5
Mung beans	12.0	87.0	1.0
Black gram	12.5	86.5	1.0
Lentils	8.0	89.9	2.1
Cowpeas	10.6	87.3	2.1
Horse gram	12.5	86.0	1.5
Kidney beans	9.7	89.3	1.0
Garden peas	10.0	88.3	1.7

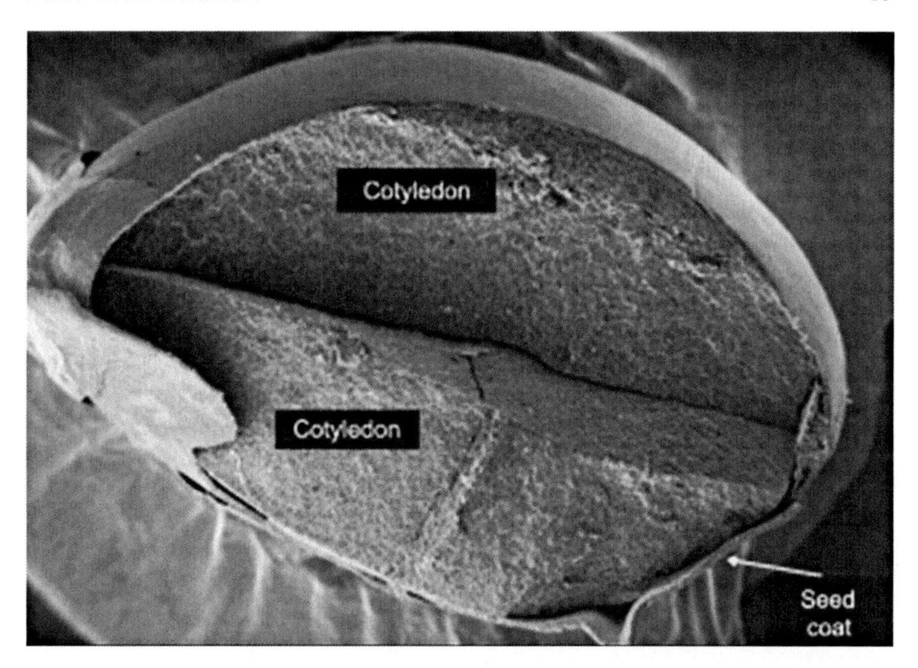

Figure 2.2 Scanning electron micrograph (26 ×) of the cross-section of a lentil pulse showing the seed coat and cotyledon (Bhattacharya *et al.*)[11]

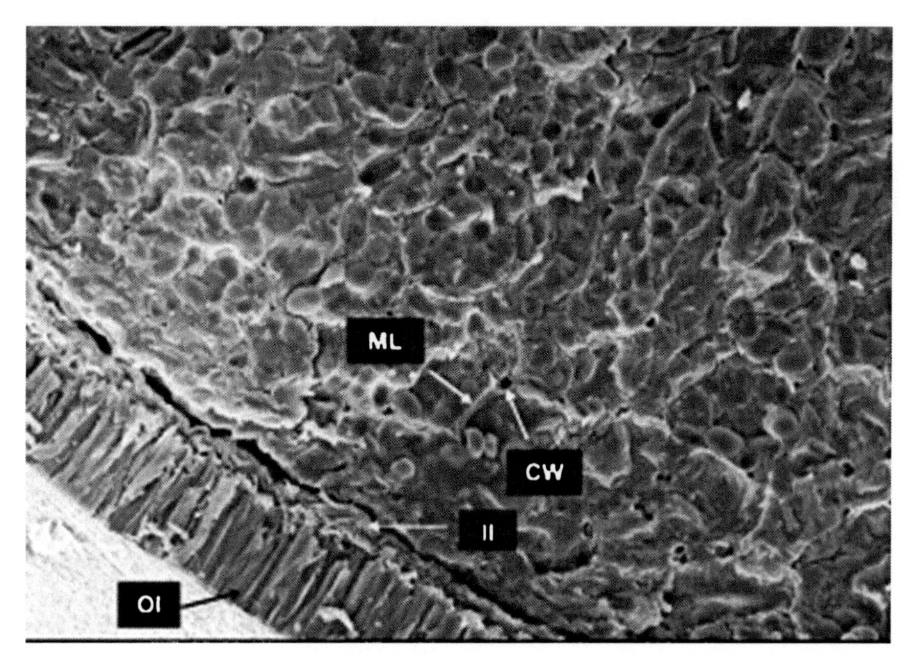

Figure 2.3 Micrograph (300 ×) showing different internal structures. CW, cell wall; ML, middle lamella; OI, outer integument; II, inner integument (Bhattacharya *et al.*)[11]

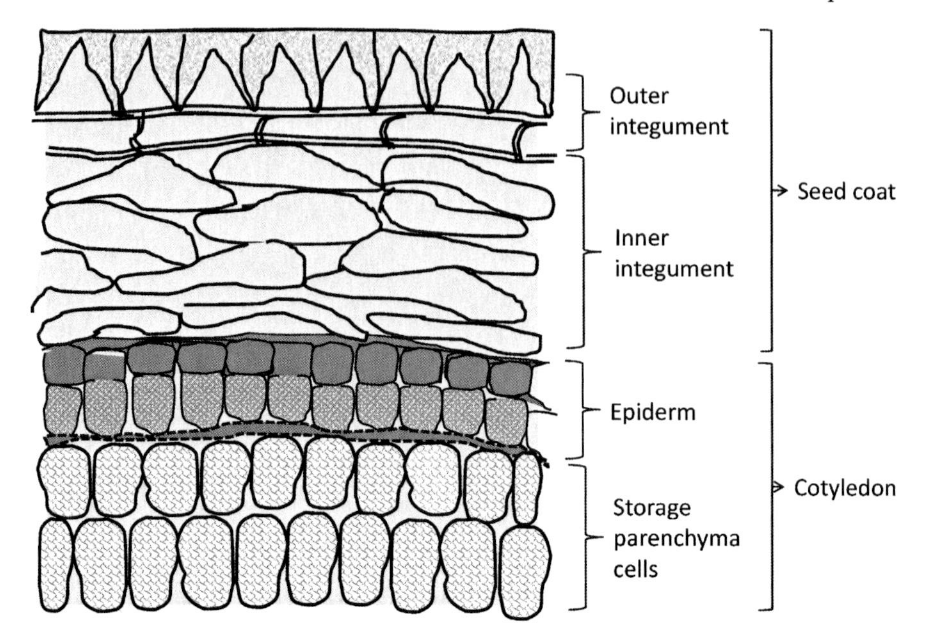

Figure 2.4 Schematic of cellular arrangement within a pulse grain

remains is the embryonic structure. This consists of two cotyledons (or seed leaves) and a short axis above and below them commonly known as the embryonic axis (radical and plumule). Apart from the germ, cotyledons and seed coat, other external structural features observed on pulse grains are discussed in the following section.

2.3.1. Hilum

The hilum is a scar on the seed coat left by the funiculus, which is the point of the attachment in the pod. The position of the hilum varies. In general, it is located near the middle edge where the seed breaks away from the stalk. The size and shape of the hilum varies from species to species, ranging from round to oblong, oval or elliptical. Low-magnification SEM indicates that the hilum exhibits a spongy-like amorphous microstructure.[12] This type of structure was observed for black beans as well as for black gram and "Carioca" beans.[12–14] The amorphous structure of the hilum is characterised by the presence of cavities arranged in a honeycomb-like structure or crosslinking network that appears to project inside the hilum (Figure 2.5).[12]

2.3.2. Micropyle

The micropyle is a minute hole or small opening in the seed coat next to the hilum, located at the radical end of the hilum. It serves as an entry point for diffusion of water. The shape of the micropyle in pulses varies, ranging from circular or triangular to fork-shaped.

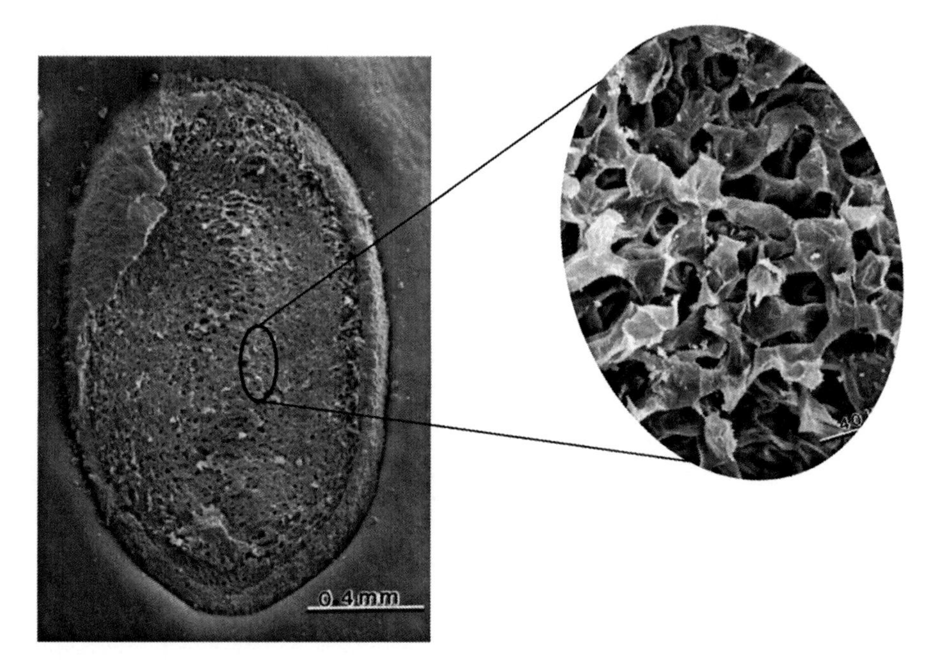

Figure 2.5 The external structures of black bean hilum and cavities are observed as a honeycomb-like structure at higher resolution. (Modified from Berrios *et al.*[12])

2.3.3. Raphe

The raphe is a sort of ridge at the side of hilum opposite to the micropyle. It is formed from the fused funiculus and represents the base of the stalk that fuses with seed coat in a mature grain.

2.3.4. Embryonic Axis

The embryonic axis includes the radicle and plumule. The radicle is the embryonic root (rudimentary root portion), whereas the plumule is the embryonic shoot (embryonic bud). The plumule is well developed in a mature seed and rests between two cotyledons. The radicle and plumule are attached to the cotyledon by a cotyledon attachment. The embryonic stem above the point of attachment is the epicotyl, whereas the embryonic stem below the point of attachment is the hypocotyl. The radicle and plumule are transport organs for nutrition to the embryo during sprouting. Both radicle and plumule play an important role during water imbibition by the seed coat.

2.3.5. Germ

The germ is the metabolically active part of the pulse grain. The embryo contains the shoot and root, which grow under proper conditions during germination. The endosperm of the pulse grain is the initial source of stored nutrition allowing growth of the seedling before the formation of leaves and the onset of photosynthesis to produce nutrients. The two cotyledons act as leaves during germination. The radicle or embryonic root has almost no protection except that provided by the seed coat. Therefore, the seed is unusually breakable, especially when it is dried and roughly treated.

2.3.6. Seed Coat

The seed coat covers the whole grain structure and is the outermost layer of the seed which protects the embryonic structure from damage caused by water absorption and microbial attack. The seed coat protects the pulse grain from the external environment and from infection during development and as such is relatively resistant to water. Seed coats have another major function: in addition to protecting of the embryo in the ripe seed, it is responsible for the supply of nutrients during seed development.[15,16] Usually, pulses have a moderately thick seed coat and pulses with thick seed coats generally have higher amounts of lipids. The outermost layer of the seed coat is usually known as the cuticle layer.

2.3.6.1. Microstructure of the Seed Coat

The basic structure of seed coat is quite similar among different species of pulses, consisting mainly of four layers: the waxy cuticle layer, epidermis, hypodermis and interior parenchyma.[17] Cross-sectional images of lupin seed coat and cowpea seeds are shown in Figures 2.6 and 2.7 respectively. The outermost layer of the seed coat acts as a barrier to water imbibition. The seed

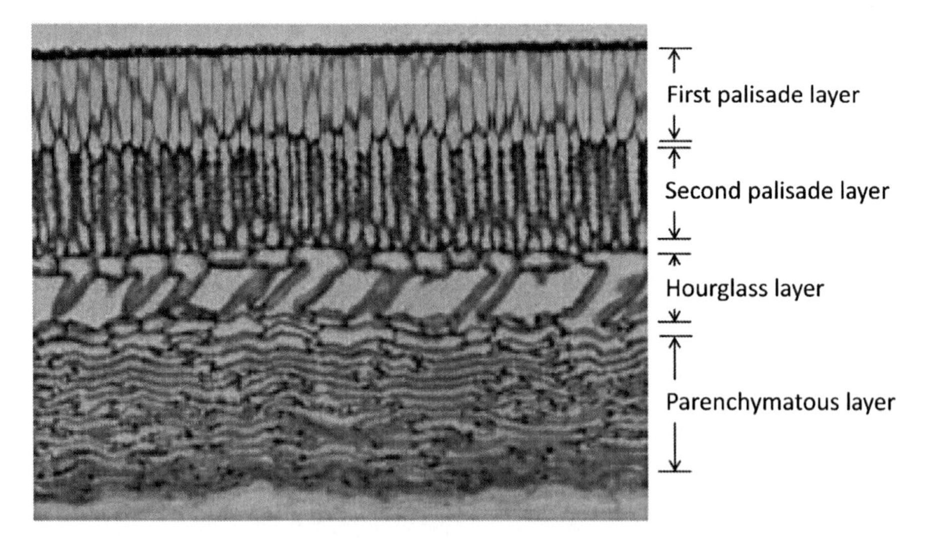

Figure 2.6 Cross-sectional images of *Lupinus angustifolius* parent genotypes obtained using light microscopy showing first palisade layer; second palisade layer; hourglass cell layer; parenchymatous layer (Clements *et al.*)[66]

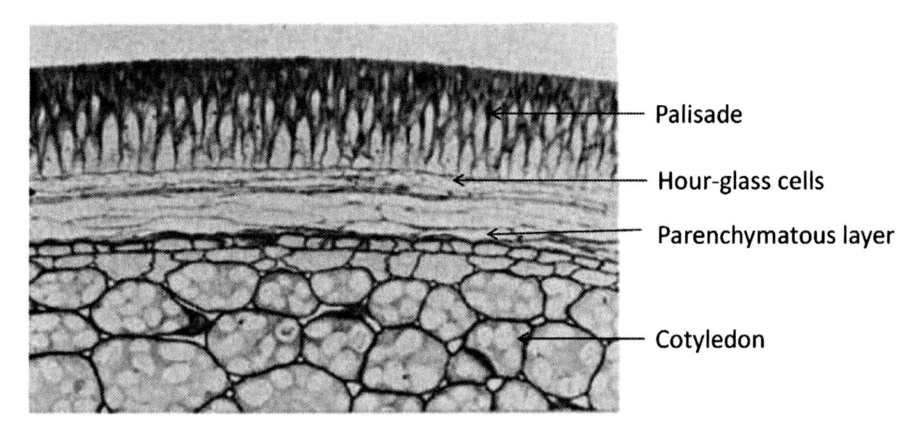

Figure 2.7 Cross-sectional image of the cowpea seed showing cotyledon; hourglass cells; parenchymatous layer; palisade cell layers (Lush & Evans, 1980).[67]

coat varies in thickness depending on the variety and species. The epidermis layer is thick walled, made up of long palisade cells (also known as malpighian cells or macrosclereids) with the long axis oriented perpendicular to the surface. The top portion or cap of the palisade cells is embedded in a suberin, which is a fatty acid found in the cell walls of the endodermis.[65] The seed coat of adzuki beans (*Vigna angularis*) has a well-organised layer of elongated palisade cells of varying thickness ranging from 40 to 60 µm depending on cultivar.[18,19] Well-organised palisade cells are associated with a relatively thick seed coat (43–59 µm). Alternatively, the epidermal cell layer may consist of an amorphous layer,[20,21] which, according to Sefa-Dedeh and Stanley,[21] consists of loosely packed cells where distinct palisade cells cannot be identified. Amorphous cells are associated with a thin seed coat (5.8– 9.9 µm).[21]

The hypodermis is made up of a single layer of cells, also known as hourglass cells, pillar cells, osteosclereids or lagenosclereids, depending on the pattern or cell wall thickness and shape. These are larger than the adjacent cell layers and are separated by wide intercellular spaces, except under the hilum where they are absent.[22] The tagmen layer consists of hourglass cells lying directly over the parenchyma cells. Hourglass cells are typically found in pulses,[16,23] with no functional role described to date. In adzuki beans, the seed coat lacks the subepidermal layer of hourglass cells and palisade cells are located directly above the parenchyma cell. Subepidermal layers of hourglass cells are present in seed coat of black beans, Mexican red beans, and blackeye peas.[24] It is worth noting that in some pulses, the hourglass cells are absent and the palisade cells overlie the mesophyll tissue. The final layer of the seed coat is the interior parenchyma usually formed of six to eight layers of thin-walled, protoplast free, tangentially elongated parenchyma cells. These are uniformly distributed throughout the whole seed coat except in the area of the hilum, where a smaller number of layers can be distinguished.

Figure 2.8 illustrates cross-sections of the pea seed coat at 30 days after anthesis showing distinct cell layers of epidermis, hypodermis, chlorenchyma, ground parenchyma, branched parenchyma and amphicribral vascular bundle. Parenchyma cells can be distinguished into three sublayers: chlorenchyma, ground parenchyma and branched parenchyma.[16] Parenchyma cells are also characterised by intercellular spaces which are filled either with air or with liquid. These intercellular spaces can be small or large depending on the species. Seed coats with branched parenchyma possess large intercellular spaces. The parenchyma is characterised by intercellular spaces, which are particularly large in the branched parenchyma. The chlorenchyma consists of large cells (up to 100 μm) that contain chloroplasts. The main function of the chloroplasts in the seed coat seems to be the transient accumulation of starch.[16,25] Cells in the second parenchyma layer contains fewer chloroplasts than the chlorenchyma; otherwise, the second parenchyma layer and the chlorenchyma are very similar. The cells of the second parenchyma layer are present in abundance, hence these are referred to as ground parenchyma. The branched parenchyma and the innermost layers of the ground parenchyma

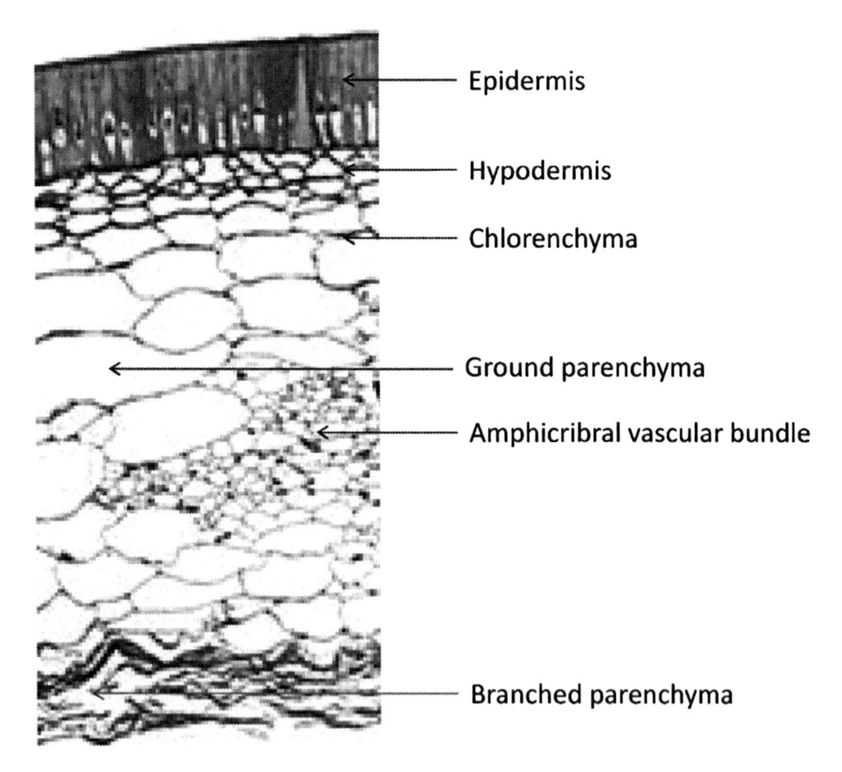

Figure 2.8 Cross-sections of the pea seed coat at 30 days after anthesis showing epidermis, hypodermis, chlorenchyma, ground parenchyma, branched parenchyma and amphicribral vascular bundle (Van Dongen *et al.*[16])

deteriorate and form a boundary layer between the living seed coat parenchyma and the cotyledons.

Immediately inside the inner parenchyma is the aleurone layer. This is the outermost layer of the endosperm, which is tightly compressed against the seed coat by the expansion of the cotyledons.[26] These are indistinguishable in the crushed layer of parenchyma above the aleurone. Figure 2.9 shows the mature soybean seed coat indicating the presence of palisade layer, hourglass cells, partially crushed parenchyma, crushed remnants of parenchyma and endothelium, aleurone and crushed remnants of endosperm.[26]

Both intra- and interspecific variation has been observed in the seed coat surface. As an example, the domesticated species of chickpea (*Cicer arietinum*) shows variable seed coat texture which varies from plate-reticulate or mamellonate plate type texture for kabuli to a double-reticulate texture for kabuli.[27] Table 2.3 indicates the variations in chickpea grain characteristics in domesticated kabuli and desi chickpea, along with SEM micrographs. Kabuli cultivars of chickpea are characterised by large, ram-shaped and cream or beige coloured seeds whereas desi cultivars have small, angular, and dark-coloured grains.

In the cowpea the external layer of the seed coat, *i.e.* the cuticle or waxy layer, may or may not be present depending on the cultivar. If it is present at all the top waxy layer is followed by epidermis, hydrodermis or hourglass cell layers, and parenchyma cell layer as shown in Figure 2.7. The epidermal layer of cowpea also consists of a palisade layer made up of well-organised, thick-walled, elongated cells with the long axis oriented perpendicular to the surface.[22] As discussed earlier, the hypodermis is located just below the

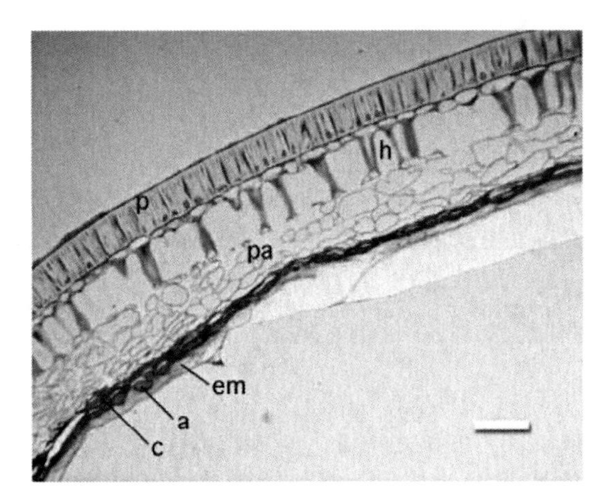

Figure 2.9 Cross-section of physiologically mature soybean seed coat (45 days after anthesis), stained with Periodic Acid Schiff and Light Green. Carbohydrates are light grey, proteins are dark grey. p, palisade layer; h, hourglass cells; pa, partially crushed parenchyma (aerenchyma); c, crushed remnants of parenchyma and endothelium; a, aleurone; em, crushed remnants of endosperm. Bar indicates 100 µm (Miller *et al.*[26]).

Table 2.3 Comparison of seed characters among desi and kabuli chickpea cultivars

Chickpea cultivar	C. arietinum *cv. kabuli*	C. arietinum *cv. desi*
Seed size* (mm, L × B)	7.59 × 5.87	7.11 × 4.76
seed colour	White–cream, cream–beige	Light brown, green, black
Shape	Cordate or circular	Cordate
Ridge on lens	Absent	Absent or present
Shape of hilum	Narrowly or widely elliptic	Narrowly or widely elliptic
seed coat texture	Double-reticulate	Plate-reticulate
SEM micrograph		

*Length × width (mm); average of five replications.
Source: Adapted from Javadi and Yamaguchi.[27]

epidermis. During the early development stages of the seed the cells in this layer are very similar to the adjacent parenchyma cells, but at later stages they differentiate into osteosclereids or "hourglass" cells. These cells obtain their final shapes after 20 days after anthesis (see Figures 2.8 and 2.10). The remaining portion of the seed coat consists of parenchyma cells. Figure 2.8

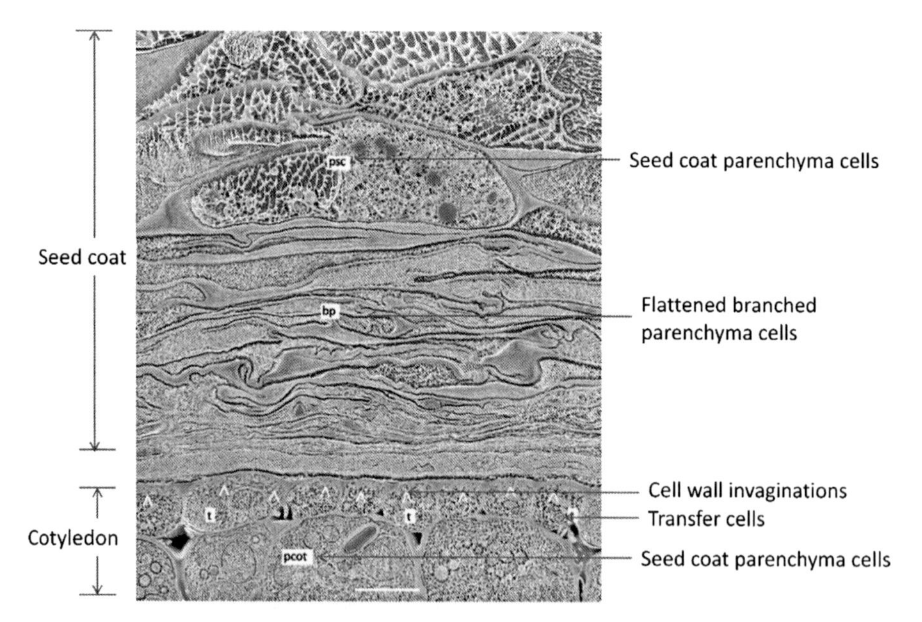

Figure 2.10 Cryo-SEM micrograph of the seed coat–cotyledon boundary approximately 20 days after anthesis, showing seed coat parenchyma (p), flattened branched parenchyma cells (bp) and transfer cells (t) of the cotyledonary epidermis. The layer between seed coat and cotyledon probably includes the wall of the embryo sac (Marinos, 1970).[68] Note that cell wall invaginations are present in the epidermal cells of the cotyledon (arrowheads) but not in seed coat parenchyma cells. Bar represents 10 μm (Van Dongen *et al.*[16])

shows a cryo-SEM micrograph of the seed coat–cotyledon boundary approximately 20 days after anthesis in pea seed, showing seed coat parenchyma and other cell layers.

2.3.7. Cotyledon

As mentioned earlier, the endosperm or cotyledon of the pulse grain is the source of stored nutrition that allows initial growth of the seedling. The cotyledon is not metabolically active and consists of numerous large parenchymatous cells without nuclei, packed with starch granules embedded in a matrix of storage proteins. The size of the parenchymatous cells ranges from 70 to 100 μm, and the most abundant structures in this region are starch. In the cotyledon, the parenchymatous cells are bounded by cell walls and middle lamellae. The middle lamella is a pectin layer, which cements the cell walls of two adjoining cells together and is mainly composed of pectic substances and protein. The parenchymatous cells are the accumulation sites for starch granules and protein bodies. The cotyledons store reserves of protein, lipid, sugars and wall polysaccharides. [28] In leguminous species the cotyledons function both as storage organs and as persistent photosynthetic structures.

2.3.7.1. Microstructure of Cotyledons

The microstructure of the cotyledons of chickpea and garden peas, both smooth and wrinkled, is shown in Figures 2.11, 2.12 and 2.13. The SEM images show that the cells in the outer layer of the cotyledon of both chickpea and garden pea are elongated and tightly packed.[1] The cells are rounder and more loosely packed in the central part of the cotyledon, with large

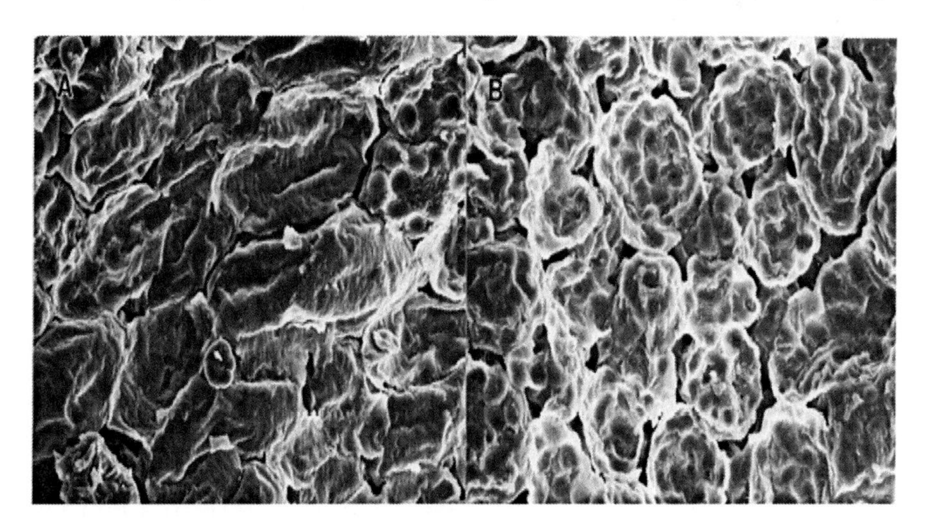

Figure 2.11 SEM of microstructure of garbanzo beans: outer layer (A) and inner layer (B) of cotyledon (Otto *et al.*[1])

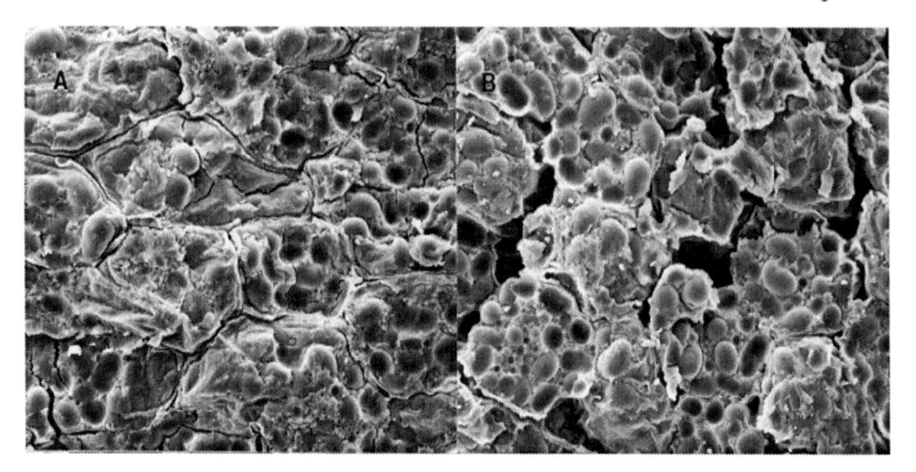

Figure 2.12 SEM of microstructure of smooth pea cv. Latah: outer layer (A) and inner layer (B) of cotyledon (Otto *et al.*[1]).

intercellular spaces, compared to the tightly packed cells in the outer layer. The structurally different outer and inner layers of pulse cotyledons behave differently during the milling of pulses.[29]

The middle lamella makes up the outer wall of the cell and is shared by adjacent cells. This follows the primary cell wall consisting of a rigid skeleton of cellulose microfibrils embedded in a gel-like matrix composed of pectic compounds, hemicellulose and glycoproteins. Finally, a secondary wall is formed after cell enlargement is complete. This is composed of cellulose, hemicellulose and lignin. The mesophyll cell walls of the cotyledons of lupins are substantially thickened with secondarily deposited reserve carbohydrates, rich in galactose and arabinose.[30,31] The presence of these wall storage materials and their fate during germination was the subject of some controversy at the turn of the century.[32–35]

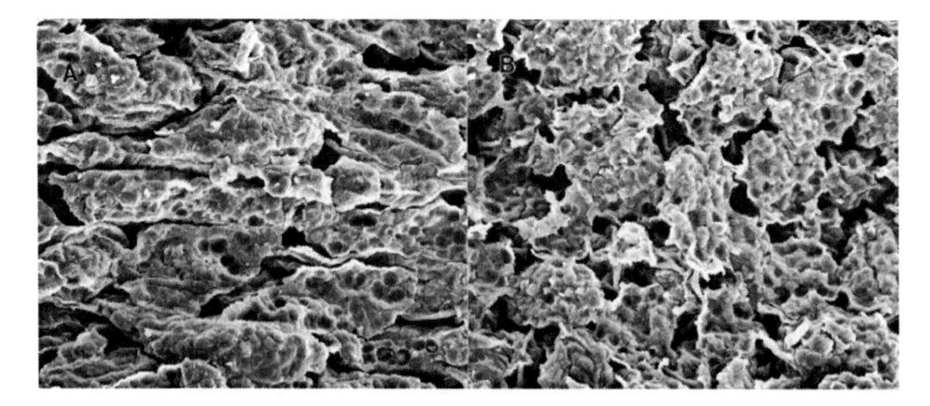

Figure 2.13 SEM of microstructure of wrinkled pea cv. Scout: outer layer (A) and inner layer (B) of cotyledon (Otto *et al.*[1]).

The parenchyma cells within the cotyledons are also known to be storage sites of carbohydrates and proteins. These cells have intercellular spaces in both compact and amorphous cotyledons. These intercellular spaces vary; a few intercellular spaces were observed in pulses with compact cotyledons whereas many intercellular spaces were seen in those with amorphous cotyledons. The parenchyma cells range in length from 60 to 120 μm with irregular shapes showing starch granules, protein bodies and a cytoplasmic matrix, and some lipid bodies mainly found along the cell wall.[36] Pulses store reserves of protein, lipid, sugars and wall polysaccharides within their fleshy cotyledons. Differential staining and light microscopy can be used to determine the state and distribution of protein, starch, lipid, and cell wall material. Figure 2.14 shows a bright field micrograph of raw bean stained with Acid Fuchsin and Toluidine Blue to differentiate protein bodies within the cells. Protein bodies are evident in the cell matrix between the starch granules and appear as small rounded structures bounded by denser-staining material. Protein bodies in the whole raw black bean (Figure 2.14) were evident in the continuous cell matrix between the starch granules and appear as small, rounded structures bounded by denser-staining material.[37] SEM examination of black beans, adzuki bean and other bean cotyledon cells.[12,13,21,38] shows that large (10–50 μm) spherical starch granules seems to be embedded in a protein matrix. Small (1–10 μm) spherical to oval individual protein bodies generally compose the protein matrix of black beans.[12,38]

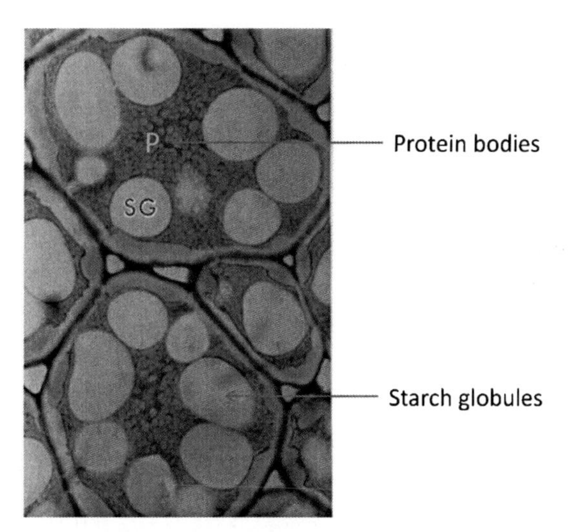

Figure 2.14 Bright field micrograph of raw black bean stained with Acid Fuchsin and Toluidine Blue O to differentiate protein bodies within the cells. Protein bodies are evident in the cell matrix between the starch granules and appear as small rounded structures bounded by denser-staining material. (Adapted from Wood *et al.*[37])

2.4. CHEMICAL CONSTITUENTS OF SEED COAT AND COTYLEDONS

The cotyledon is composed mainly of protein and carbohydrates, whereas the seed coat contains mainly fibres and phenolic compounds. The distribution of nutrients in different seed fractions, calculated as a percentage of the whole seed, shows that most of the protein, ether extract, phosphorus and iron are in the cotyledons, whereas 80–90% of crude fibre and 32–50% of calcium are in the seed coat. The seed coat of pulses is rich in minerals such as calcium, magnesium, iron, zinc, potassium and copper, as well as phytate, tannins and phenolics.[39] Pulse seed coat contains about 30–50% of the total calcium found in pulses[40] and the presence of calcium ions increases the stability and cohesion of cell walls due to their formation of insoluble complexes with pectin.[41] The presence of phenolic compounds governs the colour of the seed coat. The distribution of phenolics in pulses differs in the cotyledon and seed coats. Generally, non-flavonoid phenolic compounds are located in the cotyledon, whereas flavonoids are present in the seed coat. For example, seed coat extracts from pea exhibit antioxidant activity. This antioxidant activity is mainly due to the presence of various phenolic antioxidants such as tannins, flavonoids (flavone and flavonol glycosides) and some phenolic acids (benzoic, cinnamic acid and cinnamic acid derivatives).[42] The presence of these antioxidants in seed coat protects the grain from oxidative damage. Table 2.4 shows the proportion of phenolic compounds in the seed coat and the cotyledon of two varieties of lentils, Pardina and Castellana, grown in two different areas of the central region of Spain: Pardina 1 (Palencia), Pardina 2 (Salamanca), Castellana 1 (Salamanca) and Castellana 2 (León).

The proportion in the seed coat is low (8.2–11.4%) compared to cotyledon (88.6–91.8%). Catechins and procyanidins and, to a lesser extent, flavonols and flavones are the main contributors to the total phenolic content in these lentils, whereas the cotyledons represent a large proportion of the total grain weight, but provide only a small proportion of phenolic compounds, mainly cinnamic and benzoic compounds.[43] This indicates that the flavonoid compounds are found almost exclusively in the lentil seed coat while the non-flavonoid compounds can be found in the seed coat and the cotyledon.

Seed coat colour is an important character that determines the marketability of pulses. For example, in North America red and white varieties of dry beans are preferred, compared to Central America where red and black varieties are preferred and South America where red, yellow and white varieties are preferred. Seed coat colour is largely dependent on the presence of secondary metabolites such as tannins and anthocyanins. These pigments are located mainly in the seed coat and determine the overall hue and intensity of colour.[44] Seed coat colour is also an important quality index, along with size and shape. In many countries including Canada, USA and Australia, pulses are classified on the basis of market, industry and consumer standards. Research studies show that the variation in colour existing within pulses depends on the type and concentration of condensed tannins and anthocyanins. Seed coat colour changes during storage. The relationship between seed coat colour and the

Table 2.4 Phenolic compounds content (μg g^{-1}) and proportion (%) in the seed coat and cotyledon of two lentil varieties (Pardina and Castellana)

Compounds	*Pardina 1*	*Pardina 2*	*Castellana 1*	*Castellana 2*
Cotyledon (% of total grain weight)	88.6	90.3	91.8	91.7
Total benzoics	1.92 (8.64)	1.84 (8.60)	2.66 (34.89)	1.77(20.18)
Total cinnamics	17.89 (78.81)	18.62 (87.10)	4.63 (61.90)	6.02(68.65)
Total catechin	2.89 (12.73)	0.92 (4.3)	0.24 (3.21)	0.98(11.17)
Seed coat (% of total grain weight)	11.4	9.7	8.2	8.3
Total benzoic acids	28.39 (0.70)	32.92 (0.73)	34.70 (1.12)	37.53 (1.62)
Total cinnamic acids	19.51 (0.48)	11.78 (0.26)	20.29 (0.66)	29.45 (1.27)
Total catechins	1497.64 (36.98)	1530.11 (33.93)	1300.99 (42.00)	917.55 (39.61)
Total procyanidin dimers	2222.19 (54.88)	2305.67 (51.13)	1635.77 (52.81)	1239.05 (53.49)
Total procyanidin trimers	117.68 (0.66)	323.83 (1.78)	23.03 (1.60)	9.60 (1.82)
Total flavonols	117.68 (2.91)	323.83 (7.18)	23.03 (0.74)	9.60 (0.41)
Total flavones	136.63 (3.37)	224.83 (4.90)	33.12 (1.07)	41.29 (1.78)

Source: Adapted from Dueñas *et al.*[43]
Values in parentheses indicates proportion of phenolic compounds in percentage.

presence of tannins is well established. For example, brown cowpea seeds of contain more tannins than cream-coloured seeds,[45] and dark purple beans contain more tannins than yellow and other light-coloured beans.[46]

Seed coat cell walls contain less water-soluble polysaccharides and a higher cellulose and lignin content compared to cotyledon cell walls, which contain more water-soluble polysaccharides and less cellulose and lignin. In navy bean cultivars cellulose constitutes approximately one-third (30.61–34.55%) of the cotyledon cell walls and two-thirds of the seed coat cell walls (58.71–65.04%).[47] Similarly, lignin content is 2.5–3.0 times less in cotyledon cell walls compared to seed coat cell walls. The lignin and cellulose in seed coat cell walls govern its rigidity and contribute to the physical property of the whole grain.

2.5. CONSTITUENTS OF CELL WALLS

The cell walls of pulses contain numerous biopolymers, which govern the structure and textural properties of the grains. Some of the major biopolymers found in growing plant cell of monocots (cereals or grass family) and dicots (pulses or legume family) are listed in Table 2.5.

2.5.1. Lignins and Cellulose

Typically, a mature cell wall is made up of about 25–30% cellulose, 15–24% hemicellulose, 0.4–0.6% lignins and 28–41% pectin on a dry weight basis.

Table 2.5 Major polymers of the cell wall from a growing plant cell

Polymer	*Solubility in water*	*Prevalence in growing grasses*[c] *(%)*	*Prevalence in growing dicots*[c] *(%)*
Cellulose	Insoluble	30	30
Hemicelluloses	Soluble		
Xyloglucan	Soluble	4	25
Xylan	Soluble	30	5
Mixed glucans	Soluble	30	0
Pectin	Soluble		
Homogalacturonan	Soluble		15
Rhamnogalacturonan I	Soluble	5	15
Rhamnogalacturonan II	Soluble		5
Glycoproteins	Soluble		
Arabinogalactan proteins	Soluble	Variable	Variable
Extensin	Soluble	0.5	~ 5

[a]Solubility after extraction of polymer from the cell wall
[b]Approximate percentage on dry basis. (Growing primary walls contain ~65% water; mature cell walls have a lower water content).
[c]All pulses are dicots; cereals belong to the grass family and are monocots.
Source: Jackman and Stanley.[48]

Apart from cellulose, the other chemical constituents are fairly soluble in water. The obvious difference between chemical constituents of monocots and dicots is the presence of mixed glucans. These are rarely found in pulses, whereas cereals are good source of glucans (*e.g.* oats and barley are known sources of β-glucans). Glycoproteins are also present in pulses and constitute about 5–10% on a dry weight basis of the walls of dicotyledon cells, with carbohydrates constituting about 67% of the glycoproteins.[48] Extensins are uniformly distributed across the cell wall but do not occur in the middle lamella and forms cross-links with other cell wall polymers including pectin (Figure 2.15). These cross-links provide mechanical integrity of the wall.[48]

Variations in the chemical constituents of the cell wall and the structural integrity of the cell wall and middle lamella, along with the turgor pressure generated within cells by osmosis, govern the textural properties of pulses. The presence of pectin and formation of calcium in the plant cell wall of pulses is one of the reasons for large differences between the cooking times of pulses and cereals.

2.5.2. Pectic Substances

Pectic substances act as a cementing material for the cellulosic network, providing a structure to the primary cell walls and intercellular regions of higher plants, including pulses. The outer part of the cell wall is connected to the cell by the middle lamella, which provides a gel-like structure and serves as the intercellular bridge between the cells. Pectin is largely located in the middle

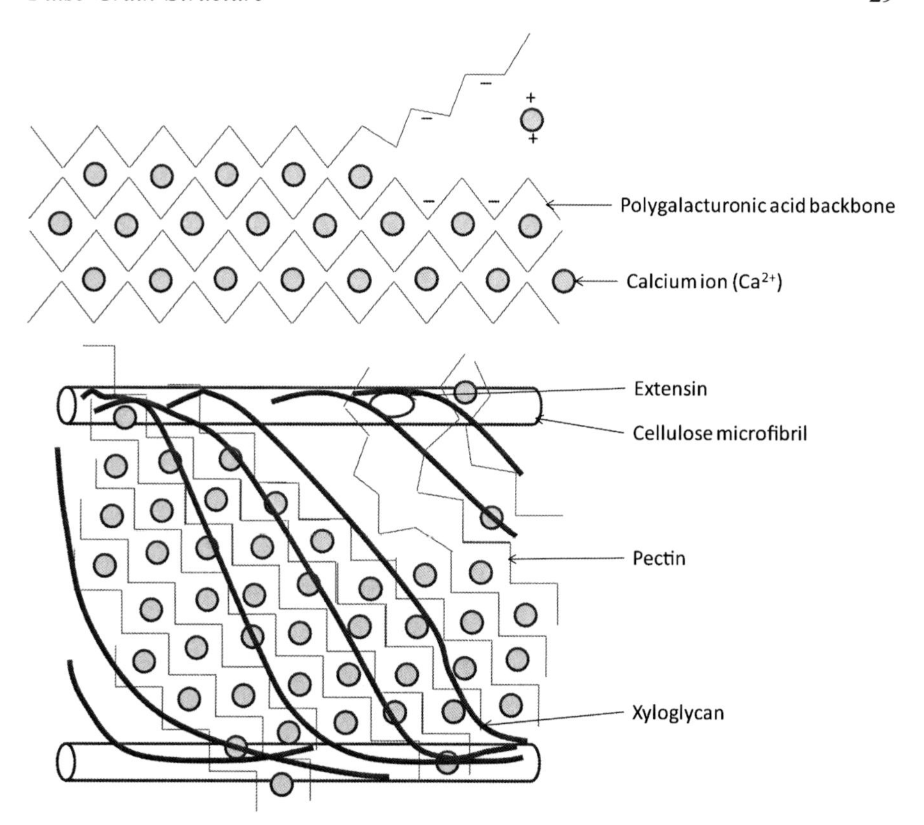

Figure 2.15 Diagrammatic representation of the primary cell wall of dicotyledonous plants. (Based on E. R. Morris *et al.*, *J. Mol. Biol.*, 1982, **155**, 507–516; F. P. C. Blamey, 2003, *Soil Sci. Plant Nutr.*, **49**, 775–783; N. C. Carpita and D. M. Gibeaut, *Plant J.*, 1993, **3**, 1–30.)

lamella, with a concentration varying from 10% to 30%.[53] The pectic acid in the middle lamella, along with calcium ions (Ca^{2+}), creates a network structure resulting in strong cohesiveness in the cell wall structure.[49] The cellulose microfibrils are embedded in a gel-like amorphous structure consist of a network of pectin and hemicellulose along with proteins.[50] Calcium can bind with unesterified pectins by forming a cross-bridge among the negatively charged carboxyl groups ($COOH^-$)of galacturonic acid.[51] Pectin also contains chemical moieties such as rhamnogalacturonan and homogalacturonan domains. The major component of the pectic substance is D-galacturonate, which is composed of methyl esters, pectin, the de-esterified pectic acid and its salts. Also included are pectate and some neutral polysaccharides lacking the galacturonan backbone.[52] Homogalacturonan is a linear polymer of (1⟶4)-α-linked galacturonic acid. Electrostatic interactions between calcium ions and carboxyl group of galacturonic acid make a sandwich-like arrangement between pectin monomers at specific sites.[41,53] The arrangement and distribution of calcium in pectin chain resembles an egg box, as shown in Figure 2.15 (Morris *et al.*).[69]

2.6. GRAIN CHARACTERISTICS AND WATER ABSORPTION

The raphe, micropyle and hilum are entry point for the diffusion of water. Differences in initial water uptake between pulses can be explained by examining the differences in the structure of the hilum, micropyle and raphe, and in the characteristics of the seed coat.[54,55] Water permeability is usually greatest in the hilum or micropyle areas, where seed coats are quite thin.[56] The hilum of pulses such as *Vicia* and *Phaseolus* species acts as hydroscopic point, which opens at low relative humidity; eventually the loss of moisture occurs and causes internal drying, whereas under high relative humidity or wet conditions these fissures are closed due to expansion of cells.[56] For example, the micropyle aperture of dry Great Northern beans (*Phaseolus vulgaris*) was observed to be >117 μm whereas, for seeds that had imbibed moisture it was 29 μm.[57] Similarly, Kyle and Randall[58] studied water diffusion at the hilum, micropyle and raphe in Great Northern and Mexican red beans (both of which are commercial market class cultivars of the common bean *Phaseolus vulgaris* L).

Chemical composition also limits grain expansion and water absorption during soaking or cooking. For example, the force required to puncture a bean is reduced by about 50% when the seed coats are removed from either soft or hard beans. Figure 2.16 shows the effect of cooking time on the puncture force for whole grains, and for cotyledons with and without a seed coat. Figure 2.16 also shows that the force required to puncture the grain depends largely on the presence of the seed coat. Water absorption by the grain is influenced by several factors, namely seed coat thickness, hilum size, seed volume, and

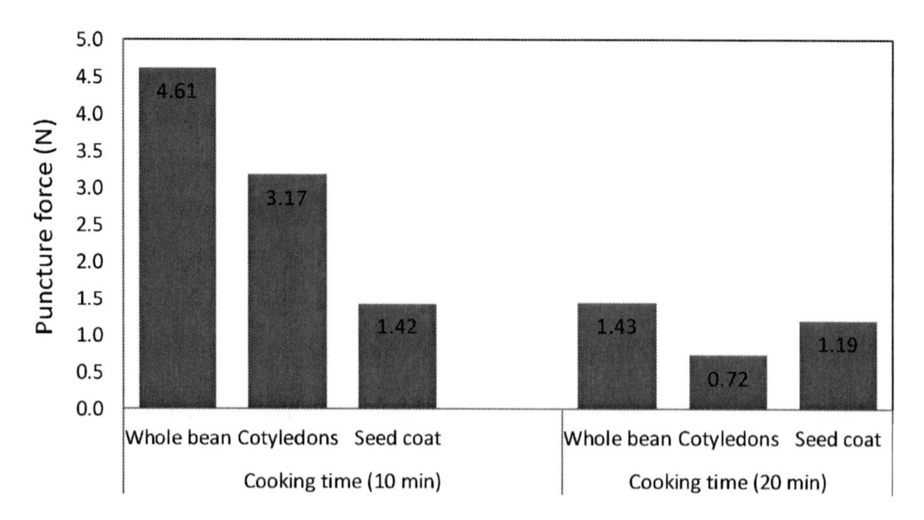

Figure 2.16 Effect of seed coat on the texture of cooked beans

colour.[20,59] The chemical constituents of seed coat, such as cellulose, hemicellulose, and lignin, also play an important role during water imbibition.

A basic understanding of calcium and pectin interaction in the plant cell wall is important in understanding the role of calcium salts (*e.g.* calcium chloride, $CaCl_2$) used in the bean canning industry to enhance the firmness of canned dry beans. The use of $CaCl_2$ results in the formation of calcium–pectin complexes, which may contribute to the firming of the seed coat and the turgidity of cell walls of the cotyledon tissue.[60–63] Pectic substances present in the cell wall result in calcium cross-links due to the formation of calcium pectate. Further, calcium cross-linkage retards water uptake and increases the firmness of the seed coat.[53,64]

REFERENCES

1. T. Otto, B. K. Baik and Z. Czuchajowska, *Cereal Chem.*, 1997, **74**, 445–451.
2. J. M. Aguilera, *J. Food Eng.*, 2005, **67**, 3–11.
3. J. Parada and J. M. Aguilera, *J. Food Sci.*, 2007, **72**, R21–R32.
4. I. Norton, S. Moore and P. Fryer, *Obesity Rev.*, 2007, **8**, 83–88.
5. J. M. Aguilera, P. J. Lillford and H. Watzke, in *Food Materials Science*, Springer, New York, 2008, pp. 3–10.
6. B. James, *Trends Food Sci. Technol.*, 2009, **20**, 114–124.
7. P. M. Falcone, A. Baiano, A. Conte, L. Mancini, G. Tromba, F. Zanini and M. A. Del Nobile, in *Advances in Food and Nutrition Research*, ed. L. T. Steve, Academic Press, New York, 2006, vol. 51, pp. 205–263.
8. C. W. Beninger, G. L. Hosfield and M. G. Nair, *Hortscience*, 1998, **33**, 328–329.
9. J. Ehlers and A. Hall, *Field Crop Res.*, 1997, **53**, 187–204.
10. S. Sefa-Dedeh and D. Stanley, *Cereal Chem.*, 1979, **56**, 379.
11. S. Bhattacharya and H. V. Narasimha, *Int. J. Food Sci. Technol.*, 2005, **40**, 213–221.
12. J. D. Berrios, B. G. Swanson and W. A. Cheong, *Scanning*, 1998, **20**, 410–417.
13. E. Joseph, S. Crites and B. Swanson, *Food Struct.*, 1993, **12**, 155–162.
14. E. Garcia, F. Lajolo and B. Swanson, *Food Struct.*, 1993, **12**, 147–154.
15. F. Boesewinkel and F. Bouman, *Seed Development and Germination*, Springer Verlag, Heidelberg, 1995, pp. 1–24.
16. J. T. Van Dongen, A. M. H. Ammerlaan, M. Wouterlood, A. C. Van Aelst and A. C. Borstlap, *Ann. Bot.*, 2003, **91**, 729–737.
17. B. G. Swanson, J. S. Hughes and H. P. Rasmussen, *Food Microstruct.*, 1985, **4**, 115–124.
18. A. Chilukuri and B. G. Swanson, *Food Struct.*, 1991, **10**, 131–135.
19. A. Engquist and B. G. Swanson, *Food Struct.*, 1992, **11**, 171–179.
20. G. N. Agbo, G. L. Hosfield, M. A. Uebersax and K. Klomparens, *Food Microstruct.*, 1987, **6**, 91–102.
21. S. Sefa-Dedeh and D. W. Stanley, *Food Technol. (Chicago)*, 1979, **33**, 77–83.

22. F. H. Souza and J. Marcos-Filho, *Rev. Brasil. Bot.*, 2001, **24**, 365–375.
23. E. J. H. Corner, *Phytomorphology*, 1951, **1**, 117–150.
24. A. M. Yousif, J. Kato and H. C. Deeth, *Food Rev. Int.*, 2007, **23**, 1–33.
25. C. Rochat and J. P. Boutin, *Physiol. Plant.*, 1992, **85**, 567–572.
26. S. S. Miller, L. A. A. Bowman, M. Gijzen and B. L. A. Miki, *Ann. Bot.*, 1999, **84**, 297–304.
27. F. Javadi and H. Yamaguchi, *Theoret. Appl. Genet.*, 2004, **109**, 317–322.
28. E. L. Hove, *J. Sci. Food. Agr.*, 1974, **25**, 851–859.
29. T. Otto, B. K. Baik and Z. Czuchajowska, *Cereal Chem.*, 1997, **74**, 141–146.
30. H. Meier and J. Reid, in *Encyclopedia of Plant Physiology*, ed. F. A. Loewus and W. Tanner, Springer Verlag, Heidelberg, 1982, Vol. 13, pp. 418–471.
31. M. L. Parker, *Protoplasma*, 1984, **120**, 233–241.
32. H. Nadelmann, *Jb. Wiss. Bot.* 1890, **21**, 609–691.
33. D. Stanley, *Food Res. Int.*, 1992, **25**, 187–192.
34. D. Stanley, *HortTechnology*, 1992, **2**, 370–378.
35. M. S. Buckeridge, H. Pessoa dos Santos and M. A. S. Tiné, *Plant Physiol. Biochem.*, 2000, **38**, 141–156.
36. K. Saio and M. Monma, *Food Struct.*, 1993, 12.
37. D. F. Wood, C. Venet and J. D. Berrios, *Scanning*, 1998, **20**, 335–344.
38. J. S. Hughes and B. G. Swanson, *Food Microstruct.*, 1985, **4**, 183–189.
39. O. C. ADEBOOYE and V. Singh, *J. Food Quality*, 2007, **30**, 1101–1120.
40. S. Deshpande and S. Damodaran, *Adv. Cereal Sci. Technol.*, 1990, 10.
41. M. Demarty, C. Morvan and M. Thellier, *Plant Cell Environ.*, 1984, **7**, 441–448.
42. A. Troszyńska, I. Estrella, M. L. López-Amóres and T. Hernández, *LWT – Food Sci. Technol.*, 2002, **35**, 158–164.
43. M. Dueñas, T. Hernández and I. Estrella, *Eur. Food Res. Technol.*, 2002, **215**, 478–483.
44. C. W. Beninger and G. L. Hosfield, *J. Agr. Food. Chem.*, 1998, **46**, 2906–2910.
45. O. Tibe and J. Amarteifio, *J. Appl.Sci. Environ. Manage.*, 2010, 11.
46. A. M. Díaz, G. V. Caldas and M. W. Blair, *Food Res. Int.*, 2010, **43**, 595–601.
47. N. Srisuma, S. Ruengsakulrach, M. A. Uebersax, M. R. Bennink and R. Hammerschmidt, *J. Agr. Food. Chem.*, 1991, **39**, 855–858.
48. R. L. Jackman and D. W. Stanley, *Trends Food Sci. Technol.*, 1995, **6**, 187–194.
49. J. P. Van Buren, ed, *Function of Pectin in Plant Tissue Structure and Firmness*, Academic Press, San Diego, CA, 1991.
50. R. G. S. Bidwell, ed, *Plant Physiology*, Macmillan, New York, 1974.
51. N. M. Steele, M. C. McCann and K. Roberts, *Plant Physiol.*, 1997, **114**, 373–381.
52. S. Banerjee and K. Chauhan, Seed Science and Technology (Netherlands), 1981, **9**, 819–822.

53. X. J. Wu, R. James and A. K. Anderson, *J. Food Process. Preserv.*, 2005, **29**, 63–74.
54. B. Klamczynska, Z. Czuchajowska and B. K. Baik, *Food Sci. Technol. Int.*, 2001, **7**, 73–81.
55. S. S. Deshpande, M. Cheryan, D. Salunkhe and B. S. Luh, *Crit. Rev. Food Sci. Nutr.*, 1986, **24**, 401–449.
56. S. Maekawa and W. J. Carpenter, *Hortscience*, 1991, **26**, 129–132.
57. J. H. Kyle, *A study of the relationship of the micropyle opening to hard seeds in Great Northern beans*. University of Idaho, 1955
58. J. H. Kyle and T. Randall, presented in part at *Proc Amer. Soc. Hort. Sci*, 1963, **83**, 461–475.
59. D. W. Stanley and J. M. Aguilera, *J. Food Biochem.*, 1985, **9**, 277–323.
60. W. Moscoso, M. Bourne and L. Hood, *J. Food Sci.*, 1984, **49**, 1577–1583.
61. M. A. Uebersax and S. Ruengsakulrach, *Abstr. Pap. Am. Chem. Soc*, 1988, 196, 75, Agfd.
62. P. Balasubramanian, A. Slinkard, R. Tyler and A. Vandenberg, *J. Sci. Food. Agr.*, 2000, **80**, 732–738.
63. A. De Lange and M. Labuschagne, *J. Sci. Food. Agr.*, 2001, **81**, 30–35.
64. M. A. Uebersax and S. Ruengsakulrach, *ACS Symp. Ser.*, 1989, **405**, 111–124.
65. A. K. Cavanagh, *A review of some aspects of the germination of Acacias*. Proceedings of the Royal Society of Victoria, 1980, **91**, 161–180.
66. J. C. Clements, A. V. Zvyagin, K. K. M. B. D. Silva, T. Wanner, D. D. Sampson, W. A. Cowling, *Plant Breeding*, 2004, **123** (3), 266–270.
67. W. M. Lush and L.T. Evans, *Field Crop Res.*, 1980, **3**, 267–286.
68. N. G. Marinos, The cytological environment of the developing embryo. *Protoplasma*, 1970, **70**, 261–279.
69. E. R. Morris, D. A. Powell, M. J. Gidley, and D. A. Rees, *J. Mol. Biol.* 1982, **155**, 507–516.

CHAPTER 3

Major Constituents of Pulses

3.1. INTRODUCTION

Pulse grains are mainly composed of carbohydrates, proteins, lipids, mineral matter and water. In addition to these, the grains contain small quantities of vitamins, enzymes and phytochemicals, which are important for human health. Pulses are consumed alone or as part of a primarily cereal-based diet. They are consumed as whole grain after suitable processing (*e.g.* baked beans), as split pulses (*e.g.* dhals) or as an ingredient (*e.g.* flour). Pulses are the main source of protein in a primarily vegetarian Indian diet. Besides proteins, pulses are also good sources of carbohydrates, vitamins, minerals, ω-3 (n-3) fatty acids and dietary fibre or non-starch polysaccharides (NSP). Pulses not only add variety to human diet, but also serve as an economical source of supplementary proteins for a large human population in developing countries. Due to the high cost and limited availability of animal proteins in the developing countries, there is an increasing interest in the use of pulses as potential sources of low-cost dietary proteins for food use.[1] Thus, pulses are recognised as important source of food protein, calories and other nutrients.[2] In many regions of the world, pulse grains are the unique supply of protein in the diet and are regarded as versatile functional ingredients or biologically active components, rather than as essential nutrients. Generally, pulses provide 3.4–3.6 kcal per gram. The chemical composition of pulses can be studied either as the classical proximate composition or the ultimate composition, as outlined in Figure 3.1. This chapter discusses the major constituents of pulses.

3.2. PROXIMATE COMPOSITION

Nutritionally pulses are high in complex carbohydrate, protein and dietary fibres; low in fat and sodium; and contain no cholesterol. In fact, cholesterol is not a constituent of foods of plant origin. There is wide variation in macro-

Pulse Chemistry and Technology
Brijesh K. Tiwari and Narpinder Singh

Published by the Royal Society of Chemistry, www.rsc.org

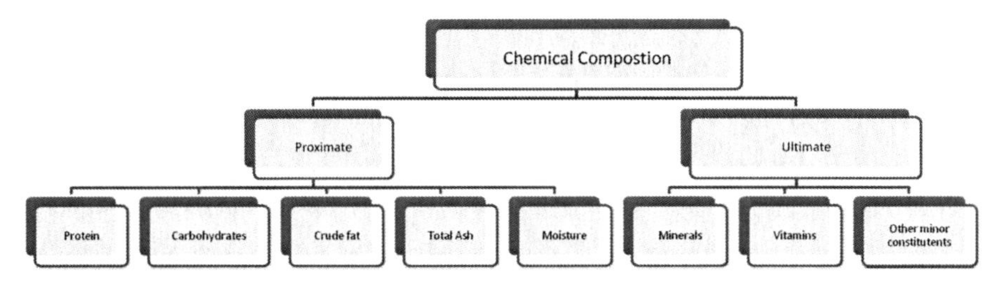

Figure 3.1 Classification of chemical compositions of pulses

and micronutrients in pulses, both inter- and intraspecies, mainly due to varietal characteristics, location, environmental and agronomic factors. In general, dried pulses contain approximately 10% moisture, 21–25% crude protein, 1–1.5% lipids, 60–65% carbohydrates, and 2.5–4% ash; an exception is chickpea, which contains about 4–5% lipids. In some pulses, such as lupin, up to 45–50% protein has been reported. Pulse proteins lack the essential amino acid methionine; however, rich in lysine, which allows these amino acids to be made available if the diet is supplemented with cereals. The proximate composition of various pulse grains is summarised in Table 3.1. Based on the proximate composition, most pulses provide 329–364 kcal/100 g. The amount of energy (measured in calories) contained in pulses can be determined from the proximate composition of pulses, *i.e.* the content of carbohydrates, proteins and fats. Carbohydrates are present in abundant amounts and constitute between 41.71% and 64.11% of the grain. Pulse carbohydrates contain monosaccharides, disaccharides, oligosaccharides and polysaccharides, but the quantity of monosaccharides is small (<1%). Starch is the primary storage carbohydrate, and constitutes a major fraction of the total carbohydrates of almost all the pulses. Pulses are recognised as an important source of protein, which ranges from 19.3% to 25.8%. Most pulses have a low fat content (0.85–6.0%), except winged bean, which has an exceptionally high fat content of 16.32% and provides high energy content (429 kcal/100 g). Ash content represents the inorganic matter and ranges between 2.48% and 4.30%. Dietary fibre ranges from 12.7% to 30.5%, and lentils contain the highest proportion. Most pulses have higher dietary fibre than wheat, maize or rice (both brown and milled).

3.2.1. Carbohydrates

Carbohydrates are chemical compounds that contain oxygen, hydrogen and carbon atoms. These may also contain some elements such as sulfur or nitrogen. Carbohydrates can be classified by the number of constituent sugar units: monosaccharides (such as glucose and fructose), disaccharides (such as sucrose and lactose), oligosaccharides and polysaccharides (such as starch and cellulose). Monosaccharides, of varying chain lengths, are represented by the general chemical formula $C_n(H_2O)_n$. Pulses have lower levels of carbohydrates than cereals, which generally contain 70–80% carbohydrates. The primary

Table 3.1 Proximate composition of different pulse grains*

Pulses	*Moisture content*	*Protein content*	*Lipid content*	*Ash content*	*Sugars content*	*Carbohydrate content*	*Dietary fibre*	*Energy (kcal/100 g)*
Kidney beans	11.75[a]	23.58[a], 22.1[h]	0.83[a],	3.83[a], 3.80[h]	2.23[a],	60.1[a], 54.3[h]	24.9[a], 24.9[j]	333[a]
Chickpeas	11.53[a]	19.30[a], 19.29[c], 17.1[d]	6.04[a], 7.11[c], 5.3[d]	2.48[a] 2.80[c]	10.70[a]	60.65[a] 60.9[d]	17.4[a], 17.4[j]	36[a]
Lentils	10.40[a],11.20[h]	25.80[a], 26.1[g]	1.06[a], 3.1[f]	2.67[a], 2.44[f]	2.03[a]	60.08[a], 56.4[g]	30.5[a],	353[a]
Mung beans	9.05[a], 9.75[b]	23.86[a], 27.5[b]	1.15[a], 1.85[b]	3.32[a], 3.76[b]	6.60[a]	62.62[a], 62.3[b]	16.3[a]	347[a]
Mungo beans	10.80[a],8.20[k]	25.21[a], 26.22[k]	1.64[a], 2.94[k]	3.36[a], 3.12[k]	–	58.99[a]	18.3[a]	341[a]
Pigeon peas	10.59[a]	21.70[a]	1.49[a]	3.45[a]	–	62.78[a]	15.0[a]	343[a]
Peas	11.27[a],0.6[e]	24.55[a], 19.3[e]	1.16[a], 0.84[e]	2.65[a], 2.7[e]	8.0[a]	60.37[a], 63.2[e]	25.5[a]	341[a]
Adzuki beans	13.44[a]	19.87[a]	0.53[a]	3.26[a]	–	62.90[a]	12.7[a]	329[a]
Black beans	11.02[a]	21.60[a], 23.6[i]	1.42[a], 0.45[i]	3.60[a], 3.51[i]	2.12[a]	62.36[a] , 67[i]	15.2[a]	341[a]
Lima beans	10.17[a]	21.46[a]	0.69[a]	4.30[a]	8.50[a]	63.38[a]	19.0[a]	338[a]
Navy beans	12.10[a]	22.33[a]	1.50[a]	3.32[a]	3.88[a]	60.75[a]	24.4[a]	337[a]
Great northern beans	10.70[a]	21.86[a]	1.14[a]	3.93[a]	2.26[a]	62.37[a]	20.2[a]	339[a]
French beans	10.77[a]	18.81[a]	2.02[a]	4.30[a]	–	64.11[a]	25.2[a]	343[a]
Winged beans	8.34[a]	29.65[a]	16.32[a]	3.98[a]	–	41.71[a]	–	409[a]
Hyacinth beans	9.38[a]	23.90[a]	1.69[a]	4.29[a]	–	60.76[a]	–	344[a]
White beans	11.32[a]	23.36[a]	0.85[a]	4.20[a]	2.11[a]	60.27[a]	15.2[a]	333[a]
Horse gram	6.8[l]	22.5[l]	1.4[l]	2.7[l]	–	66.6[l]	14.8[l]	–
Cowpeas	11.05[a], 7.4[l],	23.85[a], 24.1[l]	2.07[a], 2.3[l]	3.39[a], 2.9[l]	–	59.64[a], 63.3[l]	15.2a, 14.4[l]	343[a]

*All parameters expressed in g/100 g except for energy (kcal/100 g)

[a]USDA National Nutrient Database for Standard Reference;[35] [b]Mubarak;[36] [c]Rossi *et al.*;[37] [d]Singh and Singh;[38] [e]Alonso *et al.*;[39] [f]Singh *et al.*;[40] [g]Iqbal *et al.*;[41] [h]Costa *et al.*;[42] [i]Bravo *et al.*;[43] [j]Boye *et al.*;[44] [k]Tresina and Mohan;[45] [l]Sreerama *et al.*[46]

storage carbohydrate of pulses is starch, which constitutes a major fraction of the total carbohydrates of almost all the pulses. Some of the minor carbohydrates are discussed in Chapter 4.

3.2.1.1. Dietary Fibre

Dietary fibre is composed of plant cell wall constituents such as cellulose, hemicelluloses, polysaccharides and lignin. It is usually defined as the edible parts of plants or analogous carbohydrates that are resistant to digestion and absorption in the human small intestine, with complete or partial fermentation in the large intestine. Lentils, kidney bean, peas, navy beans, great northern bean and French bean have a dietary fibre content higher than 20%. The pulses with small grain size have higher total dietary fibre content than those with large grain size: small grains have a greater surface to volume ratio than large grains and thus a greater proportion of seed coat to cotyledon.[3] Dietary fibre is classified as either insoluble or soluble (Figure 3.2). The principal components of dietary fibre are NSP and lignin. Insoluble fibre includes cellulose, lignin and some hemicelluloses; soluble fibre includes pectins, gums and some other hemicellulose fractions. The constituents of dietary fibre, irrespective of source are often argued. However, American Association of Cereal Chemist (AACC) report published in 2001 outlines the constituents of dietary fibre as listed in Figure 3.3.

Pulses contain appreciable amounts of dietary fibre (12.7–30.5 g/100 g). Lentils, peas, French beans, kidney beans and navy beans are excellent source of dietary fibre (24.4–30 g/100 g). Lentils have relatively high dietary fibre content of 30 g/100 g whereas adzuki beans have less (12.7 g/100 g). Mung beans, pigeon peas, black beans, white beans and cowpeas have dietary fibre content in a narrow range (15–16.3 g/100 g). Cellulose is the major component of fibre in smooth and wrinkled peas, red kidney beans, navy beans, pinto beans, pink beans and blackeye peas, while in other pulses (lupin seeds, lentil, broad beans, red gram, black gram), hemicellulose is the major component. Table 3.2 shows the total insoluble and soluble dietary fibre content of some raw pulses, including seed coat (hull), obtained using enzymatic–gravimetric methods.[4] Pulses whole grains contains higher level of dietary fibres compared to flour or splits obtained after dehulling or milling. This is due to the higher level of fibre content in the seed coat than in the cotyledons. The seed coat also

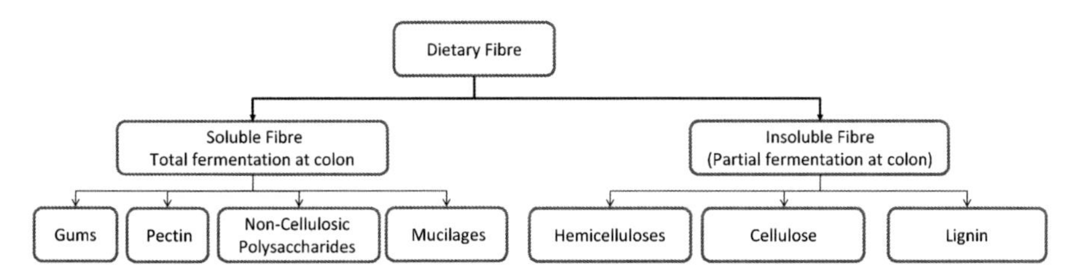

Figure 3.2 Dietary fibre of pulses

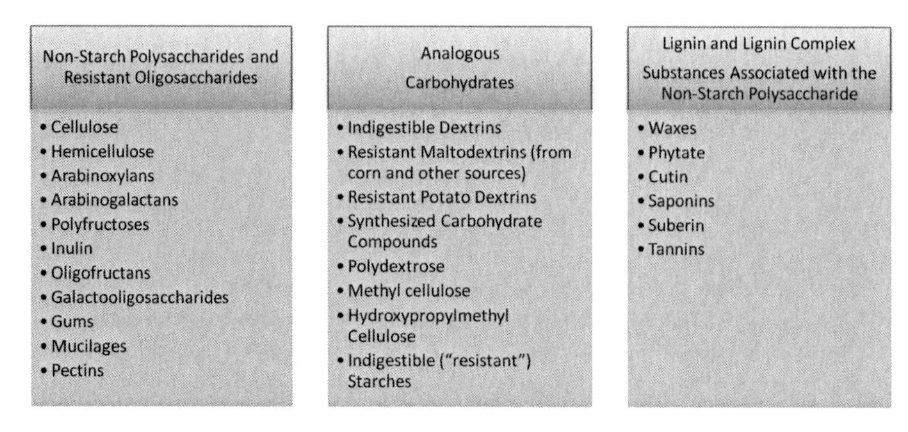

Figure 3.3 Constituents of dietary fibre (AACC, 2011)

contains a higher level of insoluble fibre fractions than the cotyledons. Often pulse fibre is also described as "inner" or "outer" fibre depending on the location (Figure 3.4). "Inner fibre" refers to cotyledon fibres, consisting of cell wall polysaccharides of varying degrees of solubility whereas "outer" refers to seed coat or hull fibres, containing largely water-insoluble polysaccharides as well as some pectin.[4,5]

3.2.1.2. Non-Starch Polysaccharides

The NSP consist of various polysaccharides including cellulose, hemicellulose, pectin, gum and mucilage. The health benefits of dietary fibres are well established in literature. Analysis of dietary fibre content in pulses should include the measurement of all components such as soluble polysaccharides, non-cellulosic polysaccharides, cellulose, and lignin and the constituent sugars of the soluble and non-cellulosic polysaccharides.[6] Soluble polysaccharides such as pectins, gums and some hemicelluloses are physiologically important

Table 3.2 Dietary fibre contents (total, insoluble, and soluble) of raw pulse seeds (g/100 g)

Pulses	*Total dietary fibre*	*Soluble fibre*	*Insoluble fibre*	*References*
Dry beans (*P. vulgaris*)	23–32	20–28	3–6	Perez-Hidalgo *et al.*; [47]Granito *et al.*; [48] Kutos *et al.*[49]
Chickpeas (*C. arietinum*)	18–22	10–18	4–8	Perez-Hidalgo *et al.*;[47] Rincon *et al.*;[50] Dalgetty and Baik[51]
Lentils (*L. culinaris*)	18–20	11–17	2–7	Perez-Hidalgo *et al.*;[47] Dalgetty and Baik[51]
Dry peas (*P. sativum*)	14–26	10–15	2–9	Borowska *et al.*; [52] Dalgetty and Baik;[51] Martín-Cabrejas *et al.*;[53] Wang *et al.*[54]

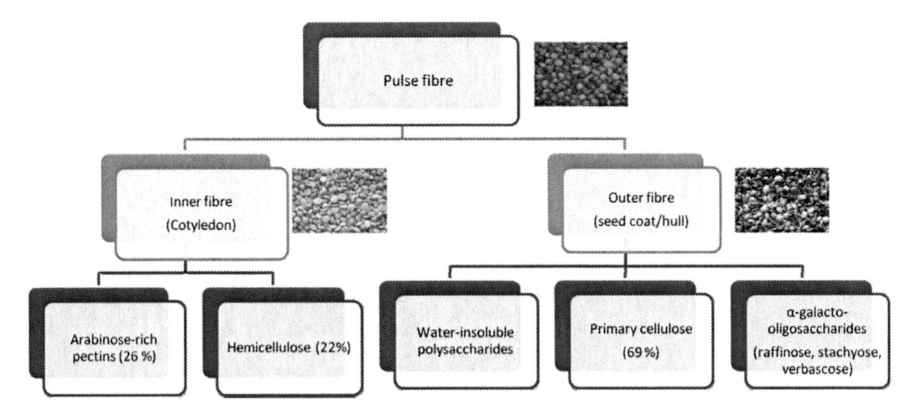

Figure 3.4 Distribution of pulse fibre within a grain

and constitute a considerable fraction of the total dietary fibre. Both soluble and insoluble fibre fractions contain a variety of sugars including arabinose, xylose, glucose, uronic acid and other sugars such as rhamnose, fucose, mannose and galactose. Of these, arabinose is the principal sugar of the soluble and insoluble fractions of pulse fibres (Table 3.3). The predominance of uronic acid reflects the presence of pectin content. Pectic substances are complex polysaccharides present in the primary cell walls and middle lamellae of all plant tissues where they play important role as hydrating agents and cementing material.[7] Pectin is the methylated ester of polygalacturonic acid, which consists of chains of 300–1000 galacturonic acid molecules joined by α-(1→4) linkages. Sugar compositional analysis of pectic substances present in the hull of field bean (*Dolichos lablab*), cowpeas (*Vigna sinensis*), and peas (*Pisum sativum*) indicates the presence of arabinose, rhamnose and xylose in addition to high amounts of uronic acid. Cellulose cannot be digested by the human digestive system, and hence passes through the gut without being absorbed. Cellulose and most types of hemicelluloses are known as insoluble fibres, whereas pectin, gum and mucilage are all soluble fibre and easily dissolve or swell when mixed with water. Insoluble fibre has a bulking action, is metabolically inert and is not fermented in the human body. It absorbs water as it passes through the digestive tract, and eases defecation. The soluble fibre is readily fermented in the colon into gases and physiologically active by-products. Pulses contain a higher amount of insoluble fibre than soluble fibre.

3.2.1.3. *Cellulose*

Cellulose is a polymer of 1000 or more glucose molecules bound together by β–1→4 linkages (Table 3.4) that cannot be broken down (*i.e.* digested) by endogenous enzymes present in the human gut. Symbiotic gastrointestinal anaerobes can release the energy of cellulose through microbial fermentation and the production of volatile fatty acids, although digestion may not be complete. The principal volatile fatty acids are acetic, propionic, and butyric

Table 3.3 Sugar content of dietary fibre

	Soluble fibre					*Insoluble fibre*				
	Pentoses		*Hexoses*			*Pentoses*		*Hexoses*		
Pulses	*Arabinose*	*Xylose*	*Glucose*	*Other*	*Uronic acid*	*Arabinose*	*Xylose*	*Glucose*	*Other*	*Uronic acid*
Garbanzo beans	44	1	28	13	15	60	6	5	12	17
Green beans	15	2	5	51	28	16	11	3	45	25
Kidney beans	43	16	10	12	18	41	18	3	21	17
Lima beans	49	4	14	16	17	37	18	4	27	14
Navy beans	41	8	8	23	21	44	17	3	17	18
Pinto beans	41	7	7	28	16	44	14	2	20	20
White beans	39	9	16	17	19	49	12	7	13	18
Lentils	21	8	35	15	21	48	14	6	15	17
Blackeye peas	45	5	19	13	18	40	22	3	19	16
Green peas	54	5	6	12	24	46	12	6	8	28

Others include rhamnose, fucose, mannose and galactose

Table 3.4 Chemical classification of dietary fibre

	Chemical constituents		
Fibre	*Main chain*	*Side chain*	*Descriptions*
CELLULOSE	Glucose	None	Main structural component of plant cell wall. Insoluble in concentrated alkali; soluble in concentrated acid
NON-CELLULOSE			
• Hemicellulose	Xylose, mannose, galactose, glucose	Arabinose, galactose, glucuronic acid	Cell wall polysaccharides which contain backbone of 1–4 linked pyranoside sugars. Vary in degree of branching and uronic acid content. Soluble in dilute alkali
• Pectic substances	Galacturonic acid	Rhamnose, arabinose, xylose, fucose	Components of primary cell wall and middle lamella vary in methyl ester content. Generally water soluble and gel-forming
• Mucilages	Galactose–mannose, glucose–mannose, arabinose–xylose, galacturonic acid–rhamnose	Galactose	Synthesised by plant secretory cells; prevent desiccation of seed endosperm. Food industry use, hydrophilic, stabiliser, *e.g.* guar
• Gums	Galactose, glucuronic acid–mannose, galacturonic acid–rhamnose	Xylose, fucose, galactose	Secreted at site of plant injury by specialised secretory cells. Food and pharmaceutical use, *e.g.* Karaya gum. Derived from algae and seaweed. Vary in uronic acid content and presence of sulphate groups. Food and pharmaceutical use, *e.g.* carrageenan, agar
• Lignin	Sinapyl alcohol, coniferyl alcohol, *p*-coumaryl alcohol	3-dimensional structure	Non-carbohydrate cell wall component. Complex cross-linked phenyl propane polymer. Insoluble in 72% H_2SO_4. Resists bacterial degradation

acids (in descending order of usual abundance) plus small and variable amounts of isobutyric, valeric and isovaleric acids.

3.2.1.4. *Hemicelluloses*

Hemicelluloses are a heterogeneous group of single and mixed polymers of arabinose, xylose, mannose, glucose, fucose, galactose and glucuronic acid, closely associated with cellulose and lignin. Examples are xyloglucans, xylans, glucomannans, arabinoxylans and glucuronoxylans. Most hemicelluloses are water insoluble, but a few will form a viscous or gel-like solution. Like cellulose, hemicelluloses cannot be digested by endogenous mammalian enzymes, although these can be partially hydrolysed under acidic conditions in the stomach. Anaerobic fermentation is required for effective use of the energy that hemicelluloses contain, and the products of fermentation are essentially the same as those of cellulose.

3.2.1.5. *Soluble Non-Starch Polysaccharides*

Soluble NSPs do not completely dissolve in water but swell to form a gel or a gummy solution. Nevertheless, they are referred to as soluble fibre. The NSPs are non-structural polysaccharides, some of which serve as plant energy reserves, and are not as digestible as starch, although fermented quite completely by ruminal and intestinal bacteria. Included among the non-starch plant energy reserves are fructans, mannans and galactans. Fructans (also known as fructosans, and including inulin) are polymers of fructose that are stored in grasses and composites,[8] as well as in parts of some food crops.[9] Fructans are broken down in an acid environment,[8] so passage through the acid stomach may result in release of some fructose monomers that can be absorbed in the small intestine.[9] Mannans are polymers of mannose found in seaweeds, algae, nuts and seeds.[10,11] Galactans are polymers of galactose found in seaweeds and algae, and with pectin in fruit pulps.[12]

3.2.1.6. *Pectic Substances*

Pectic substances are heterogeneous polysaccharides, characteristically containing galacturonic acid, rhamnose, galactose, and arabinose bound by α–1→4 linkages,[13] which cannot be digested by endogenous mammalian enzymes. Like cellulose and hemicelluloses, however, pectic substances can undergo partial or complete microbial fermentation or degradation.[14,15]

3.2.1.7. *Gums and Mucilages*

Gums and mucilages are related to pectic substances, with which these constituents share the property of swelling in water. Gums include β-glucans (soluble relatives of cellulose found in cereals, especially oats and barley), xyloglucans and mannoglucans (Table 3.4). Gums appear in plant exudates mainly as a result of physiological or pathological disturbances that induce breakdown of cell walls and cell contents. Mucilages occur in the seed coats of various pulses.

3.2.2. Protein

Proteins are complex, organic compounds composed of many amino acids linked together through peptide bonds. Proteins obtained from pulses are not only a source of constructive compounds and energy as the constituent amino acids; they also play bioactive roles and/or can be the precursors of biologically active peptides with various physiological functions. Pulses contain relatively high amount of protein and are an indispensable source of dietary protein. The quality of a protein is judged on the basis of the amino acids it contains. Traditionally, nutritionists classify amino acids as essential or non-essential, but Figure 3.5 classifies amino acids into three categories.[16–18] The essential amino acids cannot be synthesised by the human body and must be supplied by food, whereas non-essential amino acids can be synthesised in human body. The third category, the conditional amino acids, includes the amino acids that are usually not essential except during stress or illness. The National Research Council[19]classified amino acids as dispensable, non-dispensable, completely dispensable and precursors of completely dispensable amino acids.[19] Conditionally indispensable amino acids are defined as requiring a dietary source when endogenous synthesis of amino acids cannot meet the metabolic need. Dispensable amino acids can be synthesised in the body either from other amino acids or from complex nitrogenous metabolites, whereas conditionally dispensable amino acids can be synthesised under special pathophysiological conditions.[19,20]

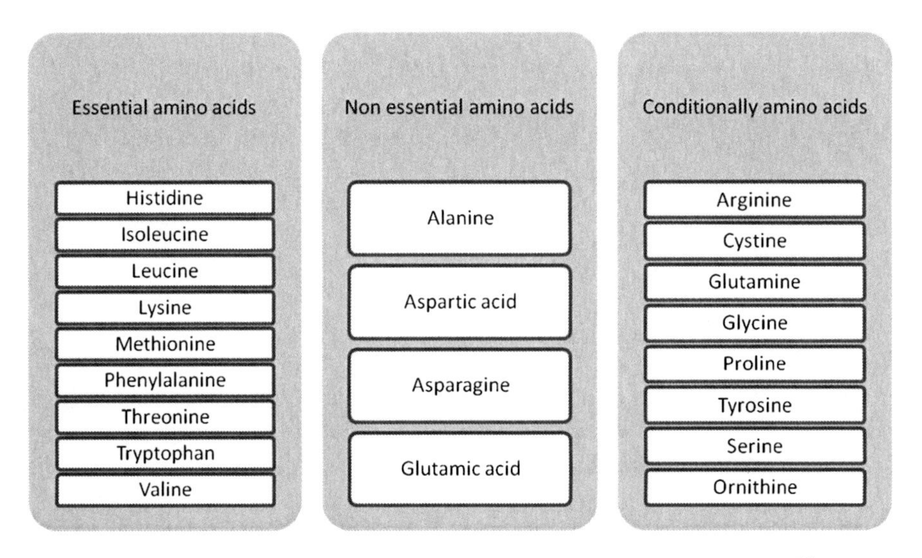

Figure 3.5 Nutritional classification of proteins (based on Trumbo *et al.*;[16] Escott-Stump;[17] ADMA.[18])

Proteins are located in the cotyledon and the embryonic axis of the seed, with small amounts present in the seed coat. The cotyledons constitute the major portion of the pulse grain and hence make the highest contribution to protein content. Pulses are the major source of food protein for vegetarians. In many regions of the world, pulse grains are the major source of dietary protein and often represent a necessary supplement to other protein sources. Pulses, like cereals, have wide variations in protein content mainly due to genetic, environmental and agronomic factors. The outer layer of the cotyledon in some pulses (chickpeas, garden peas, pigeon pea and black gram) has higher protein content than the inner part of the cotyledon. The uneven distribution of protein in the cotyledon of peas and chickpeas is well reported.[21,22]

Osborne[23] classified proteins into five classes on the basis of solubility. Albumin and globulin, which are soluble in water and dilute salt solution, are generally classified as soluble protein. The most abundant storage proteins in pulse grains are the globulins, generally classified as 7S and 11S globulins according to their sedimentation coefficients (S). The 7S and 11S globulins of pea are known as vicilin and legumin, respectively. The 7S and 11S proteins are oligomeric, usually trimmers and hexamers respectively.[24] Both the 7S and 11S globulins liberate their constituent subunits under dissociating conditions. These polypeptide chains are naturally heterogeneous in both size and charge levels.[25–27] Pulse storage proteins are relatively low in sulfur-containing amino acids (methionine), but the amount of another essential amino acid, lysine, is much greater than in cereal grains. Therefore, pulse and cereal proteins are nutritionally complementary with respect to lysine and sulfur amino acids. The degree of mutual supplementation may also depend, however, on the contents of the second limiting amino acids, *i.e.* threonine in cereals and tryptophan in pulses. These essential amino acid deficiencies have traditionally been overcome by integrating pulse-based dishes with cereal foods (pasta, rice, bread, *etc.*). However, amino acid composition only represents the potential quality of a protein food; the bioavailability is critical for the supply of amino acids in the diet. Pulse grain proteins have a lower overall nutritional quality than animal proteins and this is attributed to low content of sulfur-containing amino acids, the compact proteolysis-resistant structure of the native seed proteins,[28] and the presence of antinutritional compounds, which may affect digestibility of proteins and other components.[29,30]

3.2.2.1. Amino Acid Composition of Pulses

The amino acid composition of different pulses is shown Tables 3.5 and 3.6. Glutamic acid and aspartic acid are the major amino acids in all the pulses. Pulses are low in methionine, cystine and tryptophan. The tryptophan and cystine content in different pulses (except winged beans) ranges between 0.185 and 0.294 g/100 g and 0.184–0.375 g/100 g, respectively. Winged beans have a higher tryptophan and cystine content (0.762 g and 0.545 g/100 g respectively) compared to other pulses. The methionine content in different pulses ranges between 0.191 and 0.367 g/100 g. Pulses are rich in lysine, which varies between

Table 3.5 Essential amino acids composition of different pulses (g/100 g) at given moisture content (g/100 g)

Pulse	*Moisture content*	*Histidine*	*Isoleucine*	*Leucine*	*Lysine*	*Methionine*	*Phenylalanine*	*Threonine*	*Tryptophan*	*Valine*
Kidney beans	11.75[a]	0.656, 0.30[b]	1.041, 0.48[b]	1.882, 0.95[b]	1.618, 0.76[b]	0.355, 0.12[b]	1.275, 0.65[b]	0.992, 0.47[b]	0.279	1.233, 0.57[b]
Chickpeas	11.53[a]	0.531	0.828	1.374	1.291	0.253	1.034	0.716	0.185	0.809
Lentils	10.40[a]	0.727	1.116	1.871	1.802	0.220	1.273	0.924	0.232	1.281
Mung beans	9.05[a]	0.695	1.008	1.847	1.664	0.286	1.443	0.782	0.260	1.237
Mungo beans	10.80[a]	0.706	1.287	2.089	1.674	0.367	1.473	0.875	0.263	1.416
Pigeon peas	10.59[a]	0.774	0.785	1.549	1.521	0.243	1.858	0.767	0.212	0.937
Peas	11.27[a]	0.597	1.014	1.760	1.772	0.251	1.132	0.872	0.275	1.159
Adzuki beans	13.44[a]	0.524	0.791	1.668	1.497	0.210	1.052	0.674	0.191	1.023
Black beans	11.02[a]	0.601	0.954	1.725	1.483	0.325	1.168	0.909	0.256	1.130
Lima beans	10.17[a]	0.656	1.129	1.850	1.438	0.271	1.236	0.927	0.254	1.291
Navy beans	12.10[a]	0.507	0.952	1.723	1.280	0.278	1.158	0.711	0.247	1.241
Great northern beans	10.70[a]	0.608	0.965	1.745	1.500	0.329	1.182	0.920	0.259	1.144
French beans	10.77[a]	0.524	0.831	1.502	1.291	0.283	1.017	0.792	0.223	0.984
Winged beans	8.34[a]	0.790	1.468	2.497	2.136	0.356	1.429	1.179	0.762	1.530
Hyacinth beans	9.38[a]	0.684	1.143	2.026	1.632	0.191	1.204	0.925	0.199	1.239
White beans	11.32[a]	0.650	1.031	1.865	1.603	0.351	1.263	0.983	0.277	1.222
Cowpea	11.05a	0.740	0.969	1.828	1.614	0.340	1.393	0.908	0.294	1.137

[a]USDA National Nutrient Database for Standard Reference;[35] [b]Montoya *et al.*[55]

Table 3.6 Non-essential and conditional amino acids composition of different pulses (g/100 g) at given moisture content (g/100 g)

Pulse	*Moisture content*	*Alanine*	*Aspartic acid*	*Arginine*	*Glutamic acid*	*Cystine*	*Glycine*	*Proline*	*Tyrosine*	*Serine*
Kidney beans	11.75[a]	0.988, 0.51[b]	2.852, 1.20[b]	1.460, 0.30[b]	3.595, 1.40[b]	0.256	0.920, 0.56[b]	1.000, 0.38[b]	0.664, 0.40[b]	1.282, 0.68[b]
Chickpeas	11.53[a]	0.828	2.270	1.819	3.375	0.259	0.803	0.797	0.479	0.973
Lentils	10.40[a]	1.078	2.855	1.994	4.002	0.338	1.050	1.078	0.689	1.190
Mung beans	9.05[a]	1.050	2.756	1.672	4.264	0.210	0.954	1.095	0.714	1.176
Pigeon peas	10.59[a]	0.972	2.146	1.299	5.031	0.250	0.802	0.955	0.538	1.028
Peas	11.27[a]	1.080	2.896	2.188	4.196	0.373	1.092	1.014	0.711	1.080
Adzuki beans	13.44[a]	1.160	2.355	1.284	3.099	0.184	0.756	0.874	0591	0.976
Black beans	11.02[a]	0.905	2.613	1.337	3.294	0.235	0.843	0.916	0.608	1.175
Lima beans	10.17[a]	1.095	2.767	1.315	3.038	0.237	0.906	0.975	0.759	1.428
Navy beans	12.10[a]	0.908	2.598	1.020	3.098	0.187	0.801	1.117	0.484	1.180
Great northern beans	10.70[a]	0.916	2.644	1.353	3.333	0.238	0.853	0.927	0.615	1.189
French beans	10.77[a]	0789	2.276	1.165	2.869	0.205	0.734	0.798	0.530	1.023
Winged beans	8.34[a]	1.040	3.187	1.886	4.010	0.545	1.140	1.924	1.457	1.235
Hyacinth beans	9.38[a]	1.067	2.821	1.755	3.880	0.279	1.028	1.162	0.853	1.315
White beans	11.32[a]	0.979	2.825	1.446	3.561	0.254	0.912	0.990	0.658	1.271
Cowpeas	11.05a	1.088	2.881	1.088	4.518	0.263	0.985	1.072	0.771	1.194

[a]USDA National Nutrient Database for Standard Reference[35]; [b]Montoya *et al.*[55]

1.291 and 2.136 g/100 g. The highest lysine content of 2.137 g/100 g has been reported for winged beans. Pulses have a higher lysine content than cereals, which are considered relatively deficient in this amino acid. The valine content amongst the pulses varies between 0.809 and 1.53 mg/100, which is higher than short-, medium-and long-grain rice.

3.2.3. Crude Fat

Lipids in pulse grains are mainly found in the embryo axis. The cotyledons and seed coat contain only small amount of lipids. The total lipid content of different pulses vary depending upon the variety, origin, location, climatic, seasonal, environmental conditions and type of soil on which they are grown. Pulse grains contain 2–21% fat with the most important fatty acid components being linoleic (21–53%) and linolenic (4–22%) acids. Common beans are an important source of free unsaturated fatty acids, accounting for 61.1% of total fatty acids. The major fatty acids are palmitic (16:0), oleic (18:1), and linoleic (18:2). Major fatty acids found in common pulses are listed in Table 3.7.

Linolenic acid (18:3) is the major unsaturated fatty acid among the unsaturated fatty acids, and is present to the extent of 43.1% in common bean.[31] Chickpeas have the highest monounsaturated fatty acid (MUFA) content (34.2 g/100 g) and kidney beans have the highest content of

Table 3.7 Major fatty acids of some common pulses

Pulses	*Major fatty acid*	*Reference*
Chickpeas (*Cicer arietinum*)	Oleic acid, linoleic acid	Exler *et al.*[56]
Great northern beans (*Phaseolus vulgaris*)	Oleic acid, linoleic acid	Sessa and Rackis [57]
Garden peas (*Pisum sativum*)	Oleic acid, linoleic acid	Exler *et al.*[56]
Horse gram (*Dolichos biflorus*)	Linoleic acid	Mahadevappa and Raina[58]
Red gram (*Cajanus cajan*)	Linoleic acid	Mahadevappa and Raina[58]
Cowpeas (*Vigna unguiculata*)	Linoleic acid, linolenic acid	Mahadevappa and Raina[58]
Kidney beans (*Phaselous vulgaris.* L)	Linoleic acid, linolenic acid	Korytnyk and Metzler[59]
Pinto beans (*Phaseolus vulgaris.*L)	Linoleic acid, linolenic acid	Korytnyk and Metzler[59]
Black gram (*Vigna mungo.*L)	Linolenic acid	Mahadevappa and Raina[58]
Green gram (*Phaseolus radiatus*)	Oleic acid, linoleic acid, linolenic acid	Artman[60]
Lentils (*Lens culinaris*)	Oleic acid, linoleic acid	Exler *et al.*[56]
Blackeye peas (*Vigna sinensis*)	Linoleic acid, linolenic acid	Korytnyk and Metzler[59]
Lathyrus beans (*Lathyrus sativus*)	Linoleic acid	Choudhury and Rahman[61]
Lima beans (*Phaseolus lunatus*)	Linoleic acid	Exler *et al.*[56]

polyunsaturated fatty acids (PUFA) (71.1 g/100 g).[32] Adzuki beans contain 2.2% fat which consists mainly of phospholipids (63.5%), triglycerides (21.2%), steryl esters (7.5%), hydrocarbons (5.1%), diacylglycerols (1.3%), free fatty acids (0.9%) and other minor components. Adzuki bean lipids have linoleic, palmitic and linolenic acid contents of 45%, 25%, and 21%, respectively, of the total lipids.[33] The lipid contents of grains of eight *Vicia* species were reported to vary between 2.5% and 3.9% with palmitic (7–23%) and stearic (15–35%) acids being the major fatty acids.[34] Neutral lipids are the predominant class in most pulses, but phospholipids and glycolipids are also present in appreciable amounts. The distribution of neutral lipids, phospholipids, and glycolipids in pulses varies from 32% to 51%, 23% to 38% and 8% to 12%, respectively (Table 3.8). Phospholipids and glycolipids are essential components of the seed membrane. Neutral lipids consist primarily of triacylglycerols, accompanied by smaller proportions of di- and monoacylglycerols, free fatty acids, sterols and sterol esters.

Pulse lipids are susceptible to oxidation, which leads to several changes in lipids as well as other food constituents. The oxidation of lipids can be classified as enzymic or non-enzymic, both of which ultimately lead to the production of hydroperoxides, which in turn undergo decomposition to yield several products such as aldehydes, ketones, esters, acids, *etc.* These secondary products, particularly aldehydes which are reactive, produce numerous undesirable compounds by reacting with other components present in the system. Reactions of particular importance are the reactions between the decomposition products and proteins and amino acids, carbohydrates, minerals and vitamins.

The importance of key nutritional constituents of pulses is unquestionable because of the presence of macro- and micronutrients. Pulses supply significant amounts of proteins and they also contain up to 60% carbohydrates (mainly

Table 3.8 Neural lipids, phospholipids and glycolipids content of different pulses

Pulses	*Neutral lipids (%)*	*Phospholipids (%)*	*Glycolipids (%)*
Cowpeas	46.88	36.82	8.98
Pigeon peas	51.05	34.37	8.84
Field beans	47.62	36.07	9.82
Red gram	51.05	34.37	8.84
Horse gram	47.74	23.73	11.41
Black gram	49.76	34.84	9.17
Pinto beans	45.21	28.08	–
Lima beans	40.6	24.83	–
Broad beans	35.67	29.49	–
Blackeye beans	41.72	29.14	–
Garden peas	35.02	37.74	–
Great northern beans	32	34	–
Small red beans	40.4	34.11	–

starch). In addition, pulses are also a good source of other major and minor compounds (polyphenols, vitamins, minerals) which have important metabolic and/or physiological effects.

REFERENCES

1. N. Wang, M. J. Lewis, J. G. Brennam and A. Westby, *Food Chem.*, 1997, **58**, 59–68.
2. J. K. Chavan and D. K. Salunkhe, *Crit. Rev. Food Sci. Nutr.*, 1986, **25**, 107–158.
3. N. Wang, J. K. Daun and L. J. Malcolmson, *J. Sci. Food Agric.*, 2003, **83**, 1228–1237.
4. S. M. Tosh and S. Yada, *Food Res. Int.*, 2010, **43**, 450–460.
5. F. Meuser, in *Advanced Dietary Fibre Technology*, ed. B. V. McCleary and, L. Prosky, ,Blackwell Science, Oxford, 2001, pp. 248–269.
6. J. Anderson and S. Bridges, *Am. J. Clin. Nutr.*, 1988, **47**, 440–447.
7. G. Mualikrishna and R. N. Tharanathan, *Food Chem.*, 1994, **50**, 87–89.
8. Smith, D. *Removing and Analyzing Total Nonstructuralcarbohydrates from Plant Tissue*. College Agr. and Life Sci. Res. Rep. # 41, Univ. Wisconsin, Madison, 1969.
9. M. Ernst and W. Feldheim, *Angew. Bot.*, 2000, ***74***, 5–9.
10. M. S. Buckeridge, H. P. Santos and M. A. S. Tiné, *Plant Physiol. Biochem.*, 2000, **38**, 141–156.
11. A. Sachslehner, G. Foidl, N. Foidl, G. Gubitz and D. Haltrich, *J. Biotechnol.*, 2000, **80**, 127–134.
12. A. Femenia, M. Garcı?a-Conesa, S. Simal and C. Rosselló, *Carbohydr. Polym.*, 1998, **35**, 169–177.
13. L. Taiz and E. Zeiger, *Plant Physiology*, 2nd ed. Sinauer Associates, Sunderland, MA, 1998.
14. J. H. Cummings, D. A. T. Southgate, W. J. Branch, H. S. Wiggins, D. J. A. Houston and D. Jenkins, *Br. J. Nutr.*, 1979, **41**, 477–485.
15. B. J. H. Stevens, R. R. Selvendran, C. E. Bayliss and R. Turner, *J. Sci. Food Agric.*, 1988, **44**, 151–166.
16. P. Trumbo, S. Schlicker, A. A. Yates and M. Poos, *J. Am. Diet. Assoc.*, 2002, **102**, 1621–1630.
17. Escott-Stump S, ed. *Nutrition and Diagnosis-Related Care*, 6th ed. Lippincott Williams and Wilkins, Philadelphia, PA, 2008.
18. *A.D.A.M. Medical Encyclopedia*, Available online at http://www.ncbi.nlm.-nih.gov/pubmedhealth/PMH0002886/.
19. National Research Council. Protein and amino acids, Chapter 10 in *Dietary Reference Intakes for Energy, Carbohydrate, Fiber, Fat, Fatty Acids, Cholesterol, Protein, and Amino Acids (Macronutrients)*. The National Academies Press, Washington, DC, 2005, pp. 589–768.
20. S. Laidlaw and J. Kopple, *Am. J. Clin. Nutr.*, 1987, **46**, 593–605.

21. R. Kosson, Z. Czuchajowska and Y. Pomeranz, *J. Agric. Food Chem.*, 1994, **42**, 96–99.
22. T. Otto, T. B. K. Baik and Z. Czuchajowska, *Cereal Chem.*, 1997, **74**, 445–451.
23. Osborne, T. B. *Proteins of the Wheat Kernel.* Publ. 84, Carnegie Inst., Washington, DC, 1907, pp. 1–119.
24. M. Duranti and C. Gius, *Field Crops Res.*, 1997, 53, 31–45.
25. J. W. S. Brown, F. A. Bliss and T. C. Hall, *Theoret. Appl. Genet.*, 1981, **60**, 251–258.
26. M. Tucci, R. Capparelli, A. Costa and R. Rao, *Theoret. Appl. Genet.*, 1991, **81**, 50–58.
27. C. Horstmann, B. Schlesier, A. Otto, S. Kostka and K. Muentz, *Theoret. Appl. Genet.*, 1993, **86**, 867–74.
28. S. S. Deshpande and S. S. Nielsen, *J. Food Sci.*, 1987, **52**, 1326–1329.
29. R. Bressani, L. G. Elias and J. E. Braham, *J. Plant Foods*, 1982, **4**, 43–55.
30. T. L. Aw and B. G. Swanson, *J. Food Sci.*, 1985, **50**, 67–71.
31. E. R. Grela and K. D. Günther, *Anim. Feed Sci. Technol.*, 1995, **52**, 325–331.
32. E. Ryan, K. Galvin, T. O'Connor, A. Maguire and N. O'Brien, *Qual. Plant. – Plant Foods Hum. Nutr.*, 2007, **62**, 85–91.
33. H. Yoshida, Y. Tomiyama, N. Yoshida, M. Saiki and Y. Mizushina, *J. Am. Oil Chem. Soc.*, 2008, **87**, 535–541.
34. N. Akpinar, A. M. Akpinar and S. Turkoglu, *Food Chem.*, 2001, **74**, 449–453.
35. *USDA National Nutrient Database for Standard Reference*, 2011. Available online at: http://www.ars.usda.gov/Services/docs.htm?docid=8964 (Access date: 25/12/2011).
36. A. E. Mubarak, *Food Chem.*, 2005, **89**, 489–495.
37. M. Rossi, I. Germondari and P. Casini, *J. Agric. Food Chem.*, 1984, **32**, 811–814.
38. U. Singh and B. Singh, *Econ. Bot.*, 1992, **46**, 310–321.
39. R. Alonso, A. Aguirre and F. Marzo. *Food Chem.*, 2000, **68**, 159–165.
40. S. Singh, H. D. Singh and K. G. Sikka, *Cereal Chem.*, 1968, **45**, 13–18.
41. A. Iqbal, I. A. Khalil, N. Ateeq and M. S. Khan, *Food Chem.*, 2006, **97**, 331–335.
42. G. E. de Almeida Costa, K. D. S. Queiroz–Monici, S. M. P. M. Reis and A. C. de Oliveira, *Food Chem.*, 2006, **94**, 327–330.
43. L. Bravo, P. Siddhuraju and F. S. Calixto, *Food Chem.*, 1999, **64**, 185–192.
44. J. I. Boye, S. Aksay, S. Roufik, S. Ribéreau, M. Mondor, E. Farnworth and S. H. Rajamohamed, *Food Res. Int.*, 2010, **43**, 537–546.
45. P. S. Tresina, V. R. Mohan, *Int. J. Food Sci. Technol.*, 2011, **46**, 1739–1746.
46. Y. N. Sreerama, V. B. Sashikala, V. M. Pratape, and V. Singh, *Food Chem.*, 2012, **131**, 462–468.
47. M. A. Perez-Hidalgo, E. Guerra-Hernandez and B. Garcia-Villanova, *J. Food Compos. Anal.*, 1997, **10**, 66–72.

48. M. Granito, J. Frias, R. Doblado, M. Guerra, M. Champ and M. C. Vidal–Valverde, *Eur. Food Res. Technol.*, 2002, **214**, 226–231.
49. T. Kutos, T. Golob, M. Kac and A. Plestenjak, *Food Chem.*, 2003, **80**, 231–235.
50. F. Rincon, B. Martinez and M. V. Ibanez, *J. Sci. Food Agric.*, 1998, **78**, 382–388.
51. D. D. Dalgetty and B. K. Baik, *Cereal Chem.*, 2003, **80**, 310–315.
52. J. Borowska, R. Zadernowski, J. Borowski and W. K. Swiecicki, *Plant Breed. Seed Sci.*, 1998, **42**, 75–85.
53. M. A. Martín-Cabrejas, N. Ariza, R. Esteban, E. Mollá, K. Waldron and F. J. López-Andréu, *J. Agric. Food Chem.*, 2003, **51**, 1254–1259.
54. N. Wang, D. W. Hatcher and E. J. Gawalko, *Food Chem.*, 2008, **111**, 132–138.
55. C. A. Montoya, J. P. Lalles, S. Beebe and P. Leterme, *Food Res. Int.*, 2010, **43**, 443–449.
56. J. Exler, R. M. Avera and J. L. Weithrauch, *J. Am. Diet. Assoc.*, 1977, **71**, 412–415.
57. D. J. Sessa and J. J. Rackis., *J. Am. Oil Chem. Soc.*, 1977, **54**, 468.
58. V. G. Mahadevappa and P. L. Raina, *J. Am. Oil Chem. Soc.*, 1978, **55**, 648.
59. W. Korytnk and E. A. Metzler, *J. Sci. Food Agric.*, 1963, **14**, 841–844.
60. N. R. Artman, *Adv. Lipid Res.*, 1969, **7**, 245–330.
61. K. Choudhury and M. M. Rahman, *J. Sci. Food Agric.*, 1973, **24**, 471–473.

CHAPTER 4

Minor Constituents of Pulses

4.1. INTRODUCTION

The nutritional value of pulses is unquestionable due to the presence of macro- and micronutrients. Pulses supply significant amounts of proteins and also contain up to 60% carbohydrates (mainly starch). In addition to proteins, carbohydrates, lipids, mineral matter, and water, pulse grains contain small quantities of vitamins, enzymes, and phytochemicals which are important for human health. As well as being a source of macronutrients and minerals, pulses also contain plant secondary metabolites that are increasingly being recognised for their potential benefits to human health.[1] A number of bioactive substances, including enzyme inhibitors, lectins, phytates, oligosaccharides, and phenolic compounds, that have a role in human metabolism have been identified.[2] Bioactive compounds present in pulses have both negative and positive impacts on human health. This chapter discusses the ultimate composition of pulses including minerals, vitamins, and other minor constituents.

4.2. MINERALS

The fact that metal ions regulate an array of physiological mechanisms with considerable specificity and selectivity, as components of enzymes and other molecular complexes, underpins to the importance of minerals in human diet, *per se*. The mineral content of pulses is summarised in Table 4.1. Pulses are good source of phosphorus (P), potassium (K), calcium (Ca) and magnesium (Mg). The mineral content of pulses varies in the range 81–440 mg /100 g for calcium, 115–283 mg /100 g for magnesium, 304–451 mg /100 g for phosphorus and 955–1724 mg /100 g for potassium. Whole pulse grains have a higher mineral content than decorticated grains. Most pulses have a higher calcium content than wheat (around 32–33 mg/100 g), yellow corn (7 mg/100 g) or

Pulse Chemistry and Technology
Brijesh K. Tiwari and Narpinder Singh

Published by the Royal Society of Chemistry, www.rsc.org

Table 4.1 Mineral composition of different pulses (mg/100 g) at given moisture content (g/100 g)

Pulse	*Moisture*	*Ca*	*Fe*	*Mg*	*P*
Kidney beans	11.75[a]	143[a], 96.3[d]	8.20[a], 8.8[d]	140[a], 176[d]	407[a], 524[d]
Chickpeas	11.53[a]	105[a], 202[b], 107[d]	6.24[a], 10.2[b], 5.5[d]	115[a]	366[a]
Lentils	10.40[a]	56[a], 77[d], 57–76[i]	7.54[a], 7.6[d], 6–9[i]	122[a], 127[d], 99–109[i]	451[a], 456[d], 360–407[i]
Mung beans	9.05[a]	132[a], 244[f]	6.74[a], 9.4[f]	189[a], 174[f]	367[a], 146[f]
Mungo beans	10.80[a]	138[a], 199[e]	7.57[a], 7.5[e]	267[a]	379[a], 313[e]
Pigeon peas	10.59[a]	130[a]	5.23[a]	183[a]	367[a]
Peas	11.27[a]	55[a], 80[c]	4.43[a], 5.2[e]	115[a], 115[c]	366[a], 460[c]
Adzuki bean	13.44[a]	66[a]	4.98[a]	127[a]	381[a]
Black beans	11.02[a]	123[a], 123 g	5.02[a], 4.61 g	171[a], 184 g	352[a], 504 g
Lima beans	10.17[a]	81[a]	7.51[a]	224[a]	385[a]
Navy beans	12.10[a]	147[a], 155[d]	5.49[a], 7.6[d]	175[a], 203[d]	407[a], 618[d]
Great northern beans	10.70[a]	175[a], 156 g	5.47[a], 2.8 g	189[a], 169 g	447[a], 603 g
French beans	10.77[a]	186[a]	3.40[a]	188[a]	304[a]
Winged beans	8.34[a]	440[a]	13.44[a]	179[a]	451[a]
Hyacinth beans	9.38[a]	130[a]	5.10[a]	283[a]	372[a]
White beans	11.32[a]	240[a], 377[h]	10.44[a], 10.7[h]	190[a],289 [h]	301[a], 204[h]
Cowpeas	11.05[a]	85[a]	9.95[a]	333[a]	438[a]

[a]Data from USDA;[58] [b]Singh and Singh;[59] [c]Alonso;[40] [d]Wang and Daun;[60] [e]Singh et al.;[61] [f]*Arinathan* et al.[62]; [g]Campos-Vega et al.;[2] [h]ElMaki et al.;[63] [i]Wang et al.[64]

white short/medium/long grain rice (3–28 mg/100 g). However, the calcium content of any of the pulses, with the exception of winged beans, is not as high as that of finger millet (300–400 mg/100 g). Another important mineral, iron (Fe), which serves as a carrier of oxygen to the tissues and is an integral part of several enzyme systems, is also present in pulses (average 5–10 mg/100 g). However, the bioavailability of this iron remains questionable, as discussed in the section on phytates later in this chapter. Selenium (Se), a microelement that has been associated to maintenance of the cytochrome P450 system, DNA repair, enzyme activation and immune system function, is also found in most pulses; the selenium content in pulses varies between 3.1 and 12.8 mg/100 g. Greater northern beans, navy beans and white beans have a higher selenium content, 11.0–12.9 mg/100 g. Chickpea, lentils, mung beans, pigeon peas, winged beans and hyacinth beans have Se content around 8.2 mg/100 g, which is moderately higher than kidney beans, adzuki beans and black beans (3.2 mg/100 g). The recommended daily allowance for selenium is 50–200 mg, which implies that pulse grains (if considered as the sole source of selenium) can serve this requirement only when consumed in excess. Pulses also contain higher potassium than cereals. Among pulses, white beans have the highest potassium content of about 1795 mg/100 g. The zinc (Zn) content of pulses varies between 1.9 and 9.3 mg/100 g. Hyacinth beans and adzuki beans have an exceptionally high zinc content (5.04 and 9.8 mg/100 g, respectively), which is higher than in

Table 4.1 Extended.

K	Na	Zn	Cu	Mn	Se
1406[a], 1778[d]	24[a]	2.79[a], 3.7[d]	0.958[a], 1[d]	1.021[a], 1.4[d]	3.2[a]
875[a]	24[a]	3.43[a], 4.4[d]	0.847[a], 1[d]	2.204[a], 3.9[d]	8.2[a]
955[a], 965[d], 843–943[i]	6[a]	4.78[a], 3.9[d], 2.63–3.79[i]	0.519[a], 1[d], 0.79–1.13[i]	1.330[a], 1.6[d], 1–1.50[i]	8.3[a]
1246[a]	15[a], 29[f]	2.68[a], 1.1[f]	0.941[a], 0.34[f]	1.035[a], 0.6[f]	8.2[a]
983[a]	38[a]	3.35[a]	0.981[a]	1.527[a]	8.2[a]
1392[a]	17[a]	2.76[a]	1.057[a]	1.791[a]	8.2[a]
981[a]	15[a], 64[c]	3.01[a], 2.6[c]	0.866[a], 1.0[c]	1.391[a], 0.93[c]	1.6[a]
1254[a]	5[a]	5.04[a]	1.094[a]	1.730[a]	3.1[a]
1483[a], 1616 g	5[a]	3.65[a], 2.66 g	0.841[a], 0.24 g	1.060[a], 1.38 g	3.2[a]
1724[a]	18[a]	2.83[a]	0.740[a]	1.672[a]	7.2[a]
1185[a], 1704[d]	5[a],	3.65[a], 3.4[d]	0.834[a], 1.1[d]	1.418[a], 1.7[d]	11.0[a]
1387[a], 1758 g	14[a]	2.31[a], 1.85 g	0.837[a]	1.423[a], 1.35 g	12.9[a]
1316[a]	18[a]	1.90[a]	0.440[a]	1.200[a]	12.9[a]
977[a]	38[a]	4.48[a]	2.880[a]	3.721[a]	8.2[a]
1235[a]	21[a]	9.30[a]	1.335[a]	1.573[a]	8.2[a]
1795[a], 1291[h]	16[a], 24[h]	3.67[a]	0.984[a]	1.796[a]	12.8[a]
1375[a]	58[a]	6.11[a]	1.059[a]	1.544[a]	9.1[a]

most cereals. The role of zinc in almost all major biochemical pathways is well established. Further, it plays multiple roles in the perpetuation of genetic material, including transcription of DNA, translation of RNA and ultimately cell division. Difference in soil types and fertilizer applications affects the mineral composition of pulses. Dehulling and germination enhance the bioavailability of calcium. The presence of tannin and phytic acid in the seed coat acting as inhibitors has been demonstrated to reduce iron absorption by chelating the iron ion. The bioavailable iron is also inversely related to phytic acid, tannin and total dietary fibre content. The increase in calcium bioavailability after germination and dehulling of pulses is attributed to the reduction in phytic acid, tannin and dietary fibre content. Several studies have confirmed a negative correlation of phytic acid and dietary fibre contents of foods with percentage of calcium bioavailability.

4.3. VITAMINS

Pulses are good source of B vitamins (thiamine, riboflavin, niacin, pyridoxamine, pyridoxal and pyridoxine) and in general, poor sources of vitamins A and C, although chickpeas, kidney beans and mung beans are good sources of vitamin C. The vitamin content of different pulses is summarised in Table 4.2.

The thiamine content of pulses is high and varies from 0.4 to 1.0 mg/100 g, whereas the riboflavin content is low and varies from 0.1 to 0.3 mg/100 g. Most pulses (except winged beans and hyacinth beans) are rich source of folates.

Table 4.2 Vitamin composition of different pulses (mg/100 g) at given moisture content (g/100 g)

Pulse	*Moisture content*	*Vitamin C,*	*Thiamine*	*Riboflavin*	*Niacin*	*Pantothenic acid*	*Vitamin B_6*	*Folate*	*Choline*
Kidney beans	11.75[a]	4.0[5a]	0.529[a], 0.56[b]	0.219[a],0.16[b]	2.060[a], 1.10[b]	0.780[a], 0.44[b]	0.397[a],0.21[b]	394[a],	
Chickpeas	11.53[a]	4.0a	0.477a, 0.49b	0.212[a,] 0.26[b]	1.541[a], 1.22[b]	1.588[a],1.02[b]	0.535a,0.38[b]	557a, 299[b]	95.2a
Lentils	10.40[a]	1.7a, 0.71[b]	0.510[a], 0.29[b]	0.106[a],0.33[b]	1.495[a], 2.57[b]	0.348[a], 0.32[b]	0.403a, 0.23[b]	204a, 138[b]	
Mung beans	9.05[a]	4.8[a]	0.621a	0.233a	2.251a	1.910a	0.382a	625a	97.9a
Mungo beans	10.80a	0.0a	0.273a	0.254a	1.447a	0.906a	0.281a	216a	
Pigeon peas	10.59a	0.0a, 0.9[c]	0.643a	0.187a	2.965a	1.266a	0.283a	456a	
Peas	11.27a	1.8a	0.726a	0.215a	2.889a	1.758a	0.174a	274a	95.5a
Adzuki beans	13.44a	0.0a	0.455a	0.220a	2.630a	1.471a	0.351a	622a	
Black beans	11.02a	0.0a	0.900a	0.193a	1.955a	0.899a	0.286a	444a	66.4a
Lima beans	10.17a	0.0a	0.507a	0.202a	1.537a	1.355a	0.512a	395a	96.7a
Navy beans	12.10a	3.85[b]	0.775a, 0.58[b]	0.164a, 0.16[b]	2.188a, 1.31[b]	0.744a, 0.31[b]	0.428a, 0.21[b]	364a, 107.99[b]	87.4a,
Great northern beans	10.70a	5.3a	0.653a	0.237a	1.955a	1.098a	0.447a	482a	
French beans	10.77a	4.6a	0.535a	0.221a	2.083a	0.789a	0.401a	399a	
Winged beans	8.34a	0.0a	1.030a	0.450a	3.090a	0.795a	0.175a	45a	
Hyacinth beans	9.38a	0.0a	1.130a	0.136a	1.610a	1.237a	0.155a	23a	
White beans	11.32a	0.0a	0.437a	0.146a	0.479a	0.732a	0.318a	388a	66.2a
Cowpeas	11.05a	1.5a	0.680a	0.170a	2.795a	1.511[a]	0.361a	639a	

Source: [a]Data from USDA;[58] [b]Wang and Daun;[60] [c]Oboh.[65]

Deficiency of folates in human beings may result in megaloblastic anaemia. The folate content in different pulses varies between 23 and 622 mg/100 g. Adzuki beans, mung beans and chickpeas have folate content between 557 and 625 μg/100 g whereas winged beans and hyacinth beans have folate content between 23–45 μg/100 g. However, the folate present in pulses is not readily available to humans.

All pulses except peas, winged beans and hyacinth beans have significantly higher vitamin B_6 (pyridoxine) than milled rice. All the pulses, except white beans, have a higher niacin content than wheat flour.

Most of the pulses contain only small amount of carotene (pro-vitamin A), which varies with colour, variety, and species. Fresh garden peas have considerable amount of vitamin A. Peas contain greater amounts of α than $\beta + \gamma$-tocopherols (10.4 and 5.7 mg/100 g, respectively) and chickpeas contain similar levels of α- and $\beta + \gamma$-tocopherols (6.9 and 5.5 mg/100 g, respectively).[3]

4.4. MINOR CARBOHYDRATES

4.4.1. Monosaccharides

Monosaccharides are the simplest forms of sugars found in pulses. Pulses generally contain only small quantity of free monosaccharides, accounting for around 1% or even less.

4.4.2. Disaccharides

Disaccharides are composed of two monosaccharide units bound together by a glycosidic bond. The binding between the two sugars results in the loss of a hydrogen atom (H) from one molecule and a hydroxyl group (OH) from the other. Pulses contain a slightly higher amount of sucrose than monosaccharides and it varies from 1% to 4%.

4.4.3. Oligosaccharides

Oligosaccharides typically consist of two to nine monosaccharide units. Pulse oligosaccharides include raffinose, stachyose, verbascose and ciceritol (Table 4.3). Raffinose is a trisaccharide composed of fructose, glucose and galactose (Figure 4.1). Stachyose is a tetrasaccharide consisting of one molecule each of fructose and glucose and two molecules of galactose (Figure 4.2), while verbascose is a pentasaccharide consisting of one molecule each of fructose and glucose, and three units of galactose molecules (Figure 4.3). Ciceritol is a pinitol digalactoside (Figure 4.4) mainly present in chickpeas, lentils and white lupin[4]. Ajugose is a hexasaccharide (Figure 4.5) found in trace quantities in some pulses such as black gram, smooth peas and wrinkled peas. Figure 4.6 shows the structural relationship between different oligosaccharides.

Table 4.3 Oligosaccharide content of some pulses (% dry matter)

Pulses	*Raffinose*	*Stachyose*	*Verbascose*	*Ciceritol*
Adzuki beanss	0.25	2.80–4.06	–	–
Black gram	Trace–0.90	0.90–2.30	3.40–3.50	–
Green gram	0.30–1.10	1.65–2.50	2.10–3.80	–
Navy beans	0.41–0.53	2.59–3.26	0.00–0.13	Traces
Pinto bean	0.43–0.63	2.95–2.97	0.00–0.15	Traces
Lupins	0·5–1·1	0·9–7·4	0·6–3·4	0.65*
Field peas	0·3–1·6	1·3–5·5	1·6–4·2	–
Lentils	0·3–1·0	1·7–3·1	0·6–3·1	1.6*
Kidney bean	<0·05–0·93	0·5–4·1	0·06–4·0	Traces
Chickpeas	0·4–1·2	2·0–3·6	0·6–4·2	2.8*
Faba beans*	0.20*	0.85*	3.05*	–

*Quemener and Brillouet.[4]

Pulses contain around 3–7% of oligosaccharides and their concentrations and type vary in different pulses. Smooth peas, wrinkled peas, black gram and red gram contain higher amounts of total oligosaccharides than most pulses. Verbascose is the major oligosaccharide present in black gram, Bengal gram, red gram, mung beans and faba beans, while stachyose is the major oligosaccharide in smooth and wrinkled peas, great northern beans, California small white beans, red kidney beans, navy beans, pinto beans, pink beans, black eye beans, Bengal gram, lentils, cowpeas and lupin seeds.[5] Oligosacharides contain α-galactosidic bonds and are often known as flatus-producing sugars. The human body lacks the enzyme (α-galactosidase) required to break these bonds; as a result these sugars remain undigested in the human intestine. Anaerobic fermentation of the undigested sugars in the gut results in production of H_2, CO_2 and traces of CH_4.[6] These gases cause abdominal discomfort due to a flatus effect, and the consumption of large amounts of disaccharides in the diet sometimes leads to diarrhoea. Oligosaccharides of lentils and chickpeas are mainly composed of raffinose,

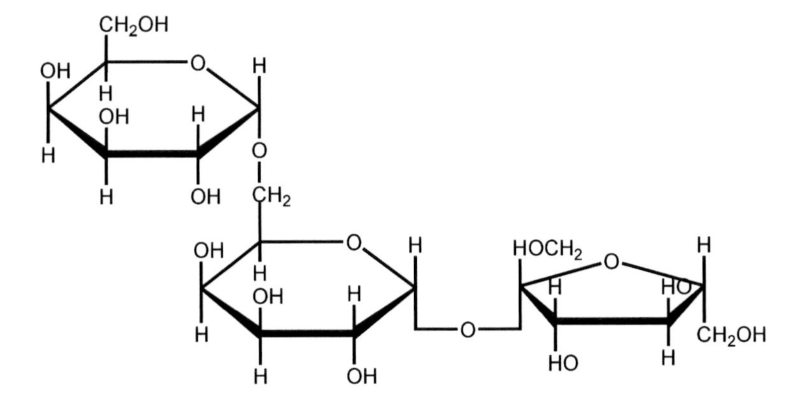

Figure 4.1 Structure of raffinose

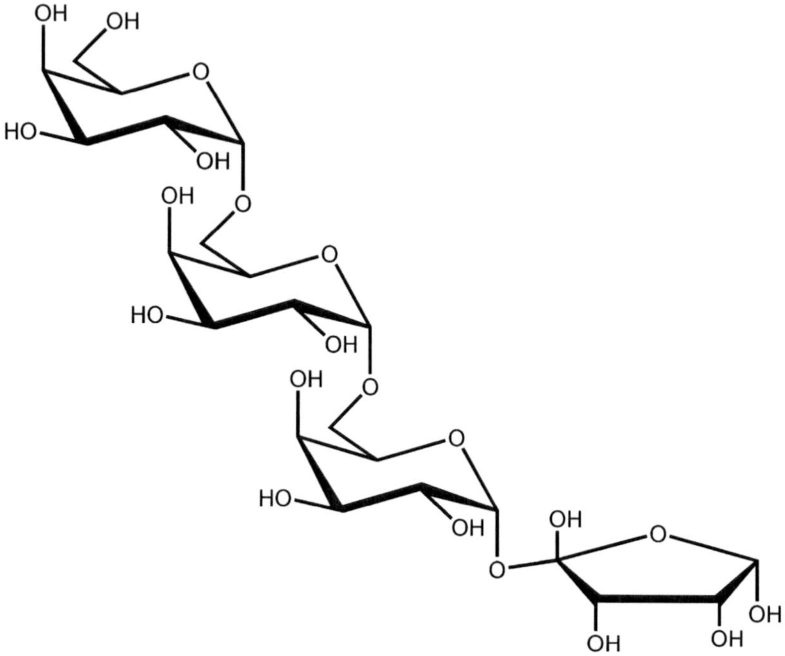

Figure 4.2 Structure of stachyose

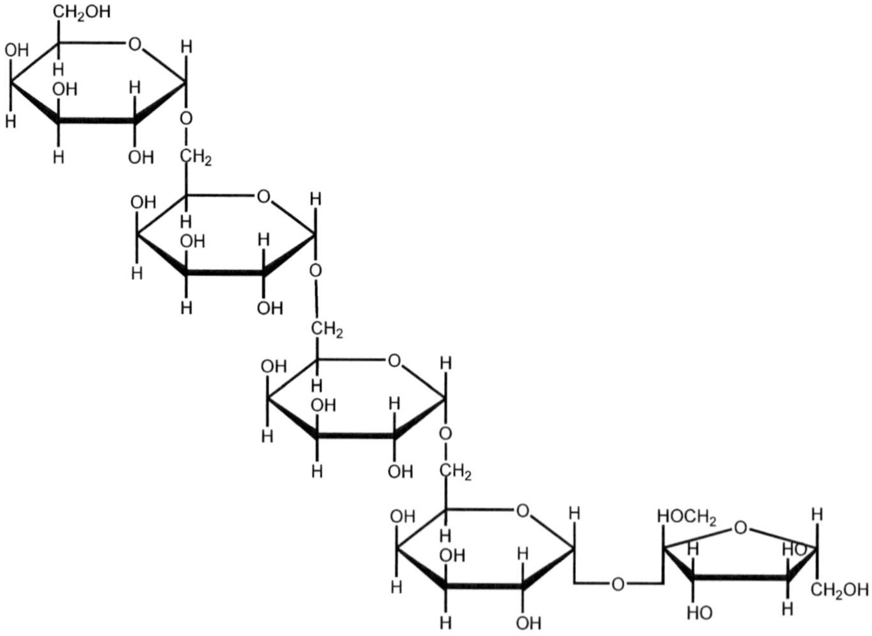

Figure 4.3 Structure of verbascose

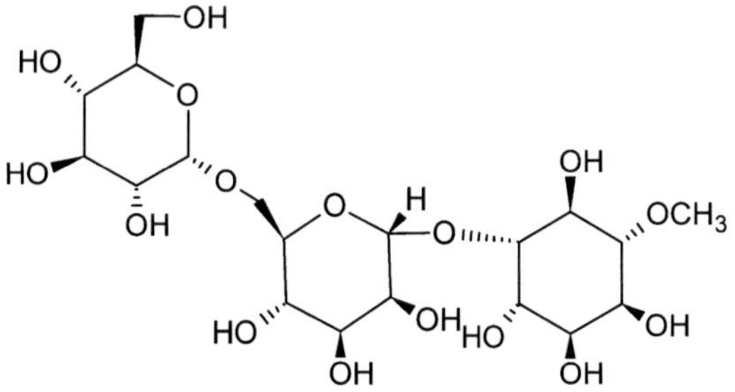

Figure 4.4 Structure of ciceritol

ciceritol, and stachyose, while those of peas are raffinose and stachyose. Verbascose was reported to be the minor oligosaccharide in lentils and peas and was absent in chickpeas and soybeans. Ciceritol is either absent or present in a non-detectable amount in peas and soybeans. The total oligosaccharide content of raw pulses has been observed to range from around 71 mg/g in yellow peas to 145 mg/g in chickpeas. Various treatments such as soaking, cooking, germination, fermentation, *etc.* causes a reduction in oligosaccharides. In black gram varieties after soaking for 16 h the levels of oligosaccharides have been reported to decrease by uo to 41.6–43.7% for raffinose, 20.6–47.6% for stachyose, 23.6–28.5% for verbascose, and 15.8–26.8% for ajugose. Whereas, cooking for 60 min can results in a decrease by up to 100% for raffinose, 55.9–76.2% for stachyose, 36.4–48.5% for verbascose, and 56.0–60.9% for ajugose in black gram varieties.

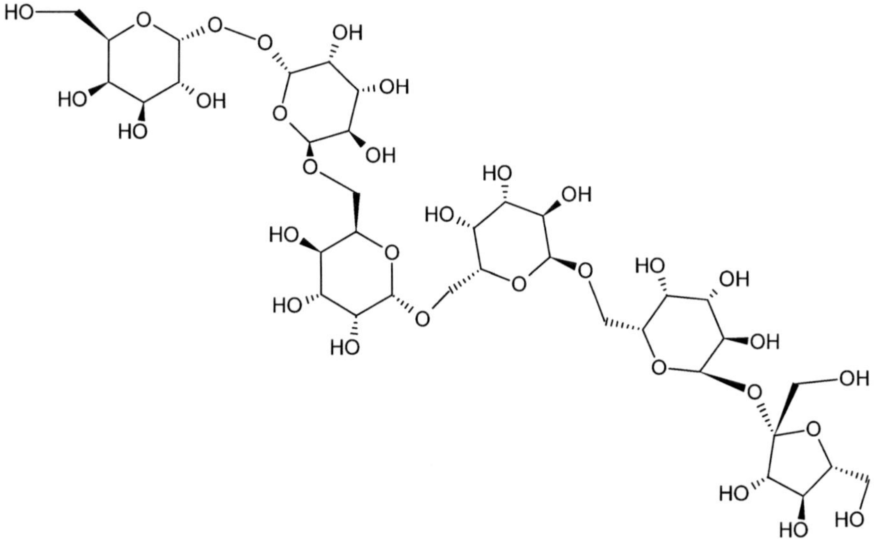

Figure 4.5 Structure of ajugose

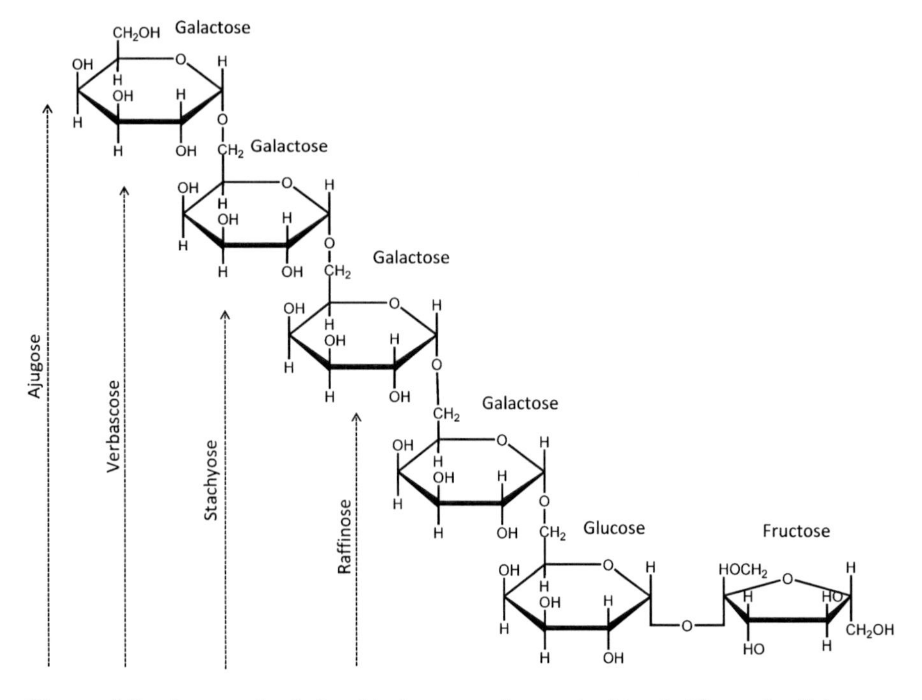

Figure 4.6 Structural relationship between oligosaccharides (raffinose family)

4.5. BIOACTIVE CONSTITUENTS

Pulses contain a number of bioactive substances including enzyme inhibitors, lectins, phytates, oligosaccharides and phenolic compounds that have roles to play in the metabolism of the humans or animals that frequently consume these foods. Their effects may be positive, negative, or both.[8] Some of these substances are considered as antinutritional factors due to their effects on the nutritional quality of the diet. Trypsin inhibitors, chymotrypsin inhibitors, phytic acid, tannins and oligosaccharides (raffinose, stachyose and verbascose) are considered antinutrients because they limit the utilisation of proteins and carbohydrates. The antinutritional factors are structurally different compounds broadly divided into two categories: proteins such as lectins and protease inhibitors, and others such as phytate, tannins or proanthocyanidins, oligosaccharides, saponins, alkaloids and haemagglutinins. Dietary polyphenols have a great diversity of structures, ranging from rather simple molecules (monomers and oligomers) to polymers. The quantity and quality of polyphenols present in plant foods can vary greatly due to factors such as plant genetics, soil composition and growing conditions, state of maturity and post-harvest conditions.[9] Tannins are polymeric flavonoids that make up a small part of the broad and diverse group of phenolic compounds produced by plants as secondary metabolites.[10,11] Tannins, present in almost all plants, are characterised as astringent, bitter polyphenolic compounds that bind to and precipitate proteins and various other organic compounds including amino acids and alkaloids. Tannin concentrations ranging from 0 to 0.7% for cowpeas, 0 to 0.2% for pigeon peas, 0.3% in adzuki beans (*Vigna angularis*), 0.1% in Japanese buckwheat, and much less or no tannin in chickpeas or green beans have been observed. The absence of tannins in certain varieties of winged beans (*Psophocarpus tetragonobolus*) has also been observed.[12] Tannins can precipitate proteins and complex with iron in the gastrointestinal lumen, reducing the absorption, digestibility and availability of these nutrients.[11,13] The profiles and quantities of polyphenols and tannins in foods are affected by processing because of their highly reactive nature, and this may affect their antioxidant activity and the nutritional value of foods.[14]

Anthocyanins are also synthesised by the flavonoid biochemical pathway but do not reduce digestibility and therefore are not considered antinutritional factors.[15] The soybean-rich diet fed to animals can lead to the development of enlarged spleen and reduced serum insulin levels as well as disruption of the normal intermediary metabolism of protein, fat and carbohydrate due to the activity of lectins.[16] Lectins are proteins that bind sugar and are present in a number of plants, especially pulses. The lectin activity in most pulses does not produce the same pathological effects as the consumption of raw soybean, although mild self-limiting ailments may arise in human populations and livestock from the consumption of some species of pulses. Peumans and Van Damme[17] reported that lectins from lentil and field pea may be harmful to humans if consumed raw, while non-toxic effects of lectins present in lentils, peas and chickpeas was reported earlier.[18] However, studies on the lectin activity of several leguminous crops demonstrated that chickpeas and pigeon peas had lectin activity below toxic levels.[19] Lectins interact with glycoprotein components of the red blood cell surface, causing agglutination; hence their alternative name, phytohemagglutenins.[20–22] Owusu-Ansah and McCurdy[22] reported that the toxic effects of lectins when ingested orally may be due to their ability to bind to specific receptor sites on the surface of intestinal epithelial cells that cause a non-specific interference with the absorption of nutrients across the intestinal wall. High-moisture heat treatment completely destroys lectins.[21]

Trypsin inhibitors are low-molecular-weight proteins capable of binding to and inactivating the digestive enzyme trypsin.[23] Protease inhibitors present in different pulses have varying degrees of susceptibility to various processing conditions. The processing technique and conditions selected to inactivate the protease antinutritional compounds may vary depending upon pulse variety or species. The variations in trypsin inhibitory activity amongst different varieties of same pulse are well known, and the decrease in trypsin inhibitory activity after a particular thermal treatment varies between varieties of the same pulse. Trypsin and chymotrypsin inhibitor activities in desi-type chickpeas have been reported to be higher than in kabuli-type chickpeas.[24]

Tannins are polyphenol compounds concentrated mainly in the hull portion of pulses. Decortication results in a more than 80% reduction of tannin content. Tannins decrease protein digestibility and solubility. They also form complexes with amino acids and reduce their availability. The reduced protein digestibility could be due to the inactivation of digestive enzymes or the formation of tannin–protein complexes.[21] Tannins interact with enzymes and non-enzyme constituents and either cause inactivation of enzymes such as trypsin and chymotrypsin or make protein insoluble. Polyphenols inhibit several enzymes including α-amylase, lipases, pectin esterases, cellulases and β-galactosidase.[25] Tannins also decrease the bioavailability of vitamins and minerals.[21] Soaking, germination and cooking decrease the tannin content in pulses; cooking causes the greatest reduction.

Enzyme inhibitors and lectins can reduce protein digestibility and nutrient absorption respectively, but effects of both diminish after cooking.[26] The consumption of raw pulse grains or flours may lead to various deleterious effects such as haemagglutination, bloating, vomiting and pancreatic enlargement, due to the activity of the antinutritional compounds. Some of the constituents that were earlier regarded as antinutrients are now considered to have some beneficial effects on human health. For example, phytic acid has antioxidant activity and protects DNA.[27,28] Phenolic compounds (flavonoids and phenolic acids) have antioxidant activity,[29–31] and galacto-oligosaccharides have some prebiotic activity.[32,33] Secondary metabolites are marketed as functional foods and nutraceuticals because of their health benefits. Efforts are also being made in different countries to develop or identify strains of common beans that are low in lectins and other antinutrional compounds.

The major polyphenolic compounds of pulses consist mainly of tannins, phenolic acids and flavonoids. Dark-coloured and highly pigmented pulses such as red kidney beans (*Phaseolus vulgaris*) and black gram (*Vigna mungo*) have higher contents of polyphenolic compounds. Condensed tannins (proanthocyanidins) have been quantified in hulls of several varieties of field beans (*Vicia faba*) and are also present in pea seeds of coloured-flower cultivars.[34] Pulses vary in total phenolic contents and antioxidant activities. Lentils have the higher phenolic, flavonoid and condensed tannin content (6.56 mg gallic acid equivalents g^{-1}, 1.30 and 5.97 mg catechin equivalents g^{-1}, respectively), as compared to red kidney beans and black beans.[35] Total phenolic content is directly associated with antioxidant activity.[36,37] Pulses with higher total phenolic content (lentils, red kidney beans, and black beans) have higher antioxidant capacity as assessed by 2,2-diphenyl-1-picryhydrazyl (DPPH) free radical scavenging, ferric reducing antioxidant power, and the oxygen radical absorbance capacity[35] The potential health benefits of common beans are attributed to the presence of secondary metabolites such as phenolic compounds that possess antioxidant properties. Ferulic acid is the most abundant phenolic acid in common beans, and intermediate levels of *p*-coumaric and sinapic acids are also present.[38] Cooking results in significant reductions in phytic acid and tannins in pulses, but prolonged cooking should be avoided because it decreases the nutritive value of pulses by reducing the levels of some essential amino acids. Pulses normally used in human nutrition are processed before consumption to reduce their levels of antinutritional factors. Table 4.7 shows the effect of processing on minor constituents.

4.5.1. Phytic Acid

Phytic acid, also known as inositol hexakisphosphate (IP6), or phytate (in salt form) is the primary storage form of both phosphate and inositol in seeds (Figure 4.7). IP6 is the major inositol phosphate form present in pulses; other forms (IP3–IP5) are present in relatively lower amount. Phosphorus in phytic acid constitutes the major portion of total phosphorus in most grains, accounting for around 50–80% of the total phosphorus. Phytic acid reduces the absorption of minerals such as phosphorus, calcium, magnesium, iron and zinc. It also adversely affects the absorption of lipids and proteins. The regular consumption of a diet rich in phytate leads to iron deficiency, which causes anaemia, and people who consume large amounts of pulse grains as a part of their daily diet may be affected by this.

The phytic acid content of grains is influenced by cultivar, climatic conditions and application of fertilizers. Generally, cooler temperatures during the growing season produce grains with a lower phytic acid content. Phytic acid is higher in foods that are grown using high-phosphate fertilizers in comparison to those grown using natural compost or manures. Different cultivars of the same grain may also have different levels of phytic acid. It has also been reported that capacity of phytic acid to chelate minerals can be used to prevent, inhibit or even cure some cancers by depriving those cells of the essential minerals (especially iron), which they need to reproduce.[39] However, the possibility of adverse effects of deficiency of essential minerals on non-cancerous cells cannot be ruled out.

Various pre-processing treatments, such as soaking, germination or fermentation, reduce the phytic acid content in some pulses. Soaking induces the activation of endogenous phytase and diffusion of the products, but the extent of reduction varies among pulse grains.[40] A correlation between phytate reduction and phytase activity has been established for germination of lentils,

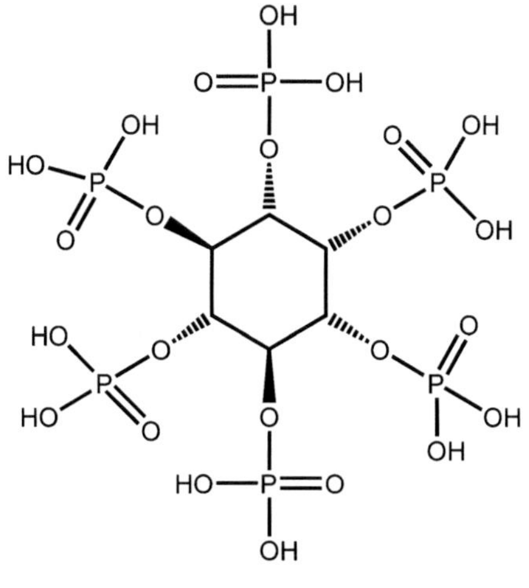

Figure 4.7 Structure of phytic acid

but in chickpeas no correlation was observed.[41] Variable effects of cooking processes on phytate content in different pulses have been reported. Several studies have reported a decrease of IP6 and a corresponding increase of IP5–IP3 due to the hydrolysis of the IP6 during the cooking.[42,43]

4.5.2. Flavonoids

The potential health benefits of common beans are attributed to the presence of secondary metabolites such as phenolic compounds that possess antioxidant properties.[44] Tannins, phenolic acids and flavonoids are some of the well-known polyphenols found in pulses. Flavonoids are widely distributed in plants, including pulses, as secondary metabolites. Flavonoids (proanthocyanins) in the seed coats of pulses impart h resistance to certain pests and weevils (e.g. *Callosobruchus maculatus*). Major classes of flavonoids include flavones, flavonols, and condensed tannins. The total flavonoid content of yellow pea is comparatively higher in chickpeas, lentils, red kidney beans and black beans. Reported content for total flavonoid expressed as gallic acid equivalent (GAE) (Table 4.4) in green peas varies from 0.08 to 0.39 mg GAE/g; black beans from 0.98 to 3.21 mg GAE/g; chickpeas from 0.18 to 3.16 mg GAE/g; lentils from 0.72 to 2.21 mg GAE/g; and red kidney beans from 0.85 to 2.93 mg GAE/g.[35] The level of total phenolic contents is higher in coloured pulses than in white and yellow ones and has been formally used as an indicator of the seed coat colour. For example, the total phenolic contents of yellow peas varies from 1.13 to 1.67 mg GAE/g; green peas from 1.04 to 1.53 mg GAE/g; black beans from 1.28 to 6.89 mg GAE/g; chickpeas from 1.54 to 1.81 mg GAE/g; lentils from 1.02 to 7.53 mg GAE/g; and red kidney beans from 1.23 to 5.90 mg GAE/g[35](Table 4.5).

4.5.3. Tannins

Tannins have been conventionally perceived as antinutritional factors or non-nutrient compounds; however, recent studies have indicated that tannins are also associated with several health benefits. Tannins commonly found in

Table 4.4 Variations in total flavonoid content (mg catechin equivalents/g) of some pulses and effect of extraction solvent (adapted from Xu and Chang[35])

Solvent	*Green peas*	*Yellow pea*	*Chickpeas*	*Lentil*	*Red kidney beans*	*Black bean*
50% acetone	0.08	0.18	0.18	1.3	2.02	2.49
80% acetone	0.39	0.32	3.16	2.01	2.71	2.96
Acidic 70% acetone	0.26	0.29	2.87	2.21	2.93	3.21
70% methanol	0.1	0.25	0.2	0.72	0.85	0.98
70% ethanol	0.19	NA	NA	1.22	1.22	1.19
100% ethanol	NA	NA	NA	1.65	NA	1.32

Table 4.5 Variations in total phenolic content (mg gallic acid equivalents/g) of some pulses and effect of extraction solvent (adapted from Xu and Chang[35])

Solvent	*Green peas*	*Yellow peas*	*Chickpeas*	*Lentils*	*Red kidney beans*	*Black beans*
50% acetone	1.53	1.67	1.81	6.56	4.98	5.04
80% acetone	1.07	1.34	1.41	6.81	5.22	5.54
Acidic 70% acetone	1.13	1.28	1.57	7.53	5.9	6.89
70% methanol	1.31	1.41	1.68	2.51	2.25	3.31
70% ethanol	1.34	1.56	1.54	2.440.12d	2.51	3.2
100% ethanol	1.04	1.13	NA	1.02	1.23	1.28

terrestrial plants and grains can be classified as hydrolysable or non-hydrolysable. Hydrolysable tannins are mainly glucose esters of gallic acid (Figure 4.8). Two types of hydrolysable tannins are known: the gallotannins, which yield only gallic acid upon hydrolysis, and the ellagitannins, which produce ellagic acid as the common degradation product.[45] Non-hydrolysable tannins, also known as condensed tannins, are polymers of flavan-3-ols (Figure 4.9). Condensed tannins are the predominant phenolic compounds in pulses. Tannins are located mainly in the testa and play an important role in the defence system of seeds that are exposed to oxidative damage by many environmental factors.[46] The concentration of condensed tannins is reported to vary depending on the type of pulse (Table 4.6): for yellow peas it varies from 0

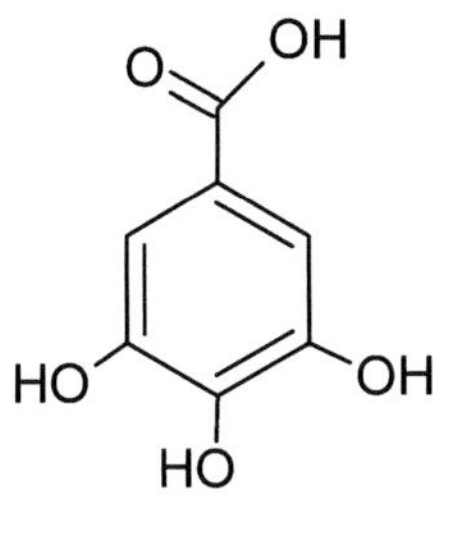

Figure 4.8 Structure of gallic acid

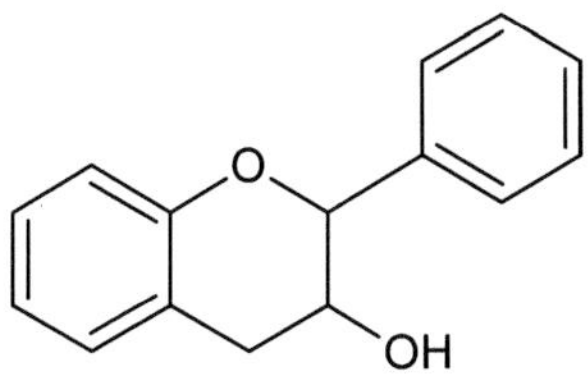

Figure 4.9 Structure of flavone-3-ols

to 1.52 mg CAE/g, green peas from 0.03 to 1.71 mg CAE/g, black beans from 0.37 to 6.74 mg CAE/g, chickpeas from traces to 1.85 mg CAE/g, lentils from 0.12 to 8.78 mg CAE/g, and red kidney beans from 0.12 to 5.53 mg CAE/g.[35] Tables 4.4–4.6 show that the variation in the phenolic composition profiles of pulses varies not only with the type but also with the methodology and solvent employed for extraction and analysis. For example, acidic 70% acetone is the most efficient solvent for the extraction of condensed tannins and ethanol is the least effective.

4.5.4. Saponins

Saponins are extremely widely distributed in the plant kingdom. They occur constitutively in a great many plant species, both wild plants and cultivated crops. Saponins may be considered a part of plants' defence systems, and as such have been included in a large group of protective molecules found in plants named "phytoanticipins" or "phytoprotectants".[47,48] However, the physiological role of saponins in plants is not yet fully understood. They occur in many plants that are consumed by humans as food, including, but not limited to soybeans, chickpeas, peanuts, mung beans, broad beans, kidney beans, lentils, garden peas, spinach, oats, fenugreek, garlic and potatoes. Chemically, saponins consist of a sugar moiety usually containing glucose, galactose, glucuronic acid, xylose, rhamnose or methylpentose, glycosidically linked to a hydrophobic aglycone (sapogenin) which may be triterpenoid (Figure 4.10a) or steroid (Figure 4.10b) in nature. The aglycone may contain one or more unsaturated C–C bonds. In pulses, and in fact, most cultivated crops the triterpenoid saponins are generally predominant, while steroid saponins are common in plants used as herbs or for their health-promoting properties.[49] Triterpenoid saponins are synthesised via the isoprenoid pathway by cyclisation of 2,3-oxidosqualene to give primarily oleanane (β-amyrin) or dammarane triterpenoid skeletons. A number of factors, such as physiological age, environmental and agronomic factors, have been shown to affect the saponin content of plants.[50]

Table 4.6 Variations in condensed tannin content (mg catechin equivalents/g) of some pulses and effect of extraction solvent (adapted from Xu and Chang[35])

Solvent	*Green peas*	*Yellow peas*	*Chickpeas*	*Lentils*	*Red kidney beans*	*Black beans*
50% acetone	0.26	0.42	1.05	5.97	3.85	3.4
80% acetone	1.71	1.52	1.85	8.78	5.53	5.29
Acidic 70% acetone	0.91	1.02	1.21	8.7	5.37	6.74
70% methanol	0.03	0	0	1.72	0.3	1.48
70% ethanol	0.05	0.03	0.05	1.18	0.54	1.57
100% ethanol	0.48	0.47	NA	0.12	0.12	0.37

Table 4.7 Effect of processing on the polyphenols, tannins, and other antinutritional factors in pulses

Pulses	*Processing method*	*Salient observation*	*Reference*
Lentils (*Lens culinaris*)	Cooking	Significant reduction in phytate content from 817.0 mg/100 g to 504 mg/100 g	Awada *et al.*[66]
Faba beans (*Vicia faba*) and kidney beans (*Phaseolus vulgaris*)	Extrusion cooking	Extrusion is the best method to remove trypsin, chymotrypsin, α-amylase inhibitors and haemagglutination activity without modifying protein content	Alonso *et al.*[40]
Faba beans (*Vicia faba*) and kidney beans (*Phaseolus vulgaris*)	Dehulling	Great reduction in condensed tannin and polyphenol levels in both beans	Alonso *et al.*[40]
Kidney bean (*Phaseolus vulgaris*)	Soaking and dehulling	Both the treatments reduced the tannin content	Deshpande *et al.*[67]
Pigeon pea (*Cajanus cajan*)	Soaking, soaking and dehulling, ordinary cooking, pressure-cooking, and germination	Phytate concentration was reduced significantly and to various extents	Duhan *et al.*[68]
Cowpeas (*Vigna sinensis*)	Pressure-cooking and solar cooking	Significantly reduced the phytic acid and polyphenol content	Pasrija and Punia[69]

Literature reveals that saponins increase on sprouting in some pulses such as mung beans and peas, but decrease in others such as moth beans, and that light availability during germination has a profound stimulating effect on the saponin content. Saponins present in pulses have three unique features:

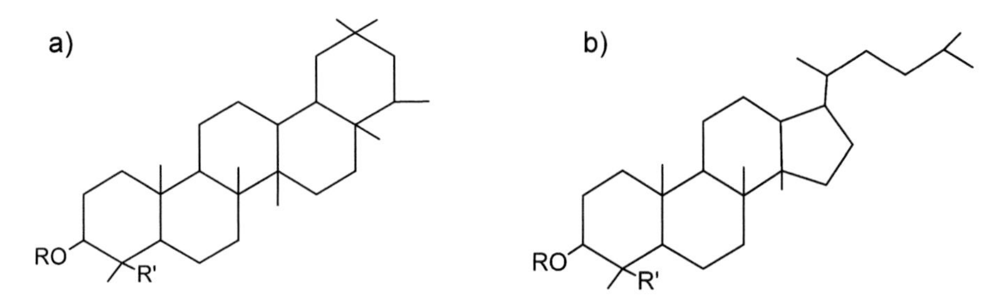

Figure 4.10 Basic structures of sapogenins: (a) a triterpenoid and (b) a steroid

- They act as bile acid sequestrants, having the ability to bind to cholesterol and pathogens entering the human body, forming molecules too large to be absorbed through the intestinal wall and helping in their removal from the body.
- They perform their function within the intestinal tract and do not enter the rest of the body, hence they are non-systemic.
- They break down and remove impacted rubber-like matter from the colon walls and act favourably on the intestinal flora, encouraging growth of beneficial bacteria and retarding harmful bacteria.

Despite these benefits, it is also worth noting that deleterious fungitoxic, haemolytic and membranolytic activities have also been ascribed to saponins. Khokhar and Chauhan[51] have reported that soaking in a solution of mixed mineral salt removes much more of the saponins in moth beans than does soaking in water (30–36% and 9–18%, respectively) and ordinary cooking of the seeds pre-soaked in water reduces the saponin level by 12–15%. Reports have also been published indicating that common domestic processing and cooking treatments reduce the saponin level of chickpeas and black gram significantly.[52] Ruiz *et al.*[53] have recorded conversion of soyasaponin VI into soyasaponin I (Figure 4.11), leaching of both saponins, and reduction of saponin level in the case of lentils upon cooking. Ruiz *et al.*[53] also did not observe significant changes in saponin content for chick peas or lentils after 6 day germination in the dark at 20 °C.

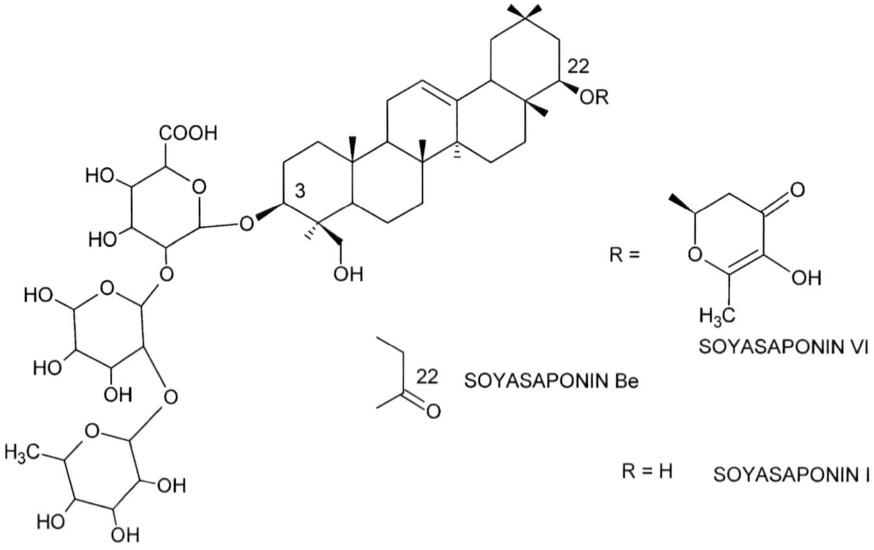

Figure 4.11 Structures of soyasaponins I, VI, and Be

Table 4.8 Phytosterols, namely β-sitosterol, campesterol, and stigmasterol content (mg/100 g) of selected pulses (adapted from Ryan *et al.*[3])

Pulse grain or seed	*β-Sitosterol (mg/100 g)*	*Campesterol (mg/100 g)*	*Stigmasterol (mg/100 g)*
Chickpeas	159.8 ± 7.1	21.4 ± 0.7	23.4 ± 0.7
Kidney beans	86.5 ± 2.6	6.5 ± 0.8	41.4 ± 1.6
Lentils	123.4 ± 4.1	15.0 ± 0.4	20.0 ± 0.6
Butter beans	85.1 ± 7.3	15.2 ± 2.9	86.2 ± 5.7

4.5.5. Phytosterols

Legumes contain a rich variety of phytochemicals, including phytosterols, natural antioxidants and bioactive carbohydrates,[54,55] which if consumed in sufficient quantities may help to reduce tumour risk.[56] Phytosterols, primarily β-sitosterol, campesterol and stigmasterol, are integral natural components of plant cell membranes that are abundant in most seeds and grains, including pulses (Table 4.8). Plant sterols inhibit the intestinal absorption of cholesterol, thereby lowering total plasma cholesterol and low-density lipoprotein (LDL) levels. Kalogeropoulos *et al.*[57] have reported that cooking leads to a decrease in the phytosterol content of legumes. However, the phytosterol content present in cooked legumes is still considerable and their benefits cannot be questioned.

REFERENCES

1. S. Rochfort and J. Panozzo, *J. Agric. Food Chem.*, 2007, **55**, 7981–7994.
2. R. Campos-Vega, G. Loarca-Pina and D. Oomah, *Food Res. Int.*, 2010, **43**, 461–482.
3. E. Ryan, K. Galvin, T. P. O'Connor, A. R. Maguire and N. M. O'Brien, *Plant Foods Hum. Nutr.*, 2007, **62**, 85–91.
4. B. Quemener and J. M. Brillouet, *Phytochemistry*, 1983, **22**, 1745–1751.
5. B. D. Oomah, A. Patras, A. Rawson, N. Singh and R. Compos-Vega, in *Pulse Foods*, ed. B. K. Tiwari, A. Gowen and B. McKenna, Academic Press, San Diego, 2011, pp. 9–55.
6. S. E. Fleming, *J. Food Sci.*, 1981, **46**, 794–798.
7. K. Girigowda, S. J. Prasanth and V. H. Mulimani, *Plant Foods Hum. Nutr.*, 2005, **60**, 173–180.
8. M. M. J. Champ, Non-nutrient bioactive substances of pulse. *Br. J. Nutr.*, 2002, **88**, S307–S319.
9. A. L. K. Faller and E. Fialho, *Food Res. Int.*, 2009, **42**, 210–215.
10. B. Winkel-Shirley, *Plant Physiol.*, 2001, **126**, 485–493.
11. A. M. Díaz, G. V. Caldas and M. W. Blair, *Food Res. Int.*, 2010, **43**, 595–601.
12. M. L. Price, A. E. Hagerman and L. G. Butler, *J. Agric. Food Chem.*, 1980, **28**, 459–461.

13. M. Brune, L. Rossander and L. Halberg, *Eur. J. Clin. Nutr.*, 1989, **43**, 547–558.
14. N. R. Dlamini, L. Dykes, L. W. Rooney, R. D. Waniska and J. R. N. Taylor, *Cereal Chem.*, 2009, **86**, 191–196.
15. O. M. Andersen and M. Jordheim, in *Flavonoids: Chemistry, Biochemistry and Applications*, ed. Ø. M. Andersen and K. R. Markham, CRC Press, Boca Raton, FL, 2006, pp. 471–553.
16. N. Sharon and H. Lis, *Glycobiology*, 2004, **14**, 53–62.
17. W. J. Peumans and E. J. Van Damme, *Plant Physiol.*, 1995, **109**, 347–352.
18. E. González de Mejía and V. I. Prisecaru, *Crit. Rev. Food Sci. Nutr.*, 2005, **45**, 425–445.
19. U. Singh, *Plant Foods Hum. Nutr.*, 1988, **38**, 251–261.
20. P. Valdeboijze, E. Bergeron, T. Gaborit and J. Delort-Laval, *Can. J. Plant Sci.*, 1980, **60**, 695–701.
21. J. K. Chavan, S. S. Kadam and D. K. Salunkhe, *CRC Crit. Rev. Food Sci. Nutr.*, 1986, **25**, 107–158.
22. Y. Owusu-Ansah and S. McCurdy, *Food Rev. Int.*, 1991, **7**, 103–134.
23. J. K. Chavan and S. S. Kadam, *CRC Crit. Rev. Food Sci. Nutr.*, 1989, **28**, 401–437.
24. U. Singh and R. Jambunathan, in *Proceedings of International Workshop on Pigeonpeas, 15–19 Dec 1980, ICRISAT, India*. Vol. 2. ICRISAT, Patancheru, A. P., India, 1981, pp. 419–425.
25. D. K. Salunkhe, S. S. Kadam and J. K. Chavan, *Postharvest Biotechnology of Food Legumes*. CRC Press, Boca Raton, FL, 1985.
26. F. M. Lajolo and M. I. Genovese. *J. Agric. Food Chem.*, 2002, **50**, 6592–6598.
27. K. Midorikawa, M. Murata, S. Oikawa, Y. Hiraku and S. Kawanishi, *Biochem. Biophys. Res. Commun.*, 2001, **288**, 552–557.
28. B. Q. Phillippy, *Adv. Food Nutr. Res.*, 2003, **45**, 1–60.
29. P.-G. Pieta, *J. Nat. Prod.*, 2000, **63**, 1035–1042.
30. P. A. Murphy and S. Hendrich, *Adv. Food Nutr. Res.*, 2002, **44**, 195–246.
31. C. T. Yeh and G. C. Yen, *J. Agric. Food Chem.*, 2003, **51**, 1474–1479.
32. P. De Boever, B. Deplancke and W. Verstraete, *J. Nutr.*, 2000, **130**, 2599–2606.
33. C. E. Rycroft, M. R. Jones, G. R. Gibson, R. A. Rastall, *Lett. Appl. Microbiol.*, 2001, **32**, 156–161.
34. S. Smulikowska, B. Pastuszewska, E. Świech, A. Ochtabińska, A. Mieczkowska, V. C. Nguyen and L. Buraczewska, *J. Anim. Feed Sci.*, 2001, **10**, 511–523.
35. B. J. Xu, S. K. C. Chang, *J. Food Sci.*, 2007, **72**, S159–S166.
36. J. M. Awika, L. W. Rooney, X. Wu, R. L. Prior and L. Cisneros-Zevallos, *J. Agric. Food Chem.*, 2003, **51**, 6657–6662.
37. R. Amarowicz, A. Troszynska, N. Barylko-Pikielna and F. Shahidi, *J. Food Lipid*, 2004, **11**, 278–286.

38. D. L. Luthria and M. A. Pastor-Corrales, *Bean Improvement Cooperative Annual Report*, 2006, **49**, 6–7.
39. J. R. Zhou and J. W. Erdman, *J. Food Sci. Nutr.*, 1995, **35**, 495–508.
40. R. Alonso, A. Aguirre and F. Marzo, *Food Chem.*, 2000, **68**, 159–165.
41. R. Greiner, M. M. Pedrosa, M. Muzquiz, G. Ayet, C. Cuadrado and C. Burbano, *Opportunities for High Quality, Healthy and Added-Value Crops to Meet European Demands*, EAAP, Wageningen1998, pp. 82–83.
42. N. Khatoon and J. Prakash, *Int. J. Food Sci. Nutr.*, 2004, **55**, 441–448.
43. Z. Rehman and W. H. Shah, *J. Food Chem.*, 2005, **91**, 327–331.
44. J. K. Lin and M. S. Weng, in *The Science of Flavonoids*, ed. E. Grotewold, Springer New York, 2006, pp. 213–238.
45. C. Andrés-Lacueva, A. Medina-Remon, R. Llorach, M. Urpi-Sarda, N. Khan, G. Chiva-Blanch, R. Zamora-Ros, M. Rotches-Ribalta and R. M. Lamuela-Raventós, *Phenolic compounds: Chemistry and occurrence in fruits and vegetables*, Wiley-Blackwell: Iowa, 2010.
46. A. Troszynska, I. Estrella, M. L. López-Amorós and T. Hernández, *Lebensm.-Wiss. Technol.*, 2002, **35**, 158–164.
47. J. P Morrissey and A. E Osbourn, *Microbiol. Mol. Biol. Rev.*, 1999, **63**, 708–724.
48. K. Haralampidis, M. Trojanowska and A. E. Osbourn, *Adv. Biochem. Eng./Biotechnol.*, 2002, **75**, 31–49.
49. G. Francis, Z. Kerem, H. P. S. Makkar and K. Becker, *Br. J. Nutr.*, 2002, **88**, 587–605.
50. Y. Yoshiki, S. Kudou, K. Okubo, *Biosci. Biotechnol. Biochem.*, 1998, **62**, 2291–2299.
51. S. Khokhar and B. M. Chauhan, *J. Food Sci.*, 1986, **51**, 591−594.
52. S. Jood, B. M. Chauhan and A. C. Kapoor, *J. Sci. Food Agric.*, 1986, **37**, 1121–1124.
53. R. G. Ruiz, K. Price, M. Rose, M. Rhodes and R. Fenwick, *Z. Lebensm. Unters. Forsch.*, 1996, **203**, 366–369.
54. R. Amarowicz and R. B. Pegg, *Eur. J. Lipid Sci. Technol.*, 2008, **110**, 865–878.
55. S. Rochfort and J. Panozzo, *J. Agric. Food Chem.*, 2007, **55**, 7981–7994.
56. J. C. Mathers, *Br. J. Nutr.*, 2002, **88**, S273–S279.
57. N. Kalogeropoulos, A. Chiou, M. Ioannou, Karathanos, T. Vaios, M. Hassapidou and N. K. Andrikopoulos, *Food Chem.*, 2010, **121**, 682–690.
58. *USDA National Nutrient Database for Standard Reference*, Release 23 (2010) (Accessed 10 July 2011).
59. U. Singh and B. Singh, *Econ. Bot.*, 2000, **46**, 310–321.
60. N. Wang and J. K Daun, *J. Sci. Food Agric.*, 2004, **84**, 1021–1029.
61. S. Singh, H. Singh and K. Sikka, *Cereal Chem.*, 1968, **45**, 13–18.
62. V. Arinathan, V. R. Mohan, A. Maruthupandian and T. Athiperumalsami, *Tropical and Subtropical Agroecosystems*, 2009, **10**, 287–294.
63. H. B. ElMaki, S. M. Abdel Rahaman, W. H. Idris, A. B. Hassan, E. E. Babiker and A. H. El Tinay, *Food Chem.*, 2007, **100**, 362–368.

64. N. Wang, D. W. Hatcher, R. Toews and E. J. Gawalko, LWT – *Food Sci. Technol.*, 2009, **42**, 842–848.
65. G. Oboh, *J. Food Biochem.*, 2006, **30**, 579–588.
66. S. Awada, A. Hady, A. B. Hassan, M. I. Ah and E. E. Babiker, *J. Food Technol.*, 2005, **3**, 523–528.
67. S. Deshpande, S. Sathe, D. Salunkhe and D. P. Cornforth, *J. Food Sci.*, 1982, **47**, 1846–1850.
68. A. Duhan, N. Khetarpaul, S. Bishno, *Nutr. Health*, 1999, **13**, 161–169.
69. M. Pasrija and D. Punia, *Nutr. Food Sci.*, 2000, **30**, 133–137.

CHAPTER 5

Pulse Proteins

5.1. INTRODUCTION

Pulses are widely recognised as an excellent source of protein. Worldwide, protein malnutrition has continued to be a major issue. The Food and Agriculture Organization (FAO) estimated that 925 million people were undernourished in 2010 compared with 1.023 million in 2009, and 98% of these undernourished people live in developing countries. In this situation, it must be emphasised that pulse proteins are extremely important to a large segment of the world's population where consumption of animal protein is limited, unaffordable or restricted for religious reasons. The consumption of pulses provides many potential health benefits, including reduced risk of cardiovascular disease, cancer, diabetes, osteoporosis, hypertension, gastrointestinal disorders, adrenal disease and reduction of LDL cholesterol.[1–4] The growing awareness of the usefulness of including pulses in the diet has increased demand for pulses throughout the world. Pulses are often combined with different locally available grains in various food formulations to improve nutrition and add variety. Pulses, in combination with other grains, can also help in overcoming the protein malnutrition problem of developing countries. The synergetic effect between pulses and cereal proteins in providing a balanced essential amino acid profile for both human and animal nutrition is well known. Pulses are relatively low in sulfur-containing amino acids (methionine and cysteine) and high in lysine. Cereals, on the other hand, are low in lysine, so including pulses in a cereal-rich diet balances the overall amino acid intake. Pulse proteins contain higher amounts of lysine, leucine, aspartic acid, glutamic acid and arginine than cereal proteins. Pulses are very often supplemented with cereals and amount of supplementation varies depending upon the amount of the limiting amino acids in cereals. The nutrient composition of different pulses varies with varieties and environmental factors such as geographic location and growing season. Pulse proteins possess many

Pulse Chemistry and Technology
Brijesh K. Tiwari and Narpinder Singh

Published by the Royal Society of Chemistry, www.rsc.org

functional properties, such as water holding, fat binding, foaming and gelation, which make them a potential raw material for the development of a wide variety of food products. This chapter focuses on the classification of pulse proteins and their digestibility.

5.2. CLASSIFICATION OF PULSE PROTEINS

Pulse proteins are classified into four groups based on their solubility.

Table 5.1 shows the protein profile of commonly consumed pulses. The globulins and albumins are two major classes of pulse proteins; globulins are the most abundant class of storage proteins in pulses. The globulins and albumins represent around 60–80% and 15–25%, respectively, of the extractable proteins of the pulse cotyledon. Albumins are important during seed development because they contain most of the enzymes and metabolic proteins such as lipoxygenase, protease inhibitors and lectins. Albumins are high in lysine and sulfur-containing amino acids. The globulins are generally classified as 7S and 11S globulins according to their sedimentation coefficients (S). The sedimentation coefficient of proteins depends on the molecular weight (larger proteins sediment faster) and shape. Unfolded or highly elongated shaped proteins create more hydrodynamic friction, and thus have smaller sedimentation coefficients than a folded, globular protein of the same molecular weight. The 7S and 11S globulins of pea are known as vicilin and legumin, respectively.[5] The "legumin-like" (or 11S) proteins are oligomers, usually hexamers devoid of carbohydrate. The "vicilin-like" (or 7S) proteins are generally isolated from seed extracts as oligomers, usually trimers of glycosylated subunits[6] with trivial names such as conglycinin from *Glycine max* (soybean) and phaseolin from *Phaseolus vulgaris*, which constitutes 40–50% of total protein,[7] has a low sulfur amino acid content and a high resistance towards proteolytic enzymes.[8,9] The legumins (11S) have acidic subunits with a molecular mass of 40 kDa and basic 20 kDa subunits. The vicilins (7S) have subunits with a molecular mass of 175–180 kDa. Many grain legumes contain both the legumin- and vicilin-like proteins although some species contain exclusively either the 11S or 7S protein complexes.[10] Under dissociating conditions, both the 7S and 11S globulins liberate their constituent subunits. These polypeptide chains are naturally heterogeneous, which is evident from both their size and their charge levels.[11–13] This is due to a combination of different factors, including the multigene origin of each storage globulin and the post-translational modifications of relatively few expression products.[14,15]

The third type of storage protein, convicilin, distinct from legumin and vicilin, was first purified from the pea.[16] This protein has a distinctively different amino acid composition and unlike the 7S vicilin, contains very little carbohydrate and its subunits have a molecular mass of 71 kDa. In its native form it has a molecular mass of 290 kDa, including an N-terminal extension.

Prolamins and glutelins are the minor proteins present in pulse grains.[17,18] The protein profile of pea proteins separated by SDS-PAGE under reducing

Table 5.1 Protein profiling of different pulses

Legume	*Type of protein*	*Proteins*	*References*
Red kidney beans	Globulins	50 kDa	Sai-Ut *et al.*[58]
Navy beans		55 kDa	
Adzuki beans		36 to >97 kDa with major polypeptides (36, 45, or 55 kDa)	
Red kidney beans	Trypsin inhibitors	132 kDa	Wati *et al.*[59]
Navy beans		118 kDa	
Adzuki beans		13 kDa	
Urdbeans (*Vigna mungo*)	Albumins	10.23–25.53 kDa	Mahajan *et al.*[60]
	Globulins	10.84–112.72 kDa	
	Prolamins	10.33–51.52 kDa	
	Glutelins	8.91–112.72kDa	
Chickpeas (kabuli genotypes)	Seed	25–108 kDa	Hameed *et al.*[61]
Peas (Pisum sativum)	Convicilin (2 subunits)	77.9 kDa, 72.4 kDa	Barac *et al.*[19]
	Legumin	63.5 kDa	
	Phosphorylase B	94.0 kDa	
	Bovine albumin	67.0 kDa	
	Ovalbumin	43.0 kDa	
	Carbonic anhydrase	30.0 kDa	
	Soybean trypsin inhibitor	20.1 kDa	
	α-laktalbumin	14.4 kDa	
	Lypoxigenase (Lox)	92.7 kDa	
	Protease inhibitor (PPI)	11.5 kDa	
Lentil cultivars	Seed	14.4–116 kDa	Yüzbaşioğlu *et al.*[62]
		35–116 kDa were polymorphic and used for identification	
Kidney beans	Galactinol synthase (cotyledon)	38 kDa co-purified with 41 kDa and 43 kDa peptides	Liu *et al.*[63]
Adzuki beans	7S globulin(3 subunits)α	55 kDa	Chen *et al.*[64]
	β_1	28 kDa	
	β_2	25 kDa	
Red kidney beans	Lectin	30 kDa	Hou *et al.*[65]
Mung beans	Legumin	360 kDa	Mendoza *et al.*[20]
	Vicilin	200 kDa	
	Basic 7S globulin	135 kDa	

and non-reducing conditions is illustrated in Figure 5.1.[19] The proteins were separated into multiple components with molecular mass ranging from 104.8 kDa to 9.8 kDa, which originated mainly from vicilin and legumin. The SDS-PAGE patterns of Tris extracts contained three major (47.3, 35.0, 28.7 kDa) and three minor (37.0, 33.3, 31.8 kDa) subunits of vicilin, as well as two subunits of convicilin (77.9 kDa, 72.4 kDa). Legumin showed four bands of acidic (40.89 kDa) and basic (22.3, 23.1 kDa) subunits. Under reducing condition three minor bands of a trimeric (non-reduced) form of legumin (63.5 kDa) were observed. Minor bands of 92.7 kDa and 11.5 kDa were identified as a lypoxigenase and a protease inhibitor, respectively. Under reducing conditions, subunits of vicilin, convicilin and legumin were dominant. Prolamins have a high concentration of proline and glutamine. Glutelins have a higher proportion of methionine and cystine than the globulins and are also of nutritional significance. The mung bean major storage protein was 8S globulin or vicilin, which made up 89% of the total globulins, and 11S and basic 7S globulins, with contents of about 7.6 and 3.4%, respectively (relative to total globulins) were the other minor globulins.[20] Mung bean vicilin was observed to be mainly composed of four polypeptides of about 60, 48, 32 and 26 kDa while the 11S globulins were composed of acidic and basic subunits of 40 and 24 kDa, respectively.[20] The native phaseolin was mainly composed of two polypeptides of about 50 kDa (major) and 27 kDa (minor).

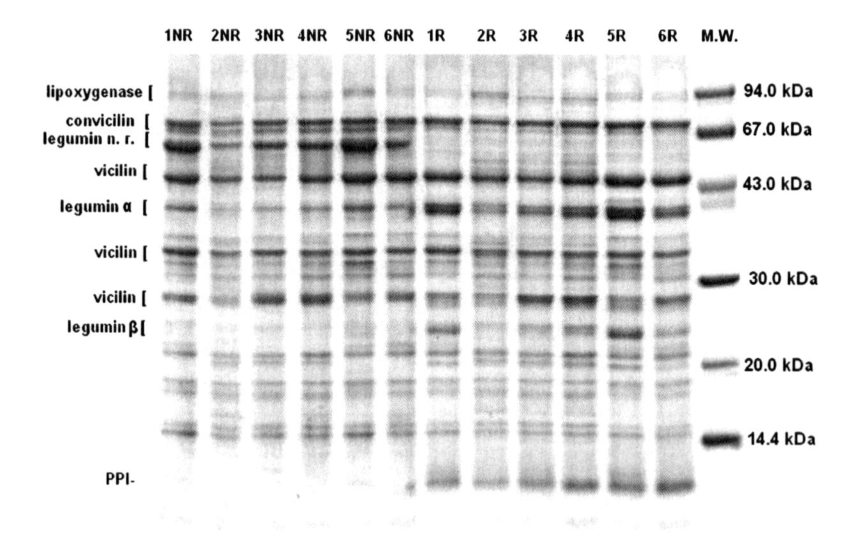

Figure 5.1 Electrophoretic patterns of pea bean proteins under reducing (R) and non-reducing (N.R) conditions. Calvedon (1R, 1NR); L1 (2R, 2NR); L2 (3R, 3NR); L3 (4R, 4NR); Maja (5R, 5NR); Miracle of America (6R, 6NR), M.W. molecular weight standards (source: Barac *et al.*[19]).

5.3. DIGESTIBILITY OF PROTEIN

Protein digestibility varies among pulse species, and is lower than for cereal proteins, milk proteins (casein) or animal proteins. This variability is due to many factors which include the intrinsic structural factors of pulse proteins and antinutritional factors (protease inhibitors, lectins, phytates, tannins, *etc.*). The differences in digestibility amongst the various pulses storage proteins are generally attributed to the difference in structural and conformational properties, such as degree of hydrophobicity and β-polypeptides of the globular proteins. Native phaseolin was observed to be extremely resistant to *in vitro* trypsin digestion.[21] Less than 50% of the original three subunits (43–51 kDa) of phaseolin were degraded after 10 min trypsin digestion to yield products in the mass region of 24–28 kDa. On the other hand, all the four major vicilin subunits (47, 43, 35, and 33 kDa) were digested within 1 min, resulting in the generation of new subunits of 27 kDa and 9–10 kDa and an increase in the bandwidth of the 20 kDa subunit. Complete proteolysis of the 29 kDa subunit of vicilin of took little under 5 min.

Heat denaturation has variable effects on protein digestibility of pulses. Denaturation of secondary, tertiary and quaternary structure of proteins occurs as a result of several factors including.

- Heat/organic chemicals: Break apart H bonds and disrupt hydrophobic attractions
- Acids/ bases: Break H bonds between polar R groups and ionic bonds
- Heavy metal ions: React with S-S bonds to form solids
- Agitation: Stretches chains until bonds break

The effects of heating on digestibility of pulse grains proteins depend on the globulin protein species. With heat treatment, proteolysis decreased in pea vicilin,[21] but increased in phaseolins of dry bean and legumins and convicilins of pea.[22] Albumins negatively affect the *in vitro* and *in vivo* protein digestibility of various pulse grains such as peas, chickpeas and dry beans in poultry and pigs.[23,24] Many antinutritional proteins such as protease inhibitors, lipoxygenase, lectins, and PA2 are albumins.[25] In the native state these albumins (particularly protease inhibitors and lectins) decreased protein digestibility by inhibiting protein hydrolysis and disturbing the absorption of peptides and amino acids.[26] Thermal treatments inactivate the antinutritional function of these albumin proteins; however, their digestibility, particularly in peas, chickpeas or dry beans, did not improve.[27,28] *In vivo* protein digestibility studies on various animals have indicated that pea albumins (PA2 and PA1) and lectins are the commonly detected protease-resistant polypeptides in peas and chickpeas.[24,28,29] The proteolytic resistance of these albumins is due to their compact structure involving disulfide bonds in protease inhibitors. Lectin structure also has some role in the digestibility of proteins. Phenolic compounds form a complex with proteins, which decreases the solubility of the proteins and makes them less susceptible to proteolytic enzymes.[30]

A multienzyme system consisting of trypsin, chymotrypsin and peptidase is used to evaluate *in vitro* protein digestibility.[31] This method is based on the measurement of change in pH of a protein suspension immediately after 10 min digestion with the multienzyme solution. The change in pH was observed to be highly correlated with the *in vivo* apparent digestibility of proteins in rats. The following regression equation was formulated from the relationship of change in pH and *in vivo* protein digestibility (Y)

$$Y = 210.464 - 18.103X$$

where X denotes the pH of the protein suspension immediately after 10 min digestion with the multienzyme solution. The most significant advantage of this method is that it require less time (<1 h) and has a high degree of sensitivity. This method is useful in evaluating the effects of trypsin inhibitor, chlorogenic acid, heat treatment, *etc.* on protein digestibility. Strong buffer salts used during the experiment may affect the measurement of protein digestibility, but a weak buffering system may not have significant effects.

Another method used for determination of protein digestibility employs pepsin and pancreatin.[32] In this method, the sample was mixed with pepsin (pH 2.0) and incubated at 37 °C for 16 h, followed by addition of pancreatin solution in borate buffer (pH 6.8). The contents were incubated for 24 h and the reaction was stopped by trichloroacetic acid. The digestibility was calculated as nitrogen in supernatant minus nitrogen in enzyme blank supernatant, expressed as percentage nitrogen in the starting material. The protein digestibility of seeds and flours of different pulses is shown in Table 5.2.

The lower digestibility of pulses is primarily due to higher dietary fibre intake, inherent structure of protein which resists proteolysis,[33–36] residual lectin activity or limited protein proteolysis of certain protein fractions. The poor digestibility of pulse proteins reduces their biological value and hence reduces their overall nutritional contribution to a diet. Nutrient digestibility decreases with increase in dietary content.[37] This is attributed to the physical entrapment or/and premature release of food particles from the stomach. Dietary fibre engulfs nutrients in a matrix, impeding the access of digestive enzymes to proteins, and hence can decrease the absorption of the digested protein.[38–42] The inhibition of the digestive enzymes in the presence of dietary fibre cannot be ruled out.[43]

Pulses are normally cooked either as whole grains or decorticated, unlike cereal grains which are usually milled/ground and then cooked. The milling of grains followed by cooking results in the rupturing of the cell wall structure and the release of proteins and starch that assist in proteolysis and amylolysis. Lectins, naturally occurring compounds that have protease and amylase inhibitory activity, also reduce the digestion and absorption of protein, but are inactivated by cooking. However, significant amounts of lectin may remain active even after cooking, depending on the cooking method,[44,45] and insufficient cooking of pulses can lead to the presence of residual amounts of these inhibitors. The regular consumption of pulses containing protease inhibitors can cause enlargement of the pancreas and secretion of more digestive enzymes to compensate for the inactivation.[46,47]

The poorer digestibility of pulses has also been attributed to the presence of carbohydrate–protein or protein–protein bonds, *i.e.* steric hindrance of proteolysis by the carbohydrate moieties of the glycoproteins.[33–36] The formation of a carbohydrate-protein bond was thought to occur due to oxidative coupling of two activated benzene rings,[48] and tyrosine residues are similarly involved in protein–protein binding.

Lower protein digestibility was also attributed to the presence of various antinutritional compounds such as polysaccharides or trypsin inhibitors. Deshpande and Nielson[49] provided evidence that the heated pulse protein is not resistant to trypsin proteolysis *in vitro*, but their research does not preclude some of the trypsin digestion products being resistant to other proteolytic enzymes necessary for absorption.

The nutritional value of a protein is generally determined in terms of amino acid profiles and digestibility. A protein with an excellent amino acid composition may have a fair nutritional value even if it has low digestibility.[50] Both the World Health Organization (WHO) and the United States Food and Drug Administration (FDA) have adopted the protein digestibility corrected amino acid score (PDCAAS) method as the official assay for evaluating protein quality. The PDCAAS value is different from the quality of protein measured by the protein efficiency ratio (PER) or the biological value (BV). The PER is based on the amino acid requirements of growing rats, which differ from those of humans. The PDCAAS method is recommended for evaluating the protein quality of foods intended for human consumption.[51] The true digestibility of protein (TDP) score considers three important parameters of protein quality evaluation,[52] *i.e.* the essential amino acid profile, digestibility, and ability to supply the essential amino acids in the amounts required by humans. Improvements in protein digestibility of raw pea seeds from 83.9% to 87.4% by extrusion and to 84.5% by soaking and cooking in water were observed.[53–55] The increase in protein digestibility by thermal treatment such as cooking, extrusion, and autoclaving has been attributed to the inactivation of antinutritional proteins such as trypsin inhibitors and lectins, and the reduction of antinutritional chemicals such as tannins, phytates, and polyphenols. The antinutritional factors remaining after the thermal treatments and the intrinsic structural factors of pulse proteins were responsible for their relatively lower digestibility as compared to milk proteins (casein). The heating brought about changes in protein conformation such as an aggregation through increased hydrophobicity and disulfide bond formation that also impaired the susceptibility of protein to proteolysis.[56–58]

Table 5.2 Pulse protein digestibility of various pulses and their products

Pulses and their product	*Sample*	*Method used*	*Protein digestibility*	*Sources*
Raw dry bean varieties	Seed	Hsu *et al.*[31]	69.0–72.2%	Nergiz *et al.*[66]
Dry bean soaked-cooking	Seed	Hsu *et al.*[31]	81.3%	Abd- El-Hady and Habiba[67]
Lima beans (B)	Flour blend (1:3, B:M)	Hsu *et al.*[31]	77.9%	Navarrete *et al.*[68]
	Extrudate blend (1:3, B:M)			
Maize (M)	Flour blend (1:1, B:M)		81.5%	
	Extrudate blend (1:1, B:M)			
			76.8%	
			81.75%	
Black beans	Raw grains	Hsu *et al.*[31]	78.9%	Sangronis *et al.*[69]
	Germinated grains		80.7%	
	Germinated-cooked grains		85.0%	
White beans	Raw grains	Hsu *et al.*[31]	77.7%	Sangronis *et al.*[69]
	Germinated grains		78.6%	
	Germinated-cooked grains		85.3%	
Vicia incana	Seeds	Hsu *et al.*[31]	78.0%	Cavada *et al.*[70]
Vicia hirsuta			86.3%	
Vicia faba			83.2%	
Chickpea	Seeds	Hsu *et al.*[31]	76.2%	Vioque *et al.*[71]
Amaranthus cruentus	Seeds	Hsu *et al.*[31]	77%	Betschart [72]
Rapeseed	Seeds	Hsu *et al.*[31]	72.8%	Larbier *et al.*[73]
Chickpeas	Protein isolates	Hsu *et al.*[31]	> 95.0%	Clemente [74]
Canavalia ensiformis,	Seeds	Hsu *et al.*[31]	74.66%	Vadivel *et al.*[75]
C. gladiata			63.39%	
Cassia floribunda			83.32%	
C. obtusifolia			74.66%	
Mucuna monosperma			78.32%	
M. pruriens var. pruriens			72.41%	

Table 5.2 Continued

Pulses and their product	*Sample*	*Method used*	*Protein digestibility*	*Sources*
M. pruriens var. utilis			72.41%	
Mung beans	Flour	Salgo *et al.*[84]	71.1%	Mubarak [76]
Kidney beans: Roba	Raw seeds	Hsu *et al.*[31]	80.7%	Shimelis *et al.*[77]
Awash			71.1%	
Beshbesh			65.6%	
Kidney beans: Roba	Processed seeds	Hsu *et al.*[31]	94.2%	Shimelis *et al.*[77]
Awash			84.1%	
Beshbesh			80.1%	
Raw peas	Seeds	Hsu *et al.*[31]	79.9–83.5%	Park *et al.*[78]
Cooked peas		Hsu *et al.*[31]	85.9–86.8%	Park *et al.*[78]
Faba beans (*Vicia faba* L.)	Flour	in vivo	86.5%	Carbonaro *et al.*[57]
Common beans (Phaseolus vulgaris L.)	Flour	in vivo	72.4%	
Common beans (*Phaseolus vulgaris*)	Flour (phaseolins)	Montoya *et al.*[79]	11–27 %	Montoya *et al.*[79]
Phaseolus vulgaris	Flour (albumins)	Mauron[83]	32.0%	Genovese *et al.*[23]
Bauhinia purpurea	Seeds	Hsu *et al.*[31]	70.0%	Vijayakumari et al[80]
Pigeon peas	Seed	Singh and Jambunathan[32]	60.4–74.4%	Chitra *et al.*[81]
Mung beans	Seed	Singh and Jambunathan[32]	67.2–72.2%	Chitra *et al.*[81]
Soybeans	Seed	Singh and Jambunathan[32]	62.7–71.6%	Chitra *et al.*[81]
Urd beans	Seed	Singh and Jambunathan[32]	55.7–63.3%	Chitra *et al.*[81]
Moth beans	Seed		58.7%	Khokhar and Chauhan[82]

REFERENCES

1. F. B. Hu, *Am. J. Clin. Nutr.*, 2003, **78**, 544–551.
2. A. Philanto and H. Korhonen, *Adv. Food Nutr. Res.*, 2003, **47**, 175–181.
3. R. N. Tharanathan and S. Mahadevamma, *Trends Food Sci. Technol.*, 2003, **14**, 507–518.
4. D. R. Jacobs and D. D. Gallaher, *Curr. Atheroscler. Rep.*, 2004, **6**, 415–423.
5. E. Derbyshire and D. J. Wright, D. Boulter, *Phytochemistry*, 1976, **15**, 3–24.
6. T. B. Osborne, *The Vegetable Proteins*, 2nd ed. Longmans, Green, New York, 1924.
7. Y. Ma and F. A. Bliss, *Crop Sci.*, 1978, **18**, 431–437.
8. U. M. Marquez and F. M. Lajolo, *J. Agric. Food Chem.*, 1981, **29**, 1068–1074.
9. A. Jivotovskaya, V. Senyuk, V. Rotari, C. Horstmann and I. Vaintraub, *J. Agric. Food Chem.*, 1996, **44**, 3768–3772.
10. W. F. Dudman and A. Millerd, *Biochem. Systemat. Ecol.*, 1975, **2**, 18–29.
11. J. W. S. Brown, F. A. Bliss and T. C. Hall, *Theoret. Appl. Genet.*, 1981, **60**, 251–259.
12. M. Tucci, R. Capparelli, A. Costa and R. Rao, *Theoret. Appl. Genet.*, 1991, **81**, 50–58.
13. C. Horstman, B. Schlesier, A. Otto, S. Kostka and K. Müntz, *Theoret. Appl. Genet.*, 1993, **86**, 867–864.
14. R. Casey, C. Domoney and N. Ellis, in *Oxford Survey of Plant Molecular and Cell Biology*, ed. B. J. Miflin, Oxford University Press, Oxford, 1986, pp. 1–95.
15. D. J. Wright, in *Developments in Food Proteins*, ed. B. J. F. Hudson, Elsevier, Amsterdam, 1986, p. 81.
16. R. R. Croy, J. A. Gatehouse, M. Tyler and D. Boulter, *Biochem. J.*, 1980, **191**, 509–516.
17. R. Gupta and S. Dhillon, *Ann. Biol.*, 1993, **9**, 71–78.
18. K. Saharan and N. Khetarpaul, *Plant Foods Hum. Nutr.*, 1994, **5**, 11–22.
19. M. Barac, S. Cabrilo, M. Pesic, S. Stanojevic, S. Zilic, O. Macej and N. Ristic, *Int. J. Mol. Sci.*, 2010, **11**, 4973–4990.
20. E. M. T. Mendoza, M. Adachi, A. E. N. Bernardo and S. Utsumi, *J. Agric. Food Chem.*, 2001, **49**, 1552–1558.
21. S. S. Deshpande and S. D. Damodaran, *Food Chem.*, 1989, **54**, 108–113.
22. C. A. Montoya, P. Leterme, N. F. Victoria, O. Toro, W. B. Souffrant, S. Beebe and J. P. Lallès, *J. Agric. Food Chem.*, 2008, **56**, 2183–2191.
23. M. I. Genovese and F. M. Lajolo, *J. Agric. Food Chem.*, 1996, **44**, 3022–3028.
24. I. Crevieu, B. Carre, A. M. Chagneau, L. Quillien, J. Gueguen and S. Berot, *J. Agric. Food Chem.*, 1997, **45**, 1295–1300.
25. J. Vioque, A. Clemente, R. Sanchez-Vioque, J. Pedroche, J. Bautista and F. Milan, *J. Agric. Food Chem.*, 1998, **46**, 3609–3613.
26. F. M. Lajolo and M. I. Genovese, *J. Agric. Food Chem.*, 2002, **50**, 6592–6598.
27. M. I. Genovese and F. M. Lajolo, *J. Agric. Food Chem.*, 1996, **44**, 3022–3028.
28. M. Le-Gall, J. Gueguen, B. Seve and L. Quillien, *J. Agric. Food Chem.*, 2005, **53**, 3057–3064.

29. M. Le-Gall, L. Quillien, B. Seve, J. Gueguen and J. P. Lalles, *J. Anim. Sci.*, 2007, **85**, 2972–2981.
30. N. R. Reddy, M. D. Pierson, S. K. Sathe and D. K. Salunkhe, *J. Am. Oil Chem. Soc.*, 1985, **62**, 541–549.
31. H. W. Hsu, D. L. Vavak, L. D. Satterlee and G. A. Miller, *J. Food Sci.*, 1977, **38**, 126–130.
32. U. Singh and R. Jambunathan, *J. Food Sci.*, 1981, **46**, 1364–1367.
33. J. H. Pazur and N. N. Arnonson, *Adv. Carbohydr. Chem. Biochem.*, 1972, **27**, 301–341.
34. V. Ganapathy, M. E. Ganapathy and F. H. Leibach, in *Textbook of Gastroenterology*, ed. T. Yamada, 3rd ed., 1999, pp. 464–477.
35. K. C. Chang and L. D. Satterlee, *J. Food Sci.*, 1981, **46**, 1368.
36. G. A. Semino, P. Restani and P. Cerletti, *J. Agric. Food Chem.*, 1985, **33**, 196.
37. C. Kies, *J. Agric. Food Chem.*, 1981, **29**, 435–440.
38. D. A. T. Southgate, *Proc. Nutr. Soc.*, 1973, **32**, 131–136.
39. L. R. Johnson., ed, *Physiology of the Gastrointestinal Tract*, Raven Press, New York, 1981.
40. B. Elsenhaus, R. Blume and W. F. Caspary, *Am. J. Clin. Nutr.*, 1981, **34**, 1837–1848.
41. G. V. Vahouny, in *Dietary Fiber in Health*, Plenum Press, New York, 1986, pp. 139–181.
42. N. Faisant, M. Champ, P. Colonna, A. Buleon, C. Molis, A. M. Langkilde, T. Schweizer, B. Flourie and J. P. Galmiche, *Eur. J. Clin. Nutr.*, 1993, **47**, 285–296.
43. B. O. Schneemen, in *Dietary Fiber in Health and Disease*, ed. G. Vahouny and D. Kritchevsky, Plenum Press, New York, 1982, pp. 73–83.
44. L. U. Thompson, *Food Res. Int.*, 1993, **26**, 131–149.
45. R. L. Rea, L. U. Thompson and D. J. A. Jenkins, *Nutr. Res.*, 1985, **5**, 919–929.
46. B. O. Schneeman and G. Dunaif, *J. Agric. Food Chem.*, 1984, **32**, 477–480.
47. Z. Madar, Y. Tencer and A. Gertler, Y. Birk, *Nutr. Metab.*, 1976, **20**, 234–242.
48. S. C. Fry. *Ann. Rev. Plant Physiol.*, 1986, **37**, 165–183.
49. S. S. Deshpande and S. S. Nielsen, *J. Food Sci.*, 1987, **52**, 1326–1329.
50. W. H. Wong and S. G. Cheung, *Aquaculture*, 2001, **193**, 123–137.
51. Food and Drug Administration, *Fed. Reg.*, **58**, 1993, 2101–2106.
52. E. C. Henley and J. M. Kester, *Food Technol.*, 1994, **48**, 74–77.
53. R. Alonso, G. Grant, P. Dewey and F. Marzo, *J. Agric. Food Chem.*, 2000, **48**, 2286–2290.
54. R. Alonso, A. Aguirrie and F. Marzo, *Food Chem.*, 2000, **68**, 159–165.
55. I. H. Han, B. G. Swanson and B-K. Baik, *Cereal Chem.*, 2007, **84**, 518–521.
56. M. Carbonaro, M. Cappelloni, S. Nicoli, M. Lucarini and E. Carnovale, *J. Agric. Food Chem.*, 1997, **45**, 3387–3394.
57. M. Carbonaro, G. Grant, M. Cappelloni and A. Pusztai, *J. Agric. Food Chem.*, 2000, **48**, 742–749.
58. S. Sai-Ut, S. Ketnawa, P. Chaiwut and S. Rawdkuen, *Asian J. Food Agro-Industry*, 2009, **2**, 493–504.

59. R. K. Wati, T. Theppakorn and S. Rawdkuen, *Asian J. Food Agro-Industry*, 2009, **2**, 245–254.
60. R. Mahajan, S. P. Malhotra and R. Singh, *Plant Foods Hum. Nutr.*, 1988, **38**, 163–173.
61. A. Hameed, T. M. Shah, B. M. Atta, N. Iqbal, M. A. Haq and H. Ali, *Pakistan J. Bot.*, 2009, **41**, 703–710.
62. E. Yüzbaşioğlu, L. Açik and S. Özcan, *Biol. Plant.*, 2008, **52**, 126–128.
63. J. J. Liu, W. Odegard and B. O. de Lumen, *Plant Physiol.*, 1995, **109**, 505–511.
64. T. H. H.Chen, L. V. Gusta, C. Tjahjadi and W. M. Breene, *J. Agric. Food Chem.*, 1984, **32**, 396–399.
65. Y. Hou, Y. Hou, L. Yanyan, G. Qin and J. Li, *J. Biomed. Biotechnol.*, 2010, **10**, 1155.
66. C. Nergiz and E. Gőkgőz, *Int. J. Food Sci. Technol.*, 2007, **42**, 868–873.
67. E. A. Abd El-Hady and R. A. Habiba, *Lebensm.-Wiss. Technol.*, 2003, **36**, 285–293.
68. C. P. Navarrete, R. González, L. Chel-Guerrero and D. Betancur-Ancona, *J. Sci. Food Agric.*, 2006, **86**, 2477–2484.
69. E. Sangronis, M. Rodríguez, R. Cava and A. Torres, *Food Res. Technol.*, 2006, **222**, 144–148.
70. E. P. Cavada, S. R. Drago, R. J. González, R. Juan, J. E. Pastor, M. Alaiz and J. Vioque, *Food Chem.*, 2011, **128**, 961–967.
71. R. Sánchez-Vioque, A. Clemente, J. Pedroche, J. Bautista and F. Millán, *J. Am. Oil Chem. Soc.*, 1999, **76**, 819–823.
72. A. A. Betschart, D. W. Irving, A. D. Shepherd, and R. M. Saunders, *J. Food Sci.*, 1981, **46**, 1181–1187.
73. Z. M. Larbier, A. M. Chagneau and M. Lessire, *Anim. Feed Sci. Technol.*, 1991, **35**, 237–246.
74. A. Clemente, J. Vioque, R. S. Vioque, J. Pedroche, J. Bautista and F. Millan, *Food Chem.*, 1999, **67**, 269–274.
75. V. Vadivel and K. Janardhanan, *Plant Foods Hum. Nutr.*, 2005, **60**, 69–75.
76. A. E. Mubarak, *Food Chem.*, 2005, **89**, 489–495.
77. E. A. Shimelis and S. K. Rakshit, *Food Chem.*, 2007, **103**, 161–172.
78. S. J. Park, T. W. Kima and B. K. Baik, *J. Sci. Food Agric.*, 2010, **90**, 1719–1725.
79. C. A. Montoya, P. Leterme, N. F. Victoria, O. Toro, W. B. Souffrant, S. Beebe and J. P. Lallès, *J. Agric. Food Chem.*, 2008, **56**, 2183–2191.
80. K. Vijayakumari, M. Pugalenthi and V. Vadive, *Food Chem.*, 2007, **103**, 968–975.
81. U. Chitra, V. Vimala, U. Singh and P. Geervani, *Plant Foods Hum. Nutr.*, 1995, **47**, 163–172.
82. S. Khokhar and B. M. Chauhan, *J. Food Sci.*, 1986, **52**, 1083–1084.
83. J. Mauron, in Proteins in Human Nutrition, Academic press, London, 1973, pp 139–154.
84. A. Salgo, K. Granzler and J. Jecasi, in *Proceedings of the international association of the cereal chemistry symposium*, eds. R. Lasztity and M. Hidvegi, Budapest, Akademiai Kiado, 1984, pp 311–321.

CHAPTER 6

Protein Isolates and Concentrates

6.1. INTRODUCTION

Currently, no global classification distinguishes between protein isolates and concentrates. However, generally a minimum protein content of 85% (dry weight basis) for classification as a protein isolate is required. Protein isolates are a more purified form of protein than protein concentrates. Pulse proteins remained underutilised by the food industry due to less information is available on their functional properties related to performance. However, the interest in utilising plant proteins in meat products has grown during the last decade. This chapter focuses on the processing technologies for fractionation, concentration and isolation of pulse proteins and their functional properties.

6.2. PROTEIN ISOLATES

The composition of protein varies depending upon the method employed in their production and raw material. The protein content of isolates produced by isoelectric precipitation and salt extraction methods ranges between 81.9–88.79% and 74.71–81.98%, respectively.[1] Protein isolates have a fat content of less than 1.0%. Isolates prepared by isoelectric precipitation and salt extraction have ash content of 3.05–5.59% and 3.57–5.33%, respectively.[1] The ash content, which represents the inorganic matter, in isolates varies depending on the strength of the acid or alkali used during isoelectric precipitation. The use of stronger acid or alkali generally results in isolates with higher salt contents. The composition of protein isolates prepared by isoelectric precipitation and salt extraction from peas, faba beans, lentils, chickpeas and soybeans is illustrated in Table 6.1. Protein isolates prepared by isoelectric precipitation from pulses, generally, have higher protein and lower carbohydrate contents than isolates prepared by salt extraction.

Pulse Chemistry and Technology
Brijesh K. Tiwari and Narpinder Singh

Published by the Royal Society of Chemistry, www.rsc.org

Table 6.1 Composition of protein isolates prepared by isoelectric precipitation and salt extraction (Adapted from Karaca *et al.*[1])

Isolates	*Method of precipitation*	*Moisture (%)*	*Protein (%)*	*Ash (%)*	*Lipids (%)*	*Carbohydrates (%)*
Pea protein isolates	Isoelectric precipitation	5.08	88.76	5.59	0.55	0.02
Faba bean protein isolates		6.46	84.14	4.03	0.39	4.98
Lentil protein isolates		5.04	81.90	3.63	0.43	9.0
Chickpea protein isolates		6.52	85.4	3.05	0.92	4.11
Soy protein isolates		4.47	87.59	2.09	0.62	5.23
Pea protein isolates	Salt extraction	9.55	81.09	5.33	0.58	3.45
Faba bean protein isolates		7.16	81.98	3.57	0.34	6.95
Lentil protein isolates		6.87	74.71	4.60	0.45	13.37
Chickpea protein isolates		6.95	81.63	3.65	0.56	7.21
Soy protein isolates		10.08	72.64	3.27	0.27	13.74

6.3. METHODS FOR PROTEIN ISOLATES

6.3.1. Air Classification

The pulses are dehulled and cotyledon portion is separated from hulls. The cotyledon portion of pulse grains is milled to very fine particles to achieve complete cellular disruption. The milling machine used for grinding the pulse grains must be capable of producing very fine grind and breaking up cells and cell fragments without severely damaging the starch granules.[2] A pin mill is preferred for milling the pulse grains to a very fine flour. The reduction of grains to fine particles is essentially required to achieve maximum separation of protein and starch fraction by air classification.[3] Finely milled flours are subjected to an air classifier in order to separate the lighter fine flour fraction, rich in protein, from the heavier coarse fraction, rich in starch. The adherent protein is derived from the membranes and stroma of the choloroplasts in which the starch granules developed.[4,5] The starch fraction consists of starch granule agglomerates embedded in a protein matrix, but further purification is carried out by repeated pin-milling and air classification.[6] Milling followed by air classification has been employed to fractionate starch- and protein-rich fractions from cowpeas, great northern beans, lentils, mung beans, navy beans, faba beans and peas.[7,8] Protein separation efficiency (PSE) is calculated as the percentage of the total flour protein recovered in the fine fraction. PSE indicates the effectiveness of air classification for fractionating pulse proteins. The fractionation efficiency of air classifiers varies with the type of classifier and composition of material. PSE decreases with the increase in moisture and fat content in the material to be fractionated. Mung beans and lentils were observed to be the most suitable pulses for air classification, while lima beans and cowpeas were the least suitable.[7] Pulses vary in milling efficiency due to difference in the hardness and thickness and structural rigidity of the cell wall and the degree of adhesion between the cell contents and the cell wall and between proteinaceous material and starch granules.[5]

6.3.2. Water Extraction

Pulse proteins can be extracted with water because they have high amounts of water-soluble proteins. The pulse grains are blended with water (1:10 grain:water ratio) under relatively low temperature conditions (around 4–5 °C). The proteins are extracted in the supernatant; the amount varies in different pulses, depending on extraction conditions and the amount of water-soluble proteins present. Repeated extractions (3–4 times) are done to increase protein recovery. The protein content of first extract is generally, higher than second and third extract. The protein content of the first extract ranged from 54% for chickpea to 67% for smooth pea.[9]

6.3.3. Salt Extraction

Protein extraction using salt extraction process/micellisation is based on the salting-in and salting-out phenomenon of food proteins. The solubility of protein depends on the salt concentration in the solution, beside other factors. Salts at low concentration stabilise the various charged groups on a protein molecule and thus attract protein into the solution and enhance its solubility. This is commonly known as *salting-in*. However, when the salt concentration is increased, a point of maximum protein solubility is usually reached; further increasing the salt concentration reduces the availability of water to solubilise protein. Finally, protein is precipitated when sufficient water molecules are not available to interact with the protein molecules. This phenomenon of protein precipitation in the presence of excess salt is known as *salting-out*. Many types of salts are employed for protein precipitation and purification through salting-out; ammonium sulfate is the most widely used because it has high solubility and is relatively inexpensive. Once the proteins are extracted using an appropriate salt solution of desired ionic strength the solution is diluted, which induces protein precipitation. The precipitated proteins are then recovered by centrifugation or filtration, and dried. Protein isolates containing 87.8% protein were produced from defatted chickpea flour suspension (10% w/v) using sodium chloride (0.5 M, pH 7.0).[10]

6.3.4. Alkaline Extraction

The pulse grains are milled into flour, which is dispersed in water (flour:water ratios between 1:5 and 1:15). A schematic diagram of the alkaline extraction and isoelectric precipitation process for production of pulse protein is shown in Figure 6.1. The pH of the slurry is adjusted to alkaline (pH 8–11) using dilute sodium hydroxide and stirred for varying periods (1–3 h) to solubilise the protein to maximum. The pH and temperature is kept constant during stirring. Extraction is usually carried out at room temperature; however, a higher temperature (up to a maximum of 50 °C) is sometimes used. Protein recovery increases with an increase in temperature, with maximum recovery at around 50 °C is because solubility is higher at this temperature. Extraction at a lower temperature increases the extraction time. The slurry is separated from

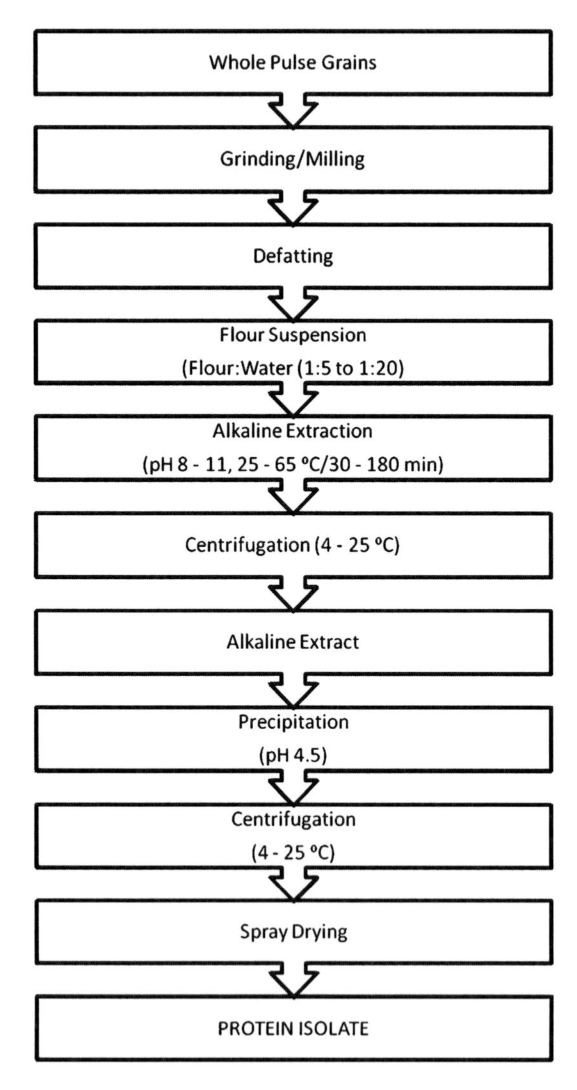

Figure 6.1 Schematic diagram of alkaline extraction and isoelectric precipitation process for production of pulse protein (Boye *et al.*[28])

insoluble material (fibre) by filtration or centrifugation. The pH of the extract is adjusted to the isoelectric point (pH 4–5) where the proteins get precipitated and are recovered by centrifugation or filtration. In a laboratory-scale wet milling process, separation of starch and protein fractions was achieved using hydrocyclones.[11] During this process, dehulled chickpea splits were milled into flour using a pin mill. The flour was defatted using isopropyl alcohol and dilute slurry (1.5%, w/w) was adjusted to pH 9.0, kept overnight and stirred for 1 h. The slurry was then subjected to a hydrocyclone to yield overflow and underflow. The overflow and underflow obtained from the first hydrocyclone were fed separately to a second hydrocyclone. The underflow from the second-pass underflow had a high starch content (89.8–99.7%), while the overflow from the second-pass underflow had a protein content ranging between 64.02% and 88.31%. The highest PSE was obtained from the overflow of the first-pass process. In the large-scale production of starch from pulses, a large number of hydrocyclones can be used to separate the protein-rich fraction from the starch-rich fraction. The hydrocyclones separate the protein- and starch-rich fractions by taking the advantage of the difference in their density. The protein-rich fraction has a lower density than the starch-rich fraction, and hence forms the overflow from the hydrocyclone during fractionation (Figure 6.2a,b). The commercial Flottweg process (Figure 6.3) involved mixing flour with water after adjusting the pH to an alkaline range (8–11) in the first tank (Figure 6.4). The slurried flour is then sent to the first decanter (Figure 6.4) centrifuge for separation of low-protein fibres from protein-rich solution. The fibre is pumped into the second tank and mixed with water. The fibre is washed of alkali and separated from the alkaline water in the second decanter, and is then dried and used as animal feed. The alkaline water is sent back to the first mixing tank. The protein-rich solution obtained from the first decanter is pumped into a third tank and adjusted to the isoelectric point (pH 4–5) to precipitate proteins. The precipitated proteins are then recovered from the mother liquor using a sedicanter (centrifugation) or filtration. The precipitates of proteins obtained from the sedicanter still contain some impurities from the mother liquor. Thus the proteins are re-slurried with water in the fourth tank and separated again using a second sedicanter (Figure 6.4). The protein precipitates are then neutralised and dried. Protein purity and recovery are dependent upon the raw material and extraction conditions such as particle size, temperature, time, flour: solvent ratio and equipment type. The centrifugation machines used, and gravitational forces employed for centrifugation, also affect the purity and extraction rate.

6.3.5. Acid Extraction

The acid extraction method is based on the same principle as alkaline extraction except that the protein extraction is done under acidic conditions. The solubility of pulse proteins is high both under very acidic conditions (*i.e.* pH <4) and alkaline conditions (*i.e.* pH <9). The low pH range is therefore used to solubilise proteins, followed by precipitation of protein at the

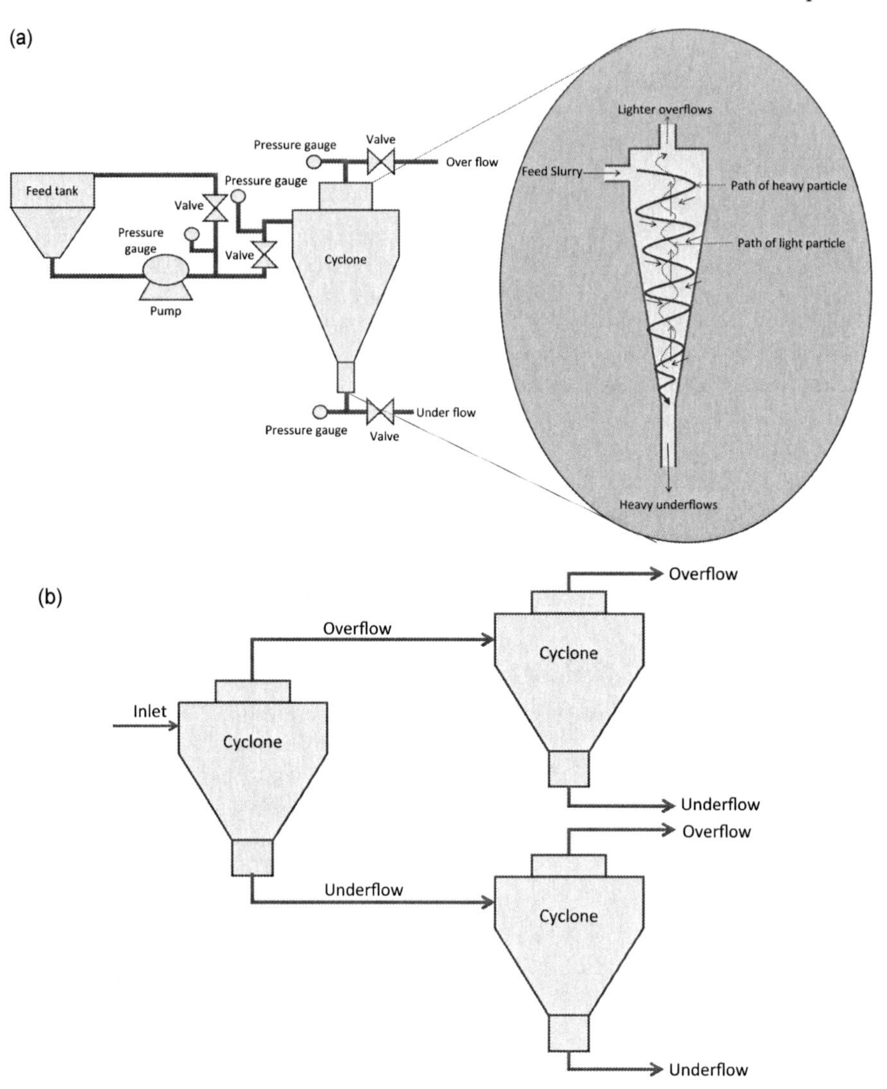

Figure 6.2 a. Schematic of hydrocyclone system (Adapted from Emami *et al.*[11]); b. Schematic of hydrocyclones process (Adapted from Emami *et al.*[11])

isoelectric point. The precipitated proteins are then recovered by centrifugation, filtration or hydrocyclones and dried. The acid extraction method is less commonly used than alkali extraction.

6.3.6. Ultrafiltration

Ultrafiltration refers to a pressure-driven separation technique in which a membrane is employed to separate different components in a fluid mixture. Ultrafiltration membranes have pore sizes less than 0.01 μm and separation of

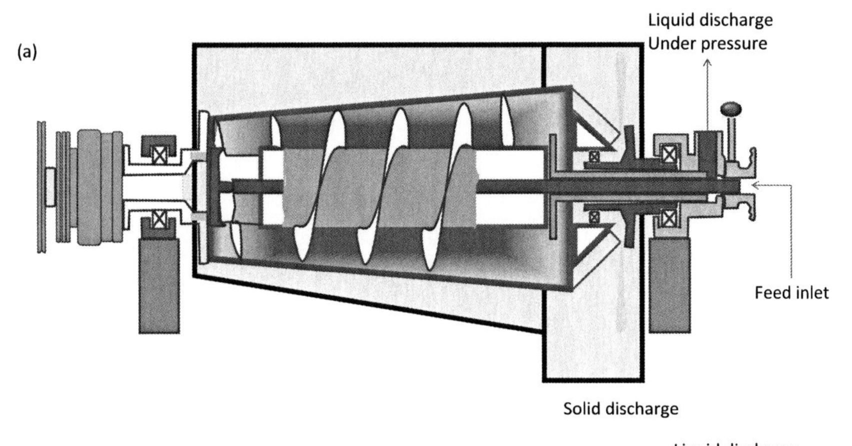

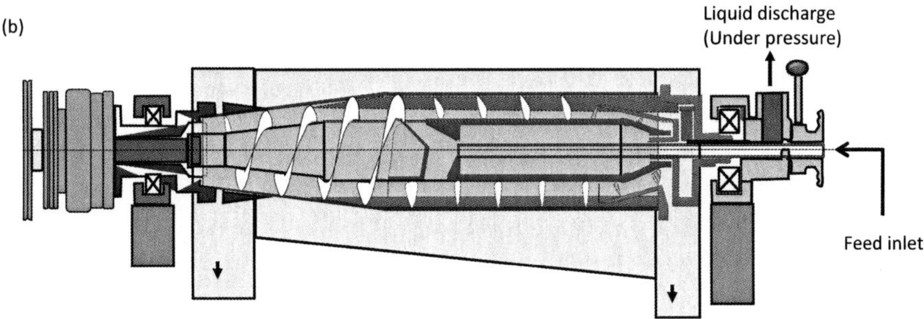

Figure 6.3 Schematic diagram of Flottweg sedicanter (a) and decanter (b)

different components in a fluid occurs based on molecular size. In this process, a pressure of 25–150 psi is employed to expel the water molecules through the membrane while the colloidal solids and salts are retained. The supernatant and protein precipitate obtained after either alkaline or acid extraction is subjected to ultrafiltration or ultrafiltration/diafiltration to concentrate the proteins. Membranes with specific molecular weight are selected depending upon the proteins to be retained. Ultrafiltration can also be used to remove undesirable components from proteins: for example, removal of antinutritional factors such as oligosaccharides or phytic acid during production of protein isolates. This technique also makes it possible to produce purified proteins with superior functional properties.

6.4. PROTEIN CONCENTRATES

Pulses are dehulled, flaked and defatted followed by the removal of soluble sugars. The removal of fat and soluble sugars increases the protein content of the product. During extraction, most of the sucrose and non-digestible oligosaccharides (stachyose, raffinose and verbascose) are removed. The removal of these undesirable oligosaccharides results in concentrates with less tendency to cause flatulence as compared to pulse flours or grains. Different

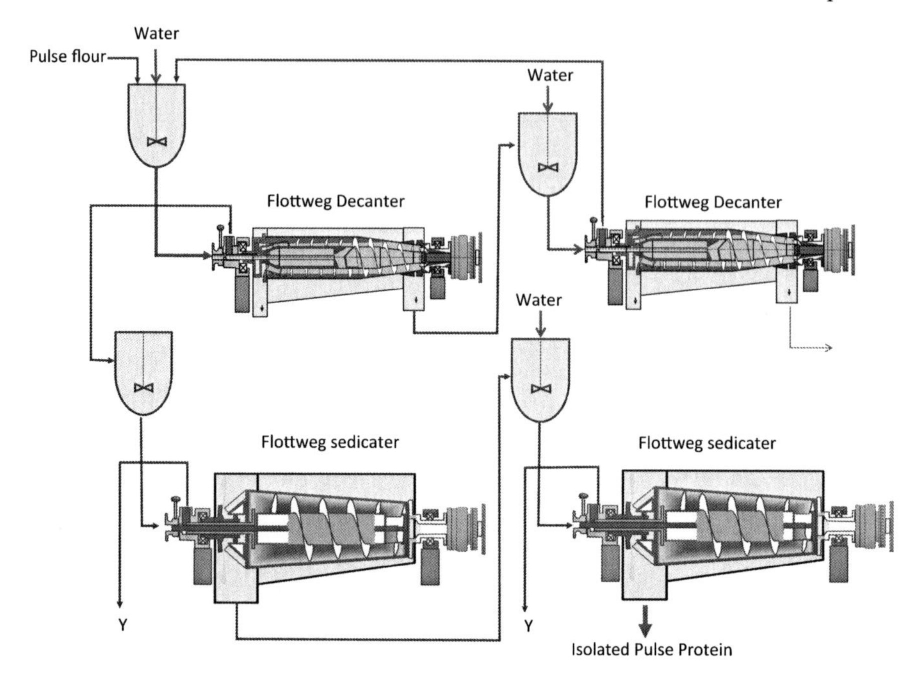

Figure 6.4 Commercial Flottweg process for production of pulse protein isolates

extraction methods are used to remove soluble sugars in the production of concentrates. The main aim is to extract the sugars without solubilising the protein portion. The yield and quality of concentrates vary with the raw material, solvent and extraction conditions used. The extent of heating during the defatting process has a significant effect on the quality and recovery of protein concentrates. Excessive heat treatments deteriorate the functional properties of protein concentrates.

6.5. METHODS FOR PROTEIN CONCENTRATES

Three processes are commonly employed for the production of protein concentrates: aqueous alcohol wash, acid wash and hot water leaching.

6.5.1. Aqueous Alcohol Process

This process employs aqueous alcohol (50–70%) as an extraction medium and is the most commonly employed process for producing protein concentrates on a commercial scale. During extraction, soluble sugars are extracted along with a small amount of soluble proteins. Proteins are denatured and become insoluble during processing. The alcohol is recovered by flash desolventising and is reused. Selecting the appropriate aqueous alcohol concentration where the proteins have minimum solubility is most important. Also, the use of excess water in the extraction medium is avoided to reduce the drying cost. More energy is required to evaporate the water at a later stage when diluted medium is used during extraction, and this adds to the cost of the final product. Protein concentrates made by this process have low solubility and poor functionality. The aqueous alcohol extraction removes some of the constituents that impart undesirable colour and flavour to protein concentrates. Hence, this process results in concentrates with better colour, less beany flavour and lower content of oligosacchrides.

6.5.2. Acid Wash Process

The majority of pulse proteins are globulins, which are insoluble in water in the region of their isoelectric point. Pulses are dehulled, flaked, defatted and washed with water near the isoelectric point (pH 4–5) that removes soluble sugars from the matrix of proteins and polysaccharide. After acid–water washing, the flakes are neutralised and dried. Some of the proteins which are soluble at pH 4–5 are also lost during processing, which reduces the protein recovery. This process yields protein concentrates with better functional properties than those produced by the aqueous alcohol process, because concentrates with higher solubility are produced.

6.5.3. Hot Water Extraction Process

Pulse proteins are easily denatured by heating and become insoluble in water. Moist heat treatments are more effective for denaturation than dry heat. The low molecular weight materials, including soluble sugars, are extracted in water at high temperature from the insoluble protein and polysaccharide matrix. Proteins extracted with this method have poor solubility and functionality.

6.5.4. Enzyme Liquefaction Process

Protein concentrate can be prepared by liquefying the starch present in flours with α-amylase and subsequently washing out the degraded starch. The grains are decorticated, milled into flour and liquefied with a thermostable α-amylase either in batch or extruder, or both (Figure 6.5). The α-amylase breaks the α-1,4 linkage of starch and liquefies. After liquefaction, the α-amylase is inactivated by heating, and then washed and centrifuged. The protein concentrate is obtained in the form of a precipitate and dried. A similar process involving enzymatic treatment of rice with a heat-stable α-amylase and a cellulase to isolate proteins up to 86% concentration was suggested by Paraman *et al.*[12] Barrows and others[13] described the production of protein concentrate from starch-containing grain or oil seed using enzymes that hydrolyse starch, maltodextrins and β-glucans. Protein-enriched flour was produced by digesting geletinised whole-grain sorghum flour with α-amylase overnight at ambient temperature.[14] Liquefaction of flour yields concentrates containing protein, fibre and lipids. Such processes employ enzymes and do

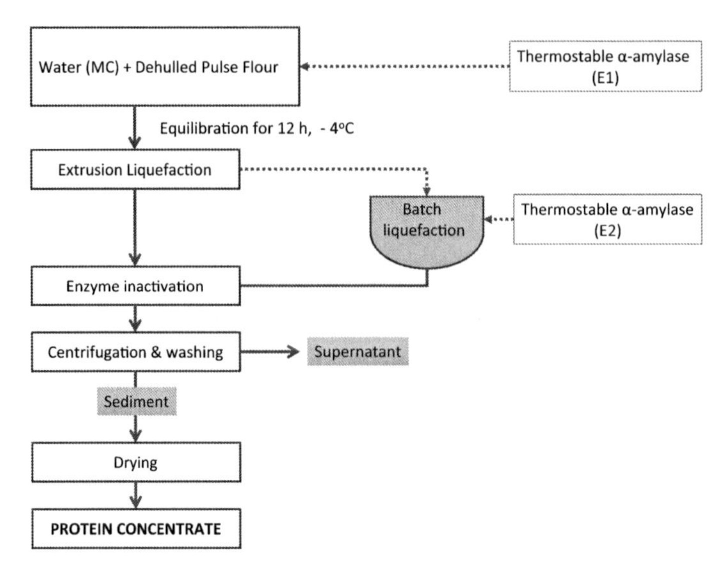

Figure 6.5 Schematic for extrusion–enzyme liquefaction process (Adapted from de Mesa-Stonestreet *et al.*[16])

not involve the use of chemicals, hence the concentrates produced are considered safe and are preferred for use in different products by the food industry. Protein concentrates with 65% protein content were produced from rice flour after treatment with a heat-stable α-amylase.[15] A method for concentrating proteins from sorghum flour utilizing a combination of extrusion and α-amylase treatment for starch liquefaction was developed by de Mesa *et al.*[16] This method involves the use of thermostable α-amylase from *Bacillus licheniformis* for starch liquefaction. In comparison to a traditional batch liquefaction process, extrusion cooking is a high-throughput process and requires less labour. The mechanical shearing during extrusion cooking degrades starch and disrupts protein bodies that limit the functionality and digestibility of proteins. The extrusion cooking liquefies starch and assists in its easy and faster removal. This process potentially results into protein concentrates of high purity.

6.6. FUNCTIONAL PROPERTIES

Functional properties of isolates and concentrates include the physicochemical properties of proteins which affect the behaviour of a food system during processing, handling, storage and consumption. The functional properties of proteins contribute quality and sensory attributes to food products. Solubility, water binding, oil binding, gelation, emulsification and foaming are the important functional properties of proteins which are measured for determining the behaviour of proteins in different food applications. Proteins from different plant sources have a wide range of properties due their heterogeneous structure and composition as well as variation in interaction with different food components (carbohydrates, lipids, ions, *etc.*) or ingredients present in a food system. It is necessary to evaluate the functional properties of proteins in order to understand their expected role in food systems and develop new products.

The processing methods such precipitation, heating, ultrafiltration, drying, *etc.* used for production of protein isolates and concentrates have variable effects on functional properties. Table 6.2 shows the variations in some functional properties of protein isolates prepared by isoelectric precipitation and salt extraction.

6.6.1. Protein Solubility

Protein solubility is an important functional property which depends on the balance of hydrophobic and hydrophilic groups on the protein surface. In other words, protein solubility is the expression of the equilibrium between the protein–solvent (hydrophilic) and the protein–protein (hydrophobic) interaction.[17] Protein–solvent interactions are generally influenced by environmental factors (*e.g.* pH, ionic strength and temperature), solvent type[18] and processing methods (*e.g.* extraction or post-extraction treatments).[19] The protein–protein interaction is facilitated through hydrophobic interactions and leads to precipitation, whereas protein–solvent interactions promote hydration and solubility.[20] Protein conformation affects solubility; the presence of polar charged amino acids near the surface increases the rate of solubilisation of the protein.

During solubility determination, different conditions are employed for solubilising the proteins. After solubilisation the quantification of protein is done by various methods. Protein solubility is measured as the amount of

Table 6.2 Solubility, emulsifying activity index and emulsion stability index of protein isolates prepared by isoelectric precipitation and salt extraction (Adapted from Karaca *et al.*[1])

Isolates	*Solubility (%)*	*Emulsifying activity index (m^2 g^{-1})*	*Emulsion stability index (min)*
Isoelectric precipitation			
Pea protein isolates	61.4	42.87	12.4
Faba bean protein isolates	89.65	44.29	69.39
Lentil protein isolates	90.73	44.51	86.79
Chickpea protein isolates	91.20	47.90	82.94
Soy protein isolates	96.53	44.20	85.97
Salt extraction			
Pea protein isolates	38.12	42.73	10.89
Faba bean protein isolates	52.5	37.11	10.97
Lentil protein isolates	89.88	37.17	11.02
Chickpea protein isolates	30.16	33.83	10.92
Soy protein isolates	96.79	43.35	25.04

protein solubilised at a particular pH under set stirring and centrifugation conditions. Proteins are stirred mechanically for a known duration, then centrifuged to separate the supernatant. The protein content in the supernatant is then determined. Protein solubility is expressed as water-soluble nitrogen or protein, nitrogen solubility index and protein dispensability index. The solubility determination is significantly affected by stirring and centrifugation conditions. Generally, solubility decreases with increase in centrifugation. The pH of the suspension is critical during measurement of protein solubility because protein solubility varies with pH. The protein solubility profile at different pH values is shown in Figure 6.6. Proteins have minimum solubility at pH 4–5 (isoelectric pH) and solubility increases with decrease or increase in pH. Morr *et al.*[21] developed a rapid and simple method for determining the solubility of food protein products. Briefly, a protein solution was prepared by dispersing the sample in sodium phosphate buffer with pH adjusted to 7, followed by overnight stirring at 4 °C. The solution was centrifuged and the nitrogen content of the supernatant was determined using the micro-Kjeldahl/Biuret method. Protein solubility (%) was calculated as the percentage ratio of the nitrogen content of the supernatant and the total nitrogen in the sample.

Thermal treatments such as moist heating, dry heating, autoclaving, boiling, extrusion cooking and drum-drying processes reduce protein solubility. The solubility of protein isolates varies with the method used for their preparation. Chickpea protein showed solubility of 72.5% and 60.4%, respectively during micellisation and isoelectric precipitation at pH 7.0.[10] Fernandez-Quintela *et al.*[22] evaluated the solubility of protein isolates from faba beans and peas with protein contents of 81.2% and 84.1%, respectively; they had the lowest solubility between pH 4 and 6 and the highest at pH 8–9. Pea protein isolates had lower solubility than faba bean and soy protein isolates at alkaline pH values. Faba bean protein isolates prepared by isoelectric precipitation had solubility of approximately 66% at both pH 3 and pH 7.[23] Pea protein isolates produced by isoelectric precipitation (61.4%) had high solubility than those produced by salt extraction (38.12%).[1] The differences in solubility of protein isolates prepared by different methods are attributed to difference in surface

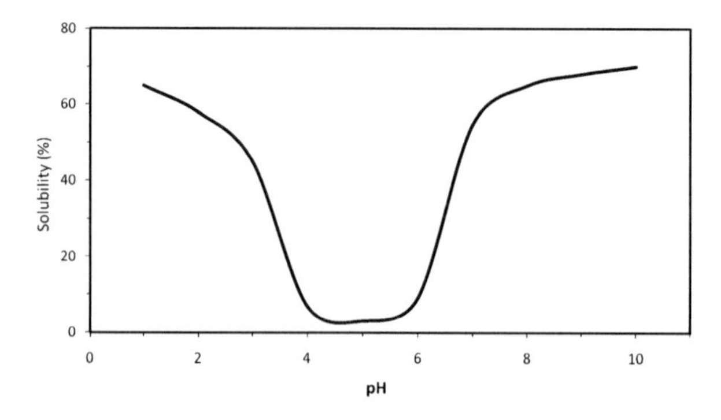

Figure 6.6 Protein solubility profile at different pH

hydrophobicity and surface charge. The solubility of protein isolates from Lentils, Faba bean and Pea were positively related with their surface charge and negatively with surface hydrophobicity.[1]

6.6.2. Protein Gelation

The protein slurry can be converted by heating to a high-viscosity progel, which sets to a gel upon cooling. The progel is irreversible, and heating the progel to a high temperature converts it to a metasol, which does not form a gel on cooling. The ability of pulse protein molecules to interact and form three-dimensional network structures, following thermally induced denaturation and molecular unfolding, is a key functional property for products of a semi-liquid or solid nature. It influences textural and flavour characteristics of the resultant products. Pulse grain proteins are globular in nature, tending to form gel network structures when their structure is only mildly disturbed following heating above a minimum unfolding temperature. Physical interactions, mainly hydrophobic and hydrogen bonding, are the key forces involved in the development of a gel network by globular proteins, while disulfide bridges can also contribute to the gel structure.[24] Processing conditions affect the gelation properties of proteins. The method of Coffmann and Garcia[25] has been used extensively in different studies to determine the least gelation concentration (LGC) of proteins. This method involves the preparation of protein suspensions in water of appropriate concentration (2–20%, w/v) in test tubes. The test tubes containing these suspensions are heated for 1 h in a boiling water bath followed by rapid cooling under running cold tap water. The test tubes are then further cooled for 2 h at 4 °C. LGC is determined as that concentration when the sample from the inverted test tube does not fall down or slip. The globulin fraction forms a gel at a concentration of around 20%. LGC of 12% for cowpea protein isolate and 10% for soy protein isolates[53] and 10% for mung bean protein isolate[25] have been observed. The method used for the preparation of isolates has greater effect on gelation behaviour of proteins than its composition. However, some studies observed no effect of protein composition.[26,27] Better gelling properties (*i.e.* lower LGC) were observed for proteins prepared by ultrafiltration than for those produced by isoelectric precipitation.[27,28] Chickpea protein isolate containing 91.5% (w/w) protein prepared by isoelectric precipitation had an LGC of 14% (w/v).[29] The gelation capacity of protein varies with the source of protein.[22] Faba bean protein isolates prepared using isoelectric precipitation had an LGC of 14% (w/v) compared to 18% and 16% (w/v), respectively for pea and soy protein isolates. Gelation decreased with germination of beans and increased with fermentation. The pH and addition of salts (sodium chloride and calcium chloride) affects the gelation properties of protein isolates.[29] Kaur and Singh[30] reported LGCs for various chickpea protein isolates ranging from 14% to 18%. Chickpea protein isolates required an LGC of 14% (w/v) to form a gel at pH 7.0 with deionised water, whereas 18% was needed at pH 3.0. Gels prepared in the presence of calcium chloride had greater strength than those prepared in sodium chloride at pH 3.0. LGCs of 10%, 8% and 12%, respectively, were observed for great northern bean flour, protein concentrate and protein isolate.[31]

6.6.3. Zeta Potential and Surface Hydrophobicity

The zeta potential (ζ) is commonly used to interpret the surface net charge of protein suspensions and their stability. The ζ of a protein solution is influenced by its ionic strength. The ζ value can be related to the stability of colloidal dispersions. The ζ indicates the degree of repulsion between adjacent, similarly charged particles in dispersion. When the potential is low, attraction exceeds repulsion, which results in the breaking up and flocculation (coagulation) of a dispersion. Colloids with high ζ (negative or positive) are electrically stabilised while those with low ζ values tend to flocculate. The ζ values generally change from a negative to a positive value with decrease in pH from alkaline (pH 10) to acidic (pH 3). The electrostatic repulsion pattern gradually change from being between negatively charged proteins to being between positive charged proteins, as a result of the gradual protonation of carboxyl groups and deprotonation of amino groups of the proteins. The surface charge of protein varies with extraction method and source of proteins. For, example, protein isolates produced by isoelectric precipitation had higher surface charge that those produced by salt extraction.[1] The differences in isolates properties between two extraction methods have been attributed to difference in composition of proteins. Isoelectric precipitated proteins comprised of mainly globulins while salt extracted proteins composed of both globulins and albumins.[27] The globulins have greater surface hydrophobicity as compared to albumins.[27]

The surface charge and hydrophobicity has a significant effect on functional properties (such as emulsifying properties) of proteins. The surface charge of proteins changes depending upon ionisation of surface groups that may be acidic or basic due to the side chains of the component amino acids. These ionisation reactions are acid–base equilibria, which depend on the pH of the solution. Tang and Sun observed pH-dependent ζ values for mung bean globulins; they reported a progressive increase in ζ from a negative value (−38.5 mW) to a positive value (38.2 mW) with pH decrease from 9.8 to 3.5.[32] The decrease of ζ under high acidic conditions may be due to the hydrolysis of glutamine and/or asparagine into glutamic acid and/or glutamic acid. The acidic/basic amino acid ratio of proteins is closely related to the surface charge (or ζ) as well as the amino acid composition. Tang and Sun[33] studied the physicochemical and conformational properties in three vicilins from *Phaseolus* legumes and suggested an association of the surface net charge of the proteins with the balance between acidic and basic amino acids, as well as the content of polar uncharged amino acids.

6.6.4. Water Absorption Capacity

Water absorption is a desirable trait in food products such as sausages, custards, puddings and dough because these products are supposed to imbibe water without dissolution of proteins, thereby attaining body thickening and viscosity.[34] The water absorption capacity (WAC) of proteins varies with the availability of polar amino acids, which are the primary sites of water interactions with proteins. The WAC of protein isolates is higher that of whole-grain flours. The process used in the production of protein isolates affects WAC. WAC capacities for pulse protein concentrates range between 0.6 and 2.7 g g^{-1}.[28] Chickpea protein isolates prepared using isoelectric precipitation (84.8% protein content) and micellisation (87.8% protein content) had WACs of 4.9 and 2.4, respectively.[10] Protein isolates from different pulses has significantly different WACs. Both the cultivar and the process conditions affect the WAC. Chickpea protein isolates had WAC between 2.3 and 3.5 g g^{-1}, lower for kabuli chickpea isolates than those from desi chickpea cultivars.[30] A faba bean protein isolate with protein content of 81.2% prepared using isoelectric precipitation had a WAC of 1.8 g g^{-1} *vs* 1.7 g g^{-1} for a pea protein isolate with 84.1% protein content prepared by same method.[22]

6.6.5. Oil Absorption Capacity

Fat or oil absorption capacity (OAC) is an important functional property of protein since it enhances the mouth feel and helps in retaining the flavour. Proteins with higher OAC are desirable in ground meat formulations, meat replacers and extenders, doughnuts, baked goods and soups.[30] Determination of OAC involves stirring samples with corn oil, centrifugation, then determination of free oil.[35] The OAC is calculated as the percentage ratio of the weight of fat absorbed by the sample and the weight of the sample. Non-polar amino acid side chains are capable of forming hydrophobic interactions with the hydrocarbon chains of lipids. Differences in OAC are generally attributed to variations in the presence of non-polar side chains.

OACs for protein isolates are higher than those of the corresponding flours. OAC is also affected by the source of proteins, *e.g.* kabuli chickpea protein isolate has higher OAC than desi chickpea protein isolate. An OAC of 2.08–3.96 g g^{-1} for chickpea protein isolates was observed.[30] Chickpea protein isolates prepared using isoelectric precipitation and micellisation, respectively, had OACs of 1.7 and 2.0 g g^{-1}.[10] Faba bean protein isolates with protein content of 81.2% prepared using isoelectric precipitation had an OAC of 1.6 g g^{-1}, higher than for pea (1.2 g g^{-1}) and soy (1.1 g g^{-1}) protein isolates.[22]

6.6.6. Emulsifying Properties

An emulsion consists of two or more completely or partially immiscible liquids, such as oil and water, where one liquid (the dispersed phase) exists in the form of droplets suspended in the other (the continuous phase). Because the surface of each droplet is an interface between hydrophobic and hydrophilic molecules, it is inherently thermodynamically unstable. The stability of an emulsion is affected by environmental conditions (temperature, pH, stresses). Food emulsions are exposed to a variety of physical, chemical and microbiological stresses during storage, which influences the emulsifying properties. Proteins are used as emulsifiers because they have the ability to be adsorbed at the oil-water interface and form layer around oil droplets. Proteins decrease interfacial tension and prevent coalescence by forming a physical barrier at the oil-water interface.[36] The emulsifying properties of proteins are measured as emulsion capacity (EC), emulsion activity index (EAI), and emulsion stability index (ESI). EAI provide an estimate of the interfacial area stabilised per unit weight of protein based on the turbidity of a diluted emulsion.[36] ESI reflects the ability of the proteins to adsorb rapidly at the water–oil interface during the formation of the emulsion, preventing flocculation and coalescence, and a high value of EAI is attributed to a highly hydrophobic nature of the protein.[37] The emulsifying properties of proteins depend on (1) the capacity to decrease interfacial energy due to adsorption at the oil–water interface; and (2) the electrostatic, structural and mechanical energy barrier caused by the interfacial layer that opposes destabilisation processes.[38] The emulsifying properties are affected by solubility,[39–41] surface charge,[42] hydrophobicity,[43] source,[23] processing conditions,[44] presence of salts[45,46] and extraction method.[10] In some studies, no relation between protein solubility and emulsifying properties has been observed.[47]

6.6.7. Foaming Properties

A foam is generally described as a colloid of many gas bubbles trapped in a liquid or solid. Foaming ability is related to the flexibility of protein molecules, which can reduce surface tension. A globular protein, which is relatively difficult to surface denature, gives low foamability,[48] but soluble proteins also contribute to foaming. The foam stability (FS) represents the percentage volume of foam remaining after a specified time compared to the initial foam volume. The FS is related to the concentration of soluble proteins, because these have the capacity to reduce surface tension at the interface between air bubbles and the surrounding liquid, and protein molecules can unfold and interact with one another to form multilayer protein films with an increased flexibility at the air–liquid interface; as a result, air bubbles become resistant to breakage and foams are stabilised.[49] Differences in foaming capacity (FC) between processed and raw bean proteins indicated that processing treatments (boiling, roasting, fermention, malting, *etc.*) decreased FC.[50] FC and FS improved in the acid and alkaline regions, and were better in both these regions than in the isoelectric region.[51] The improved FCs in the acid and alkaline regions was attributed to higher solubility of the pigeon pea proteins at these pH values. The FS of pigeon pea concentrate was found to improve as protein content and ionic strength increased.

The foaming properties of protein isolates are mainly affected by the method used for their production. Foam expansion (FE) and FS of pea protein isolates extracted by ultrafiltration were better at pH 5 than in isolates obtained by isoelectric precipitation.[52] Foam expansion (FE) and FS of chickpea protein isolates prepared using micellisation were 43.3% and 59.2%, respectively, against 47.5% and 66.6%, respectively for those prepared by isoelectric precipitation.[10] Faba bean protein isolate prepared by isoelectric precipitation had FE of 15% against 22% for soy protein isolates.[22] Faba bean protein isolates showed FS of 77% against 94% and 77%, respectively for pea protein and soy protein isolates. Different studies have shown different results for foaming properties due to variations in the purity of the isolates and conditions used during testing.

6.6.8. Thermal Properties

Thermal properties of proteins are useful for designing appropriate heat-processing techniques for the application and incorporation of proteins in food products where heat-induced gelation is essential for structure and emulsion stability.[53] The thermal stability of the proteins means their resistance to aggregation in response to heating. Thermal properties help in determining the stability of proteins during thermal processing. During heating, the native state of protein change to the denatured state, accompanied by the disruption of intermolecular bonding that results in the unfolding and aggregation of protein molecules. Protein denaturation involves conformational changes without any change in the amino acid sequence (primary structure). Denaturation does not break the peptide bonds, but it disrupts the normal alpha-helix and beta-sheets in a protein and uncoils them into random shapes. Differential scanning calorimetry (DSC) is often employed to evaluate the thermal properties of proteins and is considered reliable for monitoring thermal transformations during their denaturation. DSC thermograms provide an overall view of the stability of proteins under specified heating conditions. The change from the native to denatured state during heating is a phenomenon that requires a significant amount of heat, which appear as an endothermic peak in DSC thermograms. A thermogram for field pea and kidney bean proteins is illustrated in Figure 6.7. The onset temperature (T_o), peak denaturation temperature (T_d) and heat of transition or enthalpy (ΔH) are the parameters

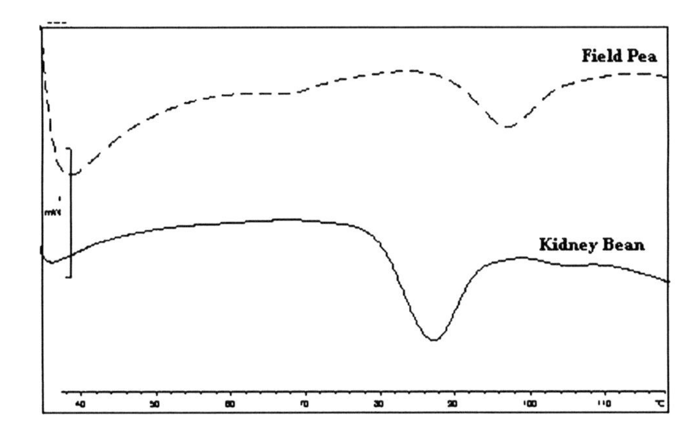

Figure 6.7 DSC thermograms of field pea and kidney bean protein isolates

obtained from the DSC thermograms. T_d is the temperature at which a transition occurs and is a measure of thermal stability under specified experimental conditions. The ΔH is measured as the area under the endothermic peak, which reflects the proportion of undenatured protein or the extent of ordered structure of a protein under specific experimental conditions.[54] Grain globulins possess T_d in the range of 83.8–107.8 °C.[55,56] T_d for bean proteins ranged from 84 °C to 91 °C, with kidney and adzuki beans having higher thermal stabilities and T_d values of 90.2 °C and 90.07 °C, respectively.[57,58] T_d for red bean globulins was 86.4 °C.[59] The thermal stability of proteins is dependent upon the balance of polar and non-polar residues. Proteins with a higher proportion of non-polar residues generally have higher T_d (*i.e.* higher heat stability). Besides the differences in structure and composition, interactions of proteins with residual salts in the isolates also have some effect on thermal stability of proteins.[60,61] Kidney bean phaseolin is much more homogeneous in polypeptide constituents and is thermally stable, with higher flexibility in quaternary conformation but lower flexibility in tertiary conformation.[32]

6.6.9. Dynamic Rheology

The changes associated with gel-forming phenomena in proteins suspension as a function of temperature are measured with dynamic rheometry. The parameters derived from rheometry to evaluate gelling behaviour of proteins are G' (storage or elastic modulus), G'' (loss modulus or viscous) and loss tangent ($\tan \delta = G''/G'$). G' is a measure of the elastic component of the network and represents the strength of the structure contributing to a three-dimensional network. G'' is a measure of the viscous component and may represent interactions which do not contribute to the three-dimensional nature of the protein molecule's denatured and exposed hydrophobic residues as a preparatory stage for gel formation. The loss tangent, $\tan \delta$, is a measure of the energy lost due to viscous flow compared to the energy stored due to elastic deformation in a single deformation cycle. During the heating of globular proteins, protein–protein interactions normally occur following denaturation. Various factors such as pH, presence of ionic salts, concentration, temperature and heating rate affect gel network formation. Higher G' indicates a stronger intermolecular network and increased interactions between protein–protein and protein–polysaccharide molecules, while low $\tan \delta$ values indicate a more elastic network.[62] Oscillatory measurements are used to determine the gelling point of proteins by several different methods. The crossover of the G' and G''[63–66] and linear extrapolation of the rapidly rising G' to the intercept with the time axis[67,68] are the methods used to determine gelling point. After the gelling point, protein aggregates are bound together into a continuous molecular structure.[69] The intercultivar and interspecies differences in gelation behaviour of globulin and legumin are well known. Pea globulin underwent heat-induced gelation whereas legumin did not gel under the same conditions.[70] O'Kane *et al.*[71–73] indicated that both pea vicilin and legumin could form gels. G' and G'' progressively increase up to a certain temperature around 80–82 °C due to the gradual development of the network structure. In the initial stage of the gelation process, the viscous behaviour of the system ($G'' > G'$) predominates. During the final stage, the elastic behaviour predominates ($G' > G''$), when protein molecules have aggregated and cross-linked resulting into formation of a three-dimensional network. After the maximum set temperature is reached the cooling phase is started. During this phase, the G' and G'' values continue to increase steadily due to continuous cross-linking as well as the slow formation and rearrangement of the network structure. The formation and stability of protein gels are attributed to the formation of hydrophobic interactions and hydrogen bonds.[74] Tan δ is an important indicator to distinguish gel formation; it decreases gradually until 80–85 °C, and remains constant and low throughout the remaining heating and cooling phases. Generally a stable gel is formed at the beginning of the cooling phase, and continuous cooling causes a slight improvement in both the G' and G'' components of the network. Figure 6.8 shows the changes in G' (storage modulus) and G'' (loss modulus) of kidney bean and field pea protein isolates during heating and cooling. When aggregation is suppressed before unfolding, the resulting network has lower opacity and higher elasticity than if random aggregation and denaturation occur simultaneously or if random aggregation occurs before denaturation.[75] The higher the randomness of aggregation, the more likely it is that a coagulum is obtained instead of a gel.[76]

The three-step process that is generally accepted for heat-induced gelation of globular proteins as described by Clark, Kavanagh, and Ross-Murphy[77] is summarized as follows:

(1) Denaturation of the protein with subsequent exposure of hydrophobic residues

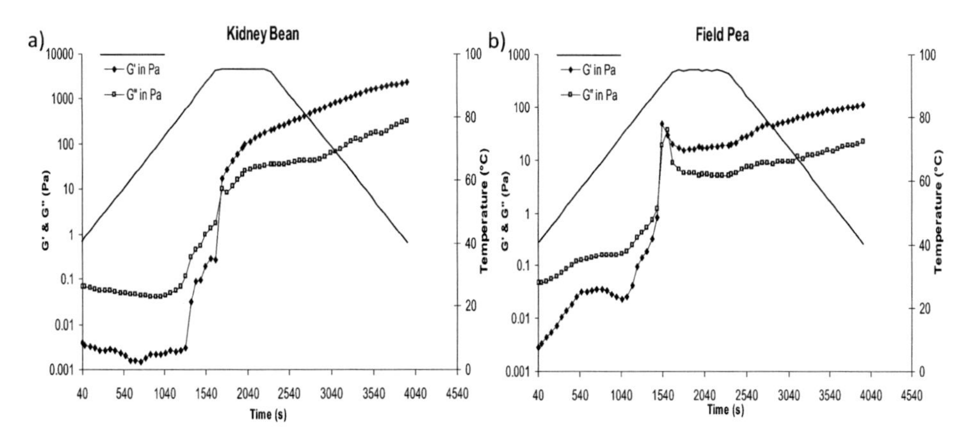

Figure 6.8 Changes in G' (storage modulus) and G'' (loss modulus) of kidney bean (a) and field pea (b) protein isolates during heating and cooling

(2) Intermolecular hydrophobic interaction of the unfolded proteins (aggregation)
(3) Agglomeration of aggregates into a network structure.

During the cooling phase, the network continues to develop and is strengthened by the formation of many short-range interactions such as hydrogen bonds.[71] The network formation by the legumin proteins in pea and soy is mainly supported by hydrophobic and hydrogen bonds, whereas disulfide bonds have minimum involvement.[72] The extent of exposure of hydrophobic residues on the protein molecular surface during heat denaturation induced gel formation. The increase in G' and G'' of gels with the increase in temperature during heating has been attributed to the presence of hydrophobic forces responsible for gelation.[78] Both hydrophobic and hydrogen bonding seems to support network formation from the pulse protein dispersion. Gelation is an important functional property of the globular proteins used to modify the texture of food products. The small strain dynamic oscillatory technique is useful in evaluating gelation properties and gel strength of protein isolates. This technique is extremely sensitive in evaluating the changes in physical structure and chemical composition of the proteins caused by heating.

REFERENCES

1. A. C. Karaca, N. Low, and M. Nickerson, *Food Res. Intl.*, 2011, **44**, 2742–2750.
2. R. W. Jones, N. W. Taylor and F. R. Senti, *Arch. Biochem. Biophys.*, 1959, **84**, 363–376.
3. R. T. Tyler and B. D. Panchuk, *Cereal Chem.*, 1982, **59**, 31–33.
4. R. D. Reichert and C. G. Young, *Cereal Chem.*, 1978, **55**, 469–472.
5. R. T. Tyler, *J. Food Sci.*, 1984, **49**, 925–930.
6. J. Gueguen, A. T. Vu and F. Schaeffer, *J. Sci. Food Agric.*, 1984, **35**, 1024–1033.
7. R. T. Tyler, C. G. Youngs and F. W. Sosulski, *Cereal Chem.*, 1981, **58**, 144–148.
8. K. Elkowicz and F. W. Sosulski, *J. Food Sci.*, 1982, **47**, 1301–1304.
9. R. Cai, B. Klamczynska and B. K. Baik, *J. Agric. Food Chem.*, 2001, **49**, 3068–3073.
10. O. Paredes-Lopez, C. Ordorica-Falomir and M. R. Olivares-Vazquez, *J. Food Sci.*, 1991, **56**, 726–729.
11. S. Emami, L. G. Tabil and R. T. Tyler, *J. Food Proc. Eng.*, 2010, **33**, 728–744.
12. I. Paraman, N. S. Hettiarachchy, C. Schaefer and M. I. Beck, *Cereal Chem.*, 2006, **83**, 663–667.
13. F. T. Barrows, C. A. Bradley, R. D. Kearns, B. D. Wasicek and R. W. Hardy, U. S. patent application 20090259018, 2009.
14. A. E. O. Elkhalifa, D. M. R. Georget, S. A. Barker and P. S. Belton, *J. Cereal Sci.*, 2009, **50**, 159–165.

15. F. F. Shih and K. Daigle, *Cereal Chem.*, 1997, **74**, 437–441.
16. N. J. de Mesa, S. Alavi and J. Gwirtz, *J. Food Eng.*, 2012, **108**, 365–375.
17. K. D. Schwenke, *Nahrung – Food*, 2001, **45**, 377–381.
18. D. J. McClements, in *Modern Biopolymer Science*, ed. S. Kasapis, I.T. Norton and J. B. Ubbink, Elsevier Academic Press, New York, 2009, pp. 129–166.
19. J. E. Kinsella, *J. Am. Oil Chem. Soc.*, 1979, **56**, 242–258.
20. S. Damodaran, in *Food Proteins: Properties and Characterization*, ed. S. Nakai and W. W. Modler, Wiley-VCH, New York, 1996, 167–234.
21. C. V. Morr, B. German, J. E. Kinsella, J. M. Regenstein, J. P. Van Buren, A. Kilara, B. A. Lewis and M. E. Mangino, *J. Food Sci.*, 1985, **50**, 1715–1721.
22. A. Fernandez-Quintela, M. T. Macarulla, A. S. Del Barrio and J. A. Martinez, *Plant Foods Hum. Nutr.*, 1997, **51**, 331–342.
23. J. R. Vose, *Cereal Chem.*, 1980, **57**, 406–410.
24. W. S. Gosal and S. B. Ross-Murphy, *Curr. Opin. Colloid Interface Sci.*, 2000, **5**, 188–194.
25. C. W. Coffman and V. V. Garcia, *J. Food Technol.*, 1977, **12**, 473–484.
26. A. Kiosseoglou, G. Doxastakis, S. Alevisopoulos and S. Kasapis, *Int. J. Food Sci. Technol.*, 1999, **34**, 253–263.
27. E. M. Papalamprou, G. I. Doxastakis, C. G. Biliaderis, V. Kiosseoglou, *Food Hydrocolloids*, 2009, **23**, 337–343.
28. J. I. Boye, S. Aksay, S. Roufik, S. Ribéreau, M. Mondor, E. Farnworth and S. H. Rajamohamed, *Food Res. Int.*, 2010, **43**, 537–546.
29. T. Zhang, B. Jiang and Z. Wang, *Food Hydrocolloids*, 2007, **21**, 280–286.
30. M. Kaur and N. Singh, *Food Chem.*, 2007, **102**, 366–374.
31. S. K. Sathe and D. E. Salunkhe, *J. Food Sci.*, 1981, **46**, 71–74.
32. C. H. Tang and X. Sun, *J. Agric. Food Chem.*, 2010, **58**, 6395–6402.
33. C. H. Tang and X. Sun, *Food Hydrocolloids*, 2011, **25**, 536–544.
34. S. Seena and K. R. Sridhar, *Food Res. Int.*, 2005, **38**, 803–814.
35. M. J. Y. Lin, F. S. Humbert, *J. Food Sci.*, 1974, **39**, 368–370.
36. K. N. Pearce and J. E. Kinsella, *J. Agric. Food Chem.*, 1978, **26**, 716–723.
37. A. Subagio, *Food Chem.*, 2006, **95**, 65–70.
38. J. R. Wagner and J. Gueguen, *J. Sci. Food Agric.*, 1999, **47**, 2181–2187.
39. U. E. Inyang and A. O. Iduh, *J. Am. Oil Chem. Soc.*, 1996, **73**, 1663–1667.
40. D. M. Ragab, E. E. Babiker and A. H. Eltinay, *Food Chem.*, 2004, **84**, 207–212.
41. N. A. El Nasri and A. H. El Tinay, *Food Chem.*, 2007, **103**, 582–589.
42. C. K. Asli, N. Low and M. Nickerson, *Food Res. Int.*, 2011, **44**, 2742–2750.
43. S. Nakai, *J. Agric. Food Chem.*, 1983, **31**, 676–679.
44. L. Sijtsma, D. Tezera, J. Hustinx and J. M. Vereijken, *Nahrung*, 1998, **42**, 215–216.
45. G. G. Palazolo, F. E. Mitidieri, J. R. Wagner, *J. Food Sci. Technol. Int.*, 2003, **9**, 409–419.
46. B. P. Singh and N. V. Queiroga, *J. Sci. Food Agric.*, 2004, **84**, 2022–2027.
47. T.-H. Mu, S.-S. Tan and Y.-L. Xue, *Food Chem.*, 2009, **112**, 1002–1005.
48. M. Kaur and N. Singh, *Food Chem.*, 2007, **102**, 366–374.

49. K. O. Adebowale and O. S. Lawal, *Food Chem.*, 2003, **83**, 237–246.
50. V. A. Obatolu, S. B. Fasoyiro and L. Ogunsunmi, *J. Food Process. Preserv.*, 2007, **31**, 240–248.
51. E. T. Akintayo, A. A. Oshodi and K. O. Esuoso, *Food Chem.*, 1999, **66**, 51–56.
52. H. Fuhrmeister and F. Meuser, *J. Food Eng.*, 2003, **56**, 119–129.
53. R. Horax, N. S. Hettiarachchy, P. Chen and M. Jalaluddin, *J. Food Sci.*, 2004, **69**, 114–118.
54. S. D. Arntfield and E. D. Murray, *Can. Inst. Food Sci. Technol. J.*, 1981, **14**, 289–294.
55. S. Gorinstein, M. Zemser, M. Friedman, W. A. Rodrigues, P. S. Martins, N. A. Vello, G. A. Tosello and O. Paredes-López, *Food Chem.*, 1996, **56**, 131–138.
56. M. F. Marcone, Y. Kakuda and R. Y. Yada, *Food Chem.*, 1998, **63**, 265–274.
57. A. M. Yousif, I. L. Bateyb, O. R. Larroqueb, B. Curtinb, F. Bekesb and H. C. Deetha, *LWT – Food Sci. Technol.*, 2003, **36**, 601–607.
58. C. H. Tang, *LWT – Food Sci Technol*, 2008, **41**, 1380–1388.
59. G. T. Meng and C. Y. Ma, *Food Chem.*, 2001, **73**, 453–460.
60. T. Arakawa and S. N. Timasheff, *Biochemistry*, 1982, **21**, 6536–6544.
61. E. D. Murray, S. D. Arntfield andf M. A. H. Ismond, *Can. Inst. Food Sci. Technol. J.*, 1985, **18**, 158–162.
62. F. O. Uruakpa, S. D. Arntfield, *LWT – Food Sci. Technol.*, 2006, **39**, 939–946.
63. H. H. Winter, *Polym. Eng. Sci.*, 1987, **27**, 1698–1702.
64. C. H. R. Friedrich and L. Heymann, *J. Rheol.*, 1988, **32**, 235–241.
65. A. H. Clark, in *Food Polymers, Gels, and Colloids*, ed. E. Dickinson, Royal Society of Chemistry, Cambridge, 1991, pp. 323–338.
66. S. Ikeda and K. Nishinari, *Food Hydrocolloids*, 2001, **15**, 401–406.
67. A. J. Steventon, L. F. Gladden and P. J. Fryer, *J. Texture Stud.*, 1991, **22**, 201–218.
68. Y. L. Hsieh, J. M. Regenstein and M. A. Rao, *J. Food Sci.*, 1993, **58**, 116–119.
69. Y. L. Hsieh and J. M. Regenstein, *J. Food Sci.*, 1992, **57**, 862–868.
70. P. S. Bora, C. J. Brekke and J. R. Powers, *J. Food Sci.*, 1994, **59**, 594–596.
71. F. E. O' Kane, R. P. Happe, J. M. Vereijken, H. Grupppen and M. A. J. S. Boekel, *J. Agric. Food Chem.*, 2004, **52**, 3149–3154.
72. F. E. O' Kane, R. P. Happe, J. M. Vereijken, H. Grupppen and M. A. J. S. Boekel, *J. Agric. Food Chem.*, 2004, **52**, 5071–5578.
73. F. E. O' Kane, J. M. Vereijken, H. Grupppen and M. A. J. S. Boekel, *J. Food Sci.*, 2005, **70**, 132–137.
74. A. T. Paulson and M. A. Tung, *J. Agric. Food Chem.*, 1989, **37**, 319–326.
75. A. Hermansson, in *Functionality and Protein Structure*, ed. A. Pour-El, ACS Symposiun Series 92, 1979, pp. 82–103.
76. M. P. Tombs, *Faraday Discuss. Chem. Soc.*, 1974, **57**, 158–164.
77. A. Clark, G. Kavanagh and S. Ross-Murphy, *Food Hydrocolloids*, 2001, **15**, 383–400.
78. S. Mleko and E. A. Foegeding, *Milchwissenschaft*, 2000, **55**, 513–516.

CHAPTER 7

Pulse Starch

7.1. INTRODUCTION

Starch, the major storage polysaccharide in pulse grains, is composed of highly branched amylopectin and sparsely branched amylose. Pulse starches generally contain a higher amount of amylose than cereal and tuber starches. Diets rich in pulses have been associated with many health benefits such as reduced calorific content, low glycaemic index, and improved heart health. These health benefits have been attributed to the high amylose content which contributes to resistant starch that, along with dietary fibre, remains undigested in the small intestine and is fermented by the colon microflora. Colonic fermentation increases the growth of beneficial bacteria leading to the production of short chain fatty acids, which have been associated with reduced risk of colon cancer. Characterisation of the structure and functional properties of starch in combination with *in vitro* and *in vivo* experiments could help in identification of characteristics of starch that are responsible for the beneficial effects on human health that are associated with pulse consumption. Pulses have been primarily looked upon as a protein source rather than as a carbohydrate source, although starch is their major component, ranging between 35% and 60%,[1] whereas the protein content of pulses varies between 14.9% and 39.4%.[2] The production of high-purity starch from certain pulse species is difficult because of the presence of highly hydrated fine fibre fractions and insoluble protein.[3] Pulse starches are isolated using wet milling and dry milling processes.[3,4] Most of the pulse starches have greater thermal and mechanical shear stability, which can make these starches an interesting ingredient for a number of food applications.

This chapter focuses on starch isolation methods, and the structure and thermal, rheological and digestibility properties of pulse starches.

Pulse Chemistry and Technology
Brijesh K. Tiwari and Narpinder Singh

Published by the Royal Society of Chemistry, www.rsc.org

7.2. STARCH ISOLATION METHODS

Starch is isolated from pulses by dry milling and wet milling process. Dry milling is used in commercial production, but wet milling method is more common in laboratories when a small quantity of high-purity starch is required. The wet milling process is used in the commercial production of starch from corn, wheat and potatoes. The method of separating starch from pulses was originally reported by Kawamura *et al.*[5] This method involves treatment with NaOH (0.2%) solution followed by washing with water, and dehydration with ethanol and water. Later, Schoch and Maywald[3] suggested the following three methods for the separation of starch from pulses:

(a) The first method, for separation of starch from mung beans, garbanzo beans and dehulled split yellow peas, involved steeping in warm water containing toluene to prevent fermentation, followed by wet grinding and repeated screening.
(b) The second method, for separation of starch from lentils, lima beans and white navy beans, involved steeping in warm water in the presence of toluene followed by re-suspension in NaOH (0.2%) solution (to dissolve most of the protein). The alkaline suspension was then screened through 220-mesh nylon to remove a portion of fine fibre and then slowly flowed down an inclined "table", which was a flat, shallow, trough of heavy-gauge stainless steel with a total slope of 0.5 inch.
(c) The third method, for separation of starch from wrinkled seeded peas, involved exhaustive alkaline steeping and washings of the isolated starch.

The separation of starch from pulse grains is difficult because of the presence of insoluble flocculent protein and fine fibre which decreases sedimentation and settles along with the starch.[3,4,6–8] The presence of highly hydratable fine fibre and the strong adherence of proteins to the starch granules in pulses make the purification of starch a cumbersome process.[3] Repeated filtration through polypropylene screens (202 and 70 μm) in combination with alkali treatment (0.02% NaOH) causes substantial reduction in the protein content in starch separated from pulses by wet milling process.[7] Amylolytic or mechanical damage to the granules during the initial isolation steps, effective deproteinisation, and losses of small granules must be avoided during starch separation.[9] The steeping of seeds in water containing 0.16% sodium hydrogen sulphite for 12 h at 50 °C, followed by grinding, sieving and centrifugation was another suggested method for separating starch from chickpeas.[10] This method involved repeated washings which yielded starch with a protein content of less than 0.8%.

The use of alkali (NaOH) is not highly desirable in industry because it needs to be neutralised and the salt thus produced has to be removed at a later stage. This adds to the production cost and makes the process less economical.

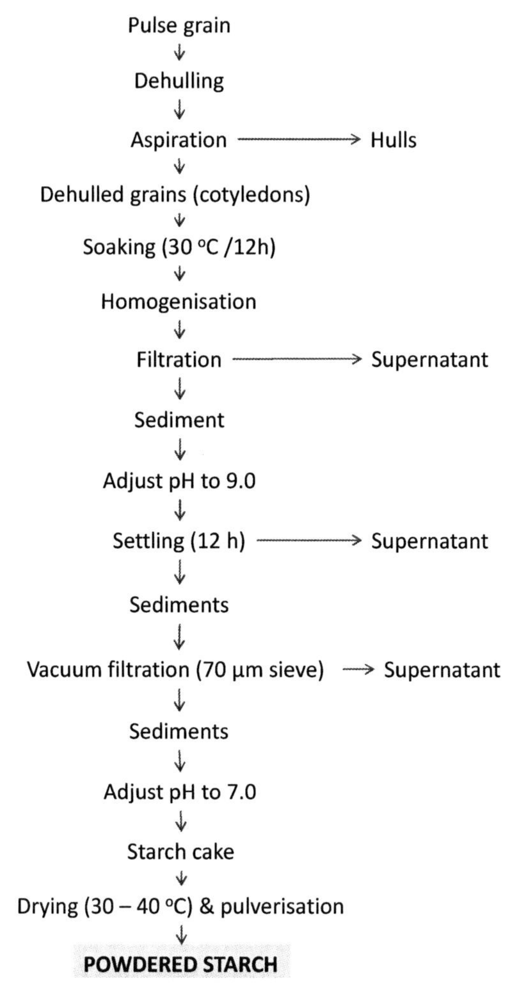

Figure 7.1 Wet milling process for production of starch (*Source*: Hoover *et al.*[27])

The wet milling process for production of starch is illustrated in Figure 7.1. Wet milling involves dehulling of pulses followed by grinding the cotyledons into fine flour. Dry milling before wet fractionation disintegrates the cell walls of cotyledons and assists in efficient separation of the embedded starch granules from the protein matrix. The flour is then homogenised in water, and protein from the sediment is extracted at pH 9.0. The sediment is washed off to remove the alkali and residual alkali neutralised to pH 7.0. The starch recovery and yield is dependent upon the extent to which the cell walls and protein matrix structure are disrupted.

The wet fractionation method for separating starch from field pea grains is illustrated in Figure 7.2.[11] This process involves dehulling of field pea seeds with an abrasive disc huller, followed by removal of the hulls from the

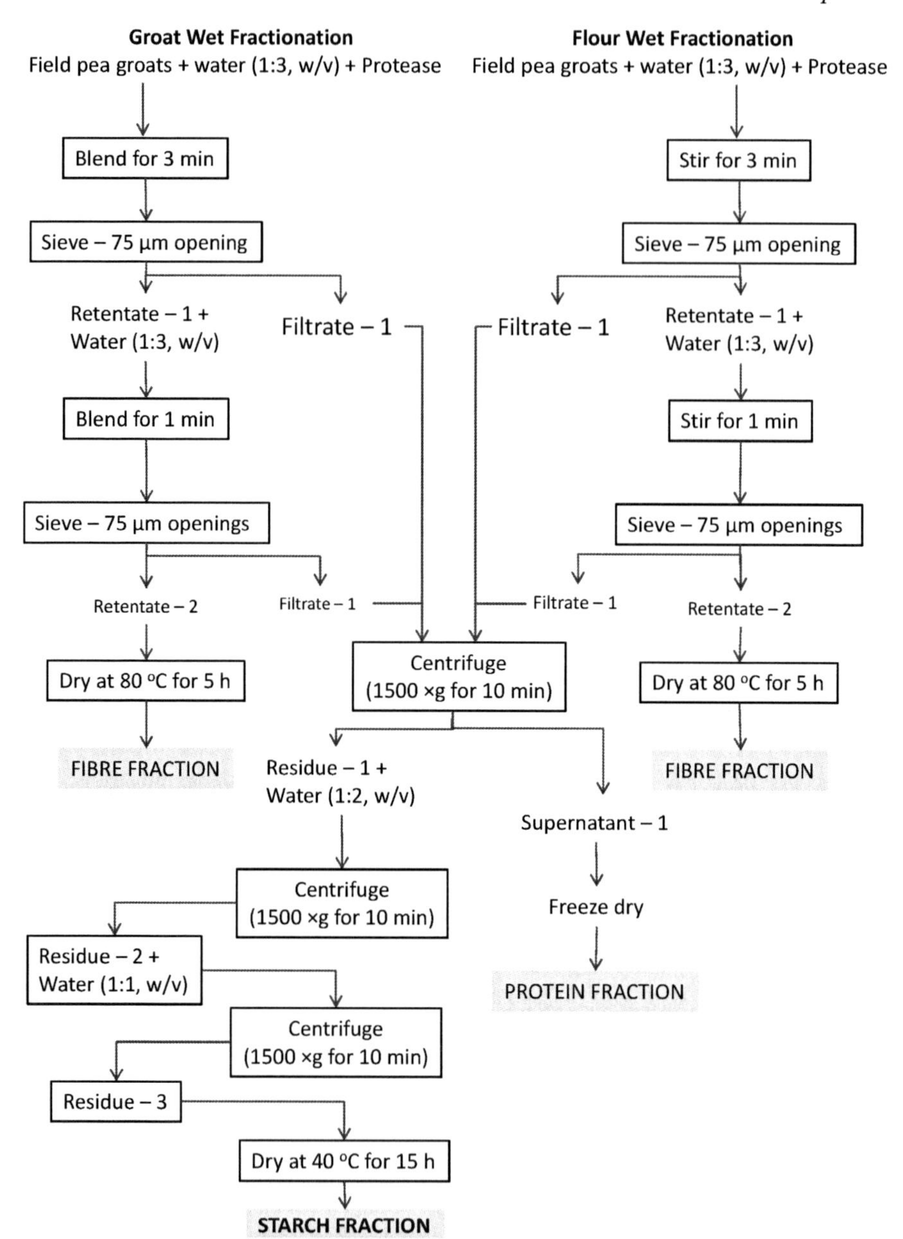

Figure 7.2 Wet fractionation of field pea (*Source*: Naguleswaran and Vasanthan[11])

cotyledons and collection of groats separately. The groats were dry milled using a cyclone mill and both groats and flours were fractionated (Figure 7.2). A comparison between groat wet fractionation (GWF) and flour wet fractionation (FWF) processes showed that the FWF process gave a higher starch yield but lower fibre and protein yields than GWF.[11] FWF resulted in higher starch yield and lower fibre and protein recovery than GWF. Starch produced from GWF had higher purity than that produced from FWF. For efficient separation and purification of starch, particle size reduction plays an important role in the fractionation.

Dry milling separation of starch involves the grinding of grains with a pin mill or hammer mill followed by air classification (Figure 7.3). A pin mill is preferred to a hammer mill for grinding. The grains are reduced to very fine particle size, which facilitates the separation of starch granules from the protein matrix.[12] The air classification process is used to separate starch fractions low in protein from the fine protein-rich fractions. Repeated milling followed by air classification is done to purify the starch. Even with repeated milling and air classification dry milling it is difficult to produce starch of the high purity that can be achieved by wet milling.[12] The dry milling process results in high starch damage. On the other hand, wet milling methods of starch separation are laborious and lengthy. In mechanical air classifiers, centrifugal force is used in addition to gravity to separate the particles by size and density within the air stream. The washing of fractions produced by dry milling with water is necessary to remove the residual protein.[4] In a study including mung beans, green lentils, great northern beans, faba beans, field peas, navy beans, baby lima beans and cowpeas a dry milling process involving pin milling and air classification was observed to be best suited for

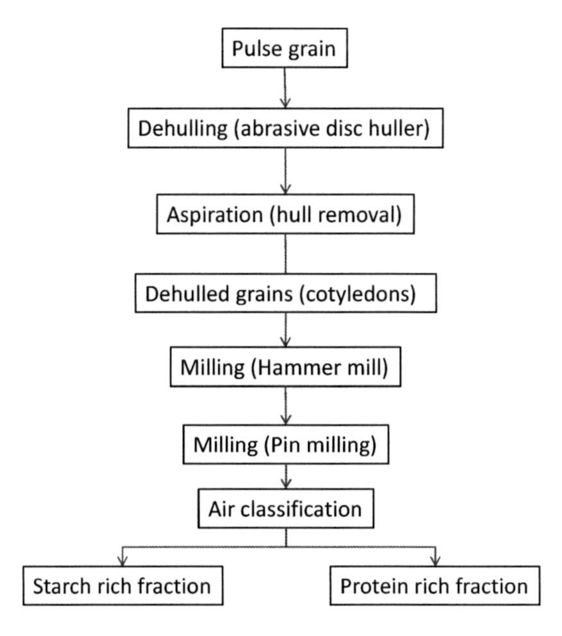

Figure 7.3 Dry milling process for separation of starch and protein-rich fractions (Hoover *et al.*[27])

fractionating the high-protein fractions from mung beans and lentils, but lima beans and cowpeas were observed to be poor choices for this method.[13]

7.3. STARCH COMPOSITION

Pulse starches consist of amylose and amylopectin. Amylose is a linear polymer composed of glucopyranose units linked through α-D-(1→4) glycosidic linkages. It is sparsely branched, with α-1,6 linkages making (approx) one branch per 1000 glucose residues.[14,15] Amylose has a molecular mass of approximately 1×10^5–1×10^6 kDa.[16–18] Amylopectin is a branched polymer with one of the highest molecular masses known among naturally occurring polymers (approximately 1×10^7–1×10^9 kDa). Pulse starches are characterised by a high amylose content (24–65%). The observed amylose content is 30.2–34.6% for black gram starches,[19,20] 28.6–34.3% for chickpea starches[19,21–22] and 38.5%, 32% and 34.2% for lentil, faba bean and field pea starches, respectively.[23] The difference in the amylose content amongst pulse starches is attributed to varietal differences,[24] the physiological state of the seed[25] or the amylose–lipid complexes formed during estimation.[26] The apparent amylose content, total amylose content and amylose complexed with lipids are shown in Table 7.1. Different pulse species have amylose complexed with native lipids in the range between 2.7% to 15.0%; pigeon pea has the lowest content. Pulse starches have a lipid content between 0.02 and 1.40, which varies with type and variety.[27] Amylose content is determined by iodine binding procedures, using amperometric, potentiometric or spectrophotometric detection methods based on iodine's ability to form a helical inclusion complex with amylose. Iodine produces a blue colour with amylose and a reddish brown colour with amylopectin. Iodine also forms a complex with long chains of amylopectin in starch, hence sometime result in overestimation of the amylose content. Phospholipids and free fatty acids form complexes with amylose and their presence in starch results in the underestimation of amylose content.

The amylopectin fraction varies in degree of polymerisation (DP) in starches from different botanical sources (Figure 7.4). The distribution of amylopectin chain DP after hydrolysis with isoamylase is determined by high-performance anion exchange chromatography equipped with a pulsed amperometric detector (HPAEC-PAD), capillary electrophoresis and mass spectrometry.[28] The amylopectin chains with DP6–12, 13–18, 19–24 and 25–30, respectively, ranged from 36.2 to 43.25%, 36.44 to 38.68%, 14.86 to 18.22% and 4.95 to 6.9% among starches separated from 20 chickpea cultivars.[29] DP6–12, DP13–24 and DP25–36 of chickpea starch ranged from 16.9 to 17.2%, 28.7 to 30.6% and 54.8 to 56.0%, respectively, as observed by Chung *et al.*[30] Amylopectin short (DP 6–10), medium (DP 11–20) and long chains (DP21–30) in rice bean germ plasm ranged from 19.85% to 28.42%, 60.69% to 65.05%, and 8.67% to 15.67%, respectively.[31]

Table 7.1 Apparent amylose content, total amylose content and amylose complexed with lipids

Pulses	*Apparent amylose content (%)*	*Total amylose content (%)*	*Amylose complexed with lipids (%)*
Kidney bean	33,[h] 27–40[m]	34–41[i]	
Chickpea, Garbanzo beans	30–34,[a] 30,[g] 27,[i] 21[j]	34–40,[a] 30–35,[i] 23[j]	9–16,[a] 9–10[j]
Lentils	27–29,[c] 32,[g] 22–23[j]	30–32,[c] 23–32,[i] 23–25[j]	10–11,[c] 6–6.5[j]
Mung bean	32,[g] 40,[f] 30[n]	33–45,[i] 45,[j] 45.3 [f]	12,[j] 12 [f]
Pigeon pea	28,[g] 28.5[k]	29.3[k]	2.7[k]
Peas	33,[g] 34.5,[h] 43–44[l]	27–46,[i] 49–50[l]	11–12[l]
Adzuki bean		45,[f] 18–35[i]	12[f]
Black beans	33–35,[c] 36,[e] 38,[g] 23–25[j]	37–39,[c] 41,[e] 27–39,[i] 27–29[j]	10–11,[c] 13,[e] 14–15[j]
Lima beans		23–32[i]	
Navy beans	26[j]	29–41,[i] 28–29[j]	8–9[j]
Great northern bean		32–41[i]	

[a]Hughes *et al.*,[107] [b]Singh *et al.*,[108] [c]Zhou *et al.*;[109] [e]Hoover and Manuel;[110] [f]Hoover *et al.*;[111] [g]Sandhu and Lim;[112] [h]Chung and Liu;[113] [i]Hoover *et al.*,[27] [j]Hoover and Ratnayake,[69] [k]Hoover;[114] [l]Ratnayake;[8] [m]Singh *et al.*;[31] [n]Li *et al.*[115]

The fine structure of amylopectin is also evaluated by debranching it using the enzyme isoamylase, followed by fractionation on gel permeation chromatography according to the wavelength range at the maximum absorption (λ_{max}) in the absorption spectra of the glucan–iodine complexes. Starch is fractionated into different fractions as listed below:

- Fraction I, representing apparent amylose content, $\lambda_{max} \geqslant 620$ nm
- Intermediate fraction (Int. Fr.), 620 nm$>\lambda_{max} \geqslant 600$ nm
- Fraction II, representing long side chains of amylopectin, 600 nm$>\lambda_{max} \geqslant 540$ nm
- Fraction III, representing long side chains of amylopectin, 540 nm$>\lambda_{max}$.[32]

Various starch fractions obtained from different pulses and cereals fractionated using gel permeation chromatography are shown in Table 7.2.

Chickpea and black gram starches contained Fr. I in the range of 34.4–35.5% and 32.9–35.9%, respectively.[33] Pigeon pea and kidney bean starches showed Fr. I of 31.8 and 35.9%, respectively. The intermediate fraction, Fr. II and Fr. III ranged between 6.0 and 8.5%, 14.7 and 15.4%, and 41.7 and 43.8%, respectively, among the chickpea starches compared to 2.7–3.0%, 16.7–18.5%, and 44.0–46.5% of similar fractions in black gram starches. A wide diversity in fine structure of starch separated from kidney bean germplasm was observed.[34] Fractions I, II and III were 27.3–40.3%, 15.6–21.4 and 37.4–47.3%, respectively in kidney bean germplasm. Pulse starches were observed to have

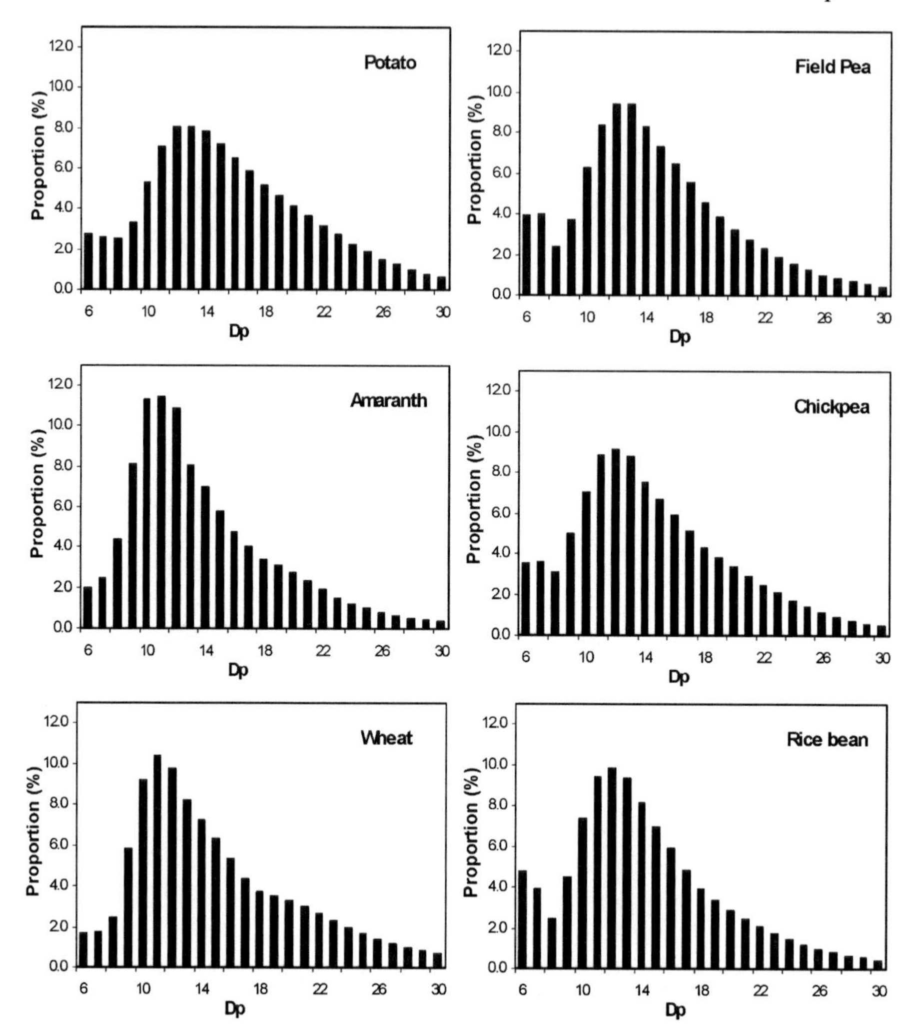

Figure 7.4 Degree of polymerisation (DP) of amylopectin in starch from different sources

lower proportion of Fr. III as compared to starch from low-amylose rice, intermediate-amylose rice and waxy maize (Table 7.2). Field pea and chickpea starches showed a lower proportion of Fr. II compared to other pulse starches while black gram and pigeon pea starch had higher proportions of these fractions. Black gram starch had the lowest content of intermediate fraction (2.5–3.0%), compared to the highest amount of the same fraction (6–8.5%) in chickpea starches. The distribution of α-1,4 chains of amylopectin (Fr. III/Fr. II) ranged between 2.5 and 3.3, lower for black gram and pigeon pea starch. Black gram and pigeon pea starches showed the highest proportion of Fr. III, followed by chickpea, kidney bean and field pea starch. Pulse starches were

Table 7.2 Isoamylase debranched starch fractions of different pulses fractionated using gel permeation chromatography[33–36]

Starch source	*Fraction I (%)*	*Intermediate fractions (%)*	*Fraction II (%)*	*Fraction III (%)*	*Fraction III/II*
Kidney beans	27.3–40.3	1.7–7	15.6–21.4	37.4–47.3	1.9–3
Chickpeas	34.4–35.5	6–8.5	14.7–15.4	41.7–43.8	2.7–3
Black gram	32.9–35.6	2.5–3.2	16.7–18.5	44–46.5	2.5–2.8
Field peas	37.9	5	13.9	43.2	3.1
Pigeon peas	31.8	4.4	18.2	45.6	2.5
Rice (high amylose, >25%)	28.3	4.9	21	45.8	2.2
Rice (intermediate amylose, 20–25%)	21.3–24.2	3.7–4.3	20.6–21.6	50.7–53.8	2.3–2.6
Rice (low amylose, 7–20%)	9.7–16.5	4.3–5	22.7–26.6	55.8–59.4	2.2–2.5
Maize (normal)	29.4–32.6	2.5–5.9	14.9–19.1	45.6–50	2.4–3.4
Maize (sugary)	41	4.1	13.4	41.5	3.1
Maize (waxy)	0–5.4	1.7–3.9	27.8–31.5	62.9–66.8	2.1–2.3

Fraction I, apparent amylose content; Intermediate fraction, mixture of amylose and long side chains of amylopectin; Fraction II, long side chains of amylopectin; Fraction III, short side chains of amylopectin.

observed to have lower Fr. II and Fr. III than rice starches. Fr. I, intermediate fraction, Fr. II and Fr. III ranges among rice starches were 9.7–28.3%, 3.7–5.0%, 20.6–26.6%, and 45.8–59.4%, respectively.[35]

Phosphorus is the non-carbohydrate constituent of starch, which significantly affects its functional properties.[37] Phosphorus is present in three major forms: phosphate monoesters, phospholipids and inorganic phosphate.[38–41] The phosphate monoesters are covalently bound to the amylopectin fraction of the starch and increase its paste clarity and viscosity, while the phospholipids result in opaque and lower-viscosity pastes.[42–44] Phosphate groups are esterified to the amylopectin fraction of starch, which contributes to the viscosity, transparency, water binding capacity and freeze–thaw stability.[44,45] The phospholipid content of the starch is proportional to the amylose content.[46,47] Phospholipids have a tendency to form a complex with amylose and long branched chains of amylopectin, which results in reduced swelling of starch. Pulse starches contain varying amount of phosphate monoester derivatives, hence have varying clarity and viscosity.[48] Wheat and rice starches have higher phospholipid content and produce starch pastes with lower transmittance as compared to that of corn and potato starches with lower phospholipid content.[37] The content and form of phosphorus in starch is influenced by growing conditions, temperature and storage.[49] Phosphate monoesters are bound to the extent of 61% and 38%, respectively, on the C-6 and C-3 of the glucose unit in starch, and possibly 1% of monoester is bound on the C-2 position.[48]

7.4. STARCH STRUCTURE

Starch is semi-crystalline in nature, with varying levels of crystallinity. The crystallinity is exclusively associated with the amylopectin, while the amorphous regions mainly represent amylose.[50,51] Table 7.3 shows the shape, size and crystallinity of some pulse starch granules. A model of the starch granule structure given by Jenkins *et al.*[52] is shown in Figure 7.5. According to this model, amylose and amylopectin are organised as alternating semi-crystalline and amorphous layers. The more and less concentric layers have alternating high and low refractive indices, densities, crystallinities and resistance to acid and enzymatic hydrolysis.[53] The dense layer in growth ring consists of ~ 16 repeats of alternating crystallites (5–6 nm) and amorphous (2–5 nm) lamellae (semi-crystalline layer).[53] Its thickness is 120–400 nm.[52] The less dense layer is largely amorphous and contains water.[53] Starch granules are therefore partially crystalline with a degree of crystallinity of 20–40%.[53] A periodicity of 9–11 nm for starches from various botanical sources measured using small angle X-ray and neutron scattering has been reported by number of researchers.[52,54,56–58] The crystalline lamellae are made up of amylopectin double helices packed in a parallel fashion while the amylopectin branching points are in amorphous regions (Figure 7.5).

The packing of amylose and amylopectin within the granules varies among the starches from different species. X-ray diffractometry is used to reveal the presence and characteristics of the crystalline structure of the starch granules. The A, B and C patterns are the different polymeric forms of the starch that differ in the packing of the amylopectin double helices. Cereal starches exhibit the typical A type, where the double helices making up the crystallites are densely packed and the structure is compact with a low water content. Tuber

Table 7.3 Shape, size, and crystallinity of pulse starch granules

Pulse	*Shape*	*Crystallinity (%)*	*Granule size (μm)*
Kidney beans	Oval, round, elliptical[a]	28–30[a]	16–42,[a] 10–30[k]
Chickpeas, garbanzo beans, Bengal gram	Oval, spherical[a,b]	23–28,[a] 18[d]	9–30[a] , 6–31[b]
Lentils	Oval, round, elliptical[a,h]	26–31,[a] 32,[c] 19,[d] 27–33[i]	6–32,[a] 2.5–25,[h] 16–17[i]
Mung beans	Oval, round[a,e]	29,[a] 22[l]	7–20,[a] 7–26[e]
Pigeon peas	Oval,elliptical,irregular[f]	33[a]	8–32[f]
Peas	Oval, round, irregular[a]	20–25,[a] 20–25[g]	2–34,[a]2.7–31.8[n]
Adzuki beans	Oval, kidney[a]		20–55[a]
Black beans	Oval, round, spherical[a,c]	17.0–21.7,[a] 32[c]	7–37[a]
Lima beans			
Navy beans	Oval, round, elliptical[a]	19–20[a]	14–32[a]
Great northern beans	Oval, round, irregular[a]		
Smooth peas	Oval, spherical, round, elliptical, irregular	18.9–36.5%	2–40
Rice beans			0.5–150

[a]Hoover *et al.*;[27] [b]Hughes *et al.*,[107] [c]Zhou *et al.*;[109] [d]Hoover and Ratnayake;[69] [e]Hoover *et al.*;[111] [f]Hoover;[114] [g]Ratnayake;[8] [h]Hoover and Manuel.[110]

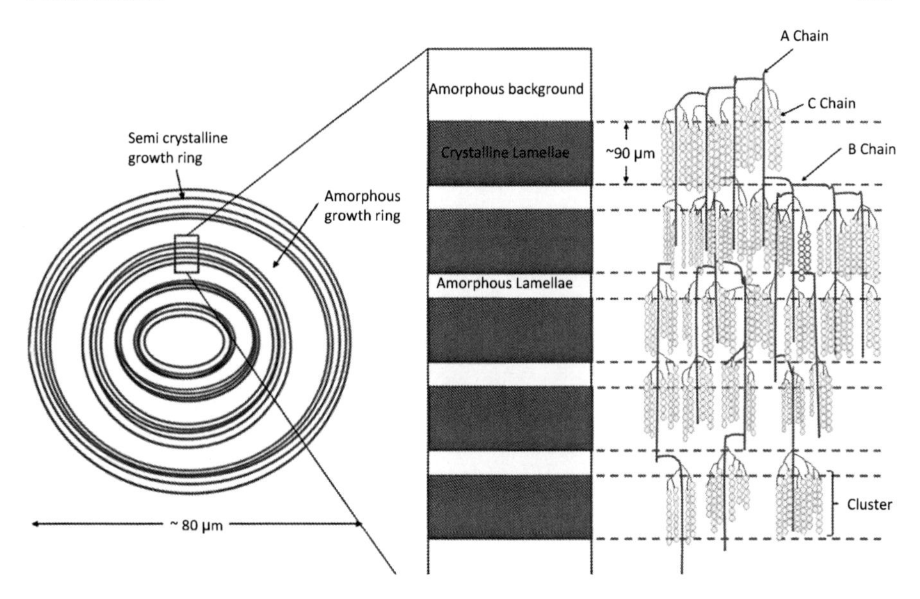

Figure 7.5 Starch granule structure model (Jenkins *et al*[52])

starches, known as B type, have less densely packed crystallites with a more open structure containing a hydrated helical core.[59,60] Pulse starch has a C-type crystalline polymorph, a mixture of A and B polymorphs, and the proportion of both of these varies in different starches.[60] The A and B types crystallise in orthogonal and hexagonal forms, respectively, having 8 and 36 water molecules per unit cell.[62] The B crystallites were proposed to be located predominantly in the centre of the granules, with the A crystallites more abundant towards the periphery.[63,64] Wrinkled pea starch has the B type, similar to high-amylose maize and tuber starches.

Starch crystallinity is usually calculated from the ratio of the area under the crystalline peaks by the total area under the X-ray diffractogram (Figure 7.6), considering the granule as a two-phase system.[65] The two methods commonly used are (1) plotting a smooth curve, which connect peak baselines on the diffractograms[66] as shown in Figure 7.7, and (2) calculating the crystallinity in comparison with an amorphous standard.[67] In both methods, the width of crystalline peaks is related to the size of crystallites: smaller crystallites give broader peaks. Both these methods were developed for cellulose, but are now well accepted in the evaluation of starches.[68] These methods do not take into account diffuse scattering from non-perfect crystalline structures, hence they underestimate the crystalline content of starches. The method described by Wakelin and Virgin[67] suggested the use of a crystalline standard; however, standards for starch were not readily available. Therefore, the approach considered for starch has been to use other crystalline standards such as quartz[69] or to assume a scale factor for the amorphous background,[70] as the scattered intensity of the amorphous sample is usually greater than that of the granular starch at certain

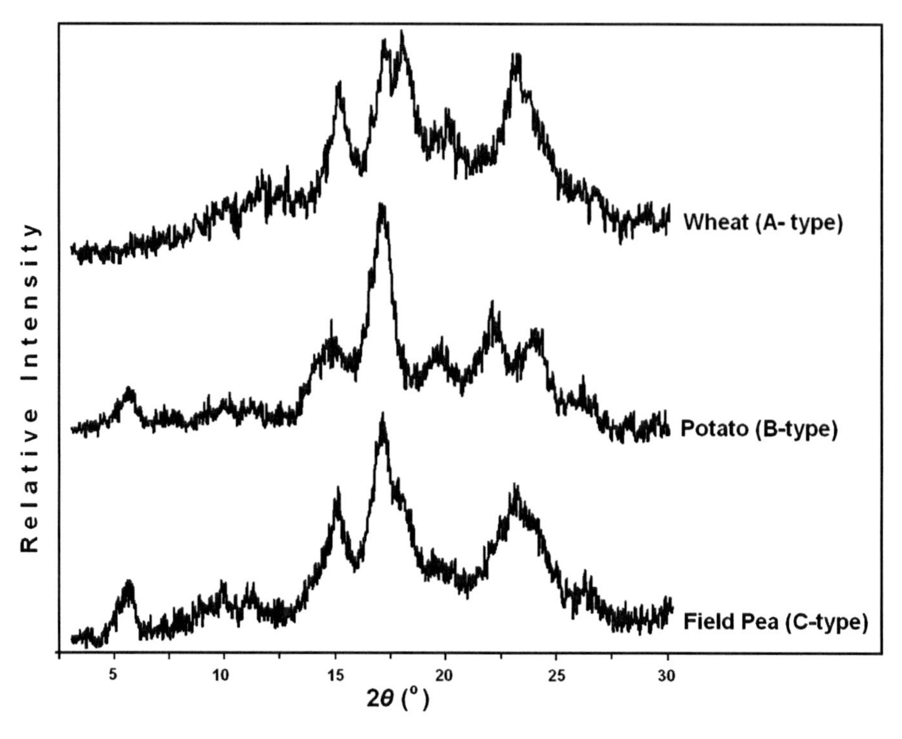

Figure 7.6 X-ray diffractograms of wheat, potato, and field pea starch

diffraction angles.[71] The peak-fitting technique to calculate crystallinity in granular starches was claimed to be better in quantifying the degree of crystalline order in these biomaterials than earlier methods.[71] The difference in relative crystallinity between pulse starches is affected by: (1) crystal size, (2) amount of crystalline regions (influenced by amylopectin content and amylopectin chain length), (3) orientation of the double helices within the crystalline domains, and (4) extent of interaction between the double helices.[69]

7.5. STARCH GRANULE SIZE

Morphological characteristics of starches from different plant sources vary with the genotype and depend on the biochemistry of the chloroplast or amyloplast, as well as the physiology of the plant.[72] The size of pulse starch granules ranges from approximately 5 to 104 μm. Starches from different pulses have a different granule size distribution. Various techniques such as light microscopy, laser light scattering, electric resistance, sieving and field flow fraction are used for evaluating starch granule distribution. Laser beam diffraction techniques are now commonly used for measuring the granule size distribution. These techniques are rapid and require only small samples. Typical unimodal, bimodal and trimodal curves for granule size distribution data obtained using laser light diffraction are shown in Figure 7.8. Most of the pulse starches are simple

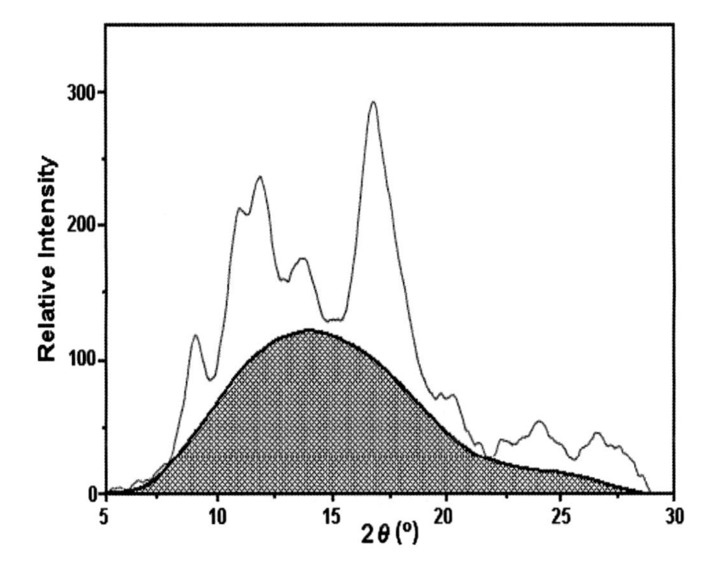

Figure 7.7 Estimation of the amorphous background of X-ray spectrum for starch using the Savitzky–Golay smoothing algorithm

granules, except wrinkled pea starch, which is a mixture of simple and compound granules.[73] The granules of pulse starches are oval, round, or spherical. Some pea starches show the presence of deeper indentations, fissures and groves in the granules.[74] Scanning electron micrographs of some native starches are shown in Figures 7.9 and Figure 7.10. The deep fissures in the starch granules are the indicator of a strong bonding between the starch and the protein matrix.[75] Chickpea starches have large oval to small spherical granules with a smooth surface without any evidence of fissures.[10] Faba bean starches exhibited numerous cracked granules, whereas the surface of black bean and

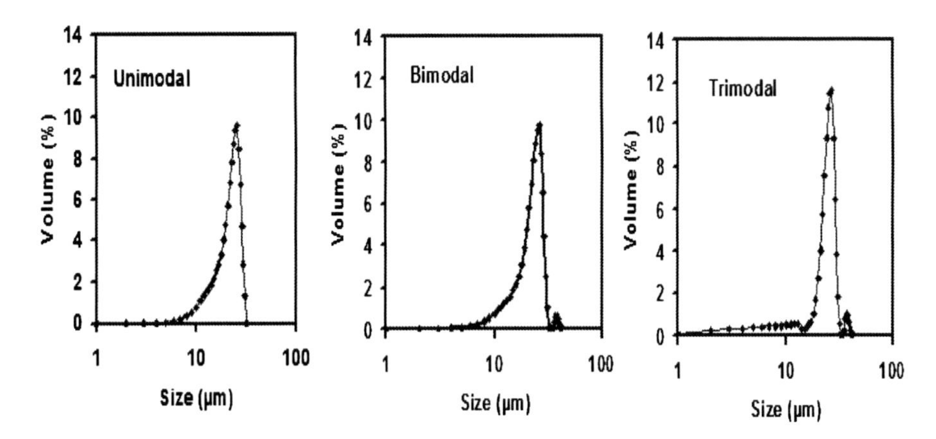

Figure 7.8 Typical unimodal, bimodal, and trimodal curves for granule size distribution (*Source*: Singh *et al.*[34])

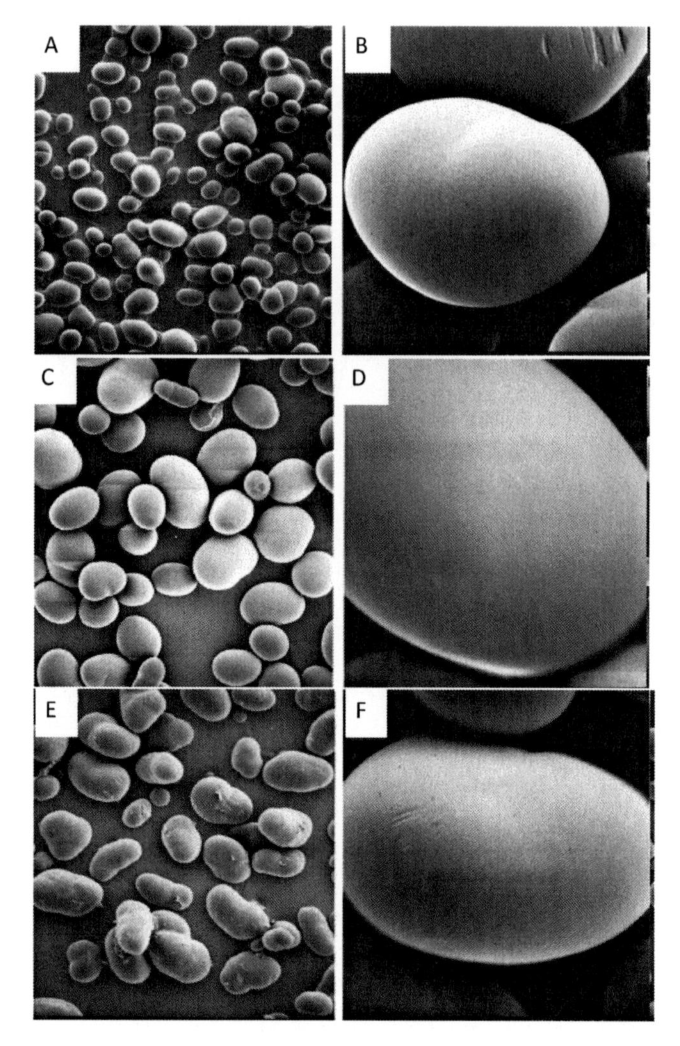

Figure 7.9 Scanning electron micrographs of native black bean (A, B), pinto bean (C, D), and lentil (E, F) starches

pinto bean starches showed no evidence of cracks or indentations.[76] The granule size distribution is an important parameter that can influence the behaviour of starch during processing.

7.6. STARCH PROPERTIES

7.6.1. Pasting Properties

Pasting refers to the changes seen in the starch on continuing heating after gelatinisation has occurred, which include further swelling and leaching of

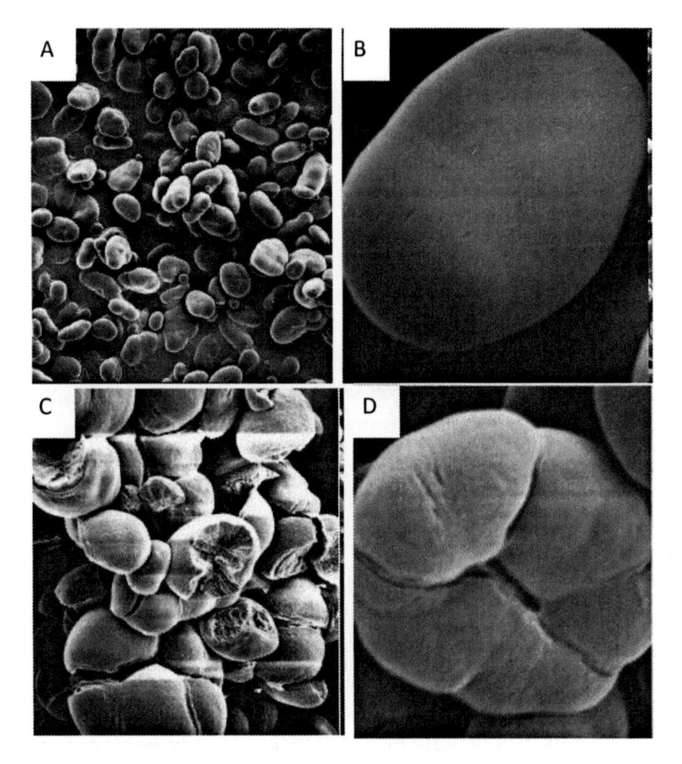

Figure 7.10 Scanning electron micrographs of native smooth (A, B) and wrinkled (C, D) pea starches

components of starch granule, and increased viscosity upon on application of shear forces.[77,78] The pasting behaviour of starches is measured under controlled increase in temperature and constant shearing conditions. During heating the starch suspension is converted into a paste, which consists of a continuous phase of solubilised amylose and/or amylopectin and a discontinuous phase of granule ghosts and fragments.[27] The Rapid Visco Analyser (Newport Scientific Pvt. Ltd, Australia) and the Brabender Viscoamylograph (Brabender GmbH & Co. Kg, Duisburg, Germany) are the instruments most commonly used for measuring the pasting behaviour of starch. During the initial stage of heating, granules absorb water and swell to several times their original size. The temperature where the initial rise in viscosity take place is an indication of the minimum temperature required to cook the starch and is termed the pasting temperature. The viscosity increases rapidly with increase in temperature and reaches a maximum value, the peak viscosity. After reaching this maximum value, viscosity decreases as a result of rupturing and fragmentation of granules during continuous stirring. This is known as breakdown viscosity, which indicates about the stability of paste towards heating under constant stirring conditions. As the paste is subsequently cooled, viscosity increases due to the aggregation of the amylose molecules.[79] The viscosity achieved upon cooling to a temperature of 50 °C is known as the final viscosity. The swelling of starch granules during heating and stirring results from the disruption of hydrogen bonds in amorphous regions. Generally, pulse starch pastes are more viscous than those of cereal starches. Some pulse starches have a higher resistance to rupture than do cereal starches. These properties of pulse starch vary with the cultivars and pulse species. For example, chickpea starches have better stability than black gram starch (Figure 7.11). A wide range of pasting behaviour among starches from different breeding lines of the same pulse species has been observed.[34] The diversity in pasting properties in starches from different kidney bean lines is illustrated in Figure 7.12.

The higher temperature causes disruptions of both amorphous and crystalline structure. The swelling is restricted by the amylose, and starches with a lower amylose content swell more than normal or high-amylose starch. Generally, pulse starches have lower breakdown viscosity, which is attributed to their high amylose content.[33] Peak viscosity and breakdown viscosity have been shown to be inversely related to amylose content in kidney bean and rice bean starches.[31,34] Amylose and lipids assist in maintaining granule integrity during heating.[47] The pasting behaviour of starches from different pulses is mainly dependent upon amylose content, but it also varies depending on amylopectin structure, phospholipid content and granule size distribution. Pasting curves of kidney bean starches varying in amylose content are shown in Figure 7.13. Starches from different sources with similar amylose content may show different swelling behaviour. Potato and tapioca starch have much larger granules than rice starch and swell more, although these starches have amylose content similar to rice starch. The size of starch granules effects the swelling properties perhaps because the larger granules have less molecular bonding, so both swell and break down at faster rate. Large starch granules generally imply higher viscosity, but make the viscosity susceptible to shear.

In spite of such differences, the more compact structure of a smaller molecule does not always mean a significant difference in gelatinisation. Rheological parameters such as peak storage modulus and loss modulus were observed to increase in the order small-, medium- and large-granule starches when subjected to temperature sweep testing.[80] The breakdown in peak storage modulus during the heating cycle and retrogradation during storage were found to be the highest for large-size and the lowest for small-size fractions. Peak viscosity, trough viscosity and final viscosity were observed to be higher in rice bean starches with a higher proportion of large granules (>30 μm) and lower in those with higher proportion of smaller granules of size <30 μm.[31]

7.6.2. Thermal Properties

When heated in the presence of excess water, starch undergoes a phase transition from order to disorder, called gelatinisation, over a temperature range characteristic of the starch source.[35] This phase transition is associated with the diffusion of water into the granule, hydration and swelling of the starch granules, uptake of heat, melting of crystalline and double helices, loss

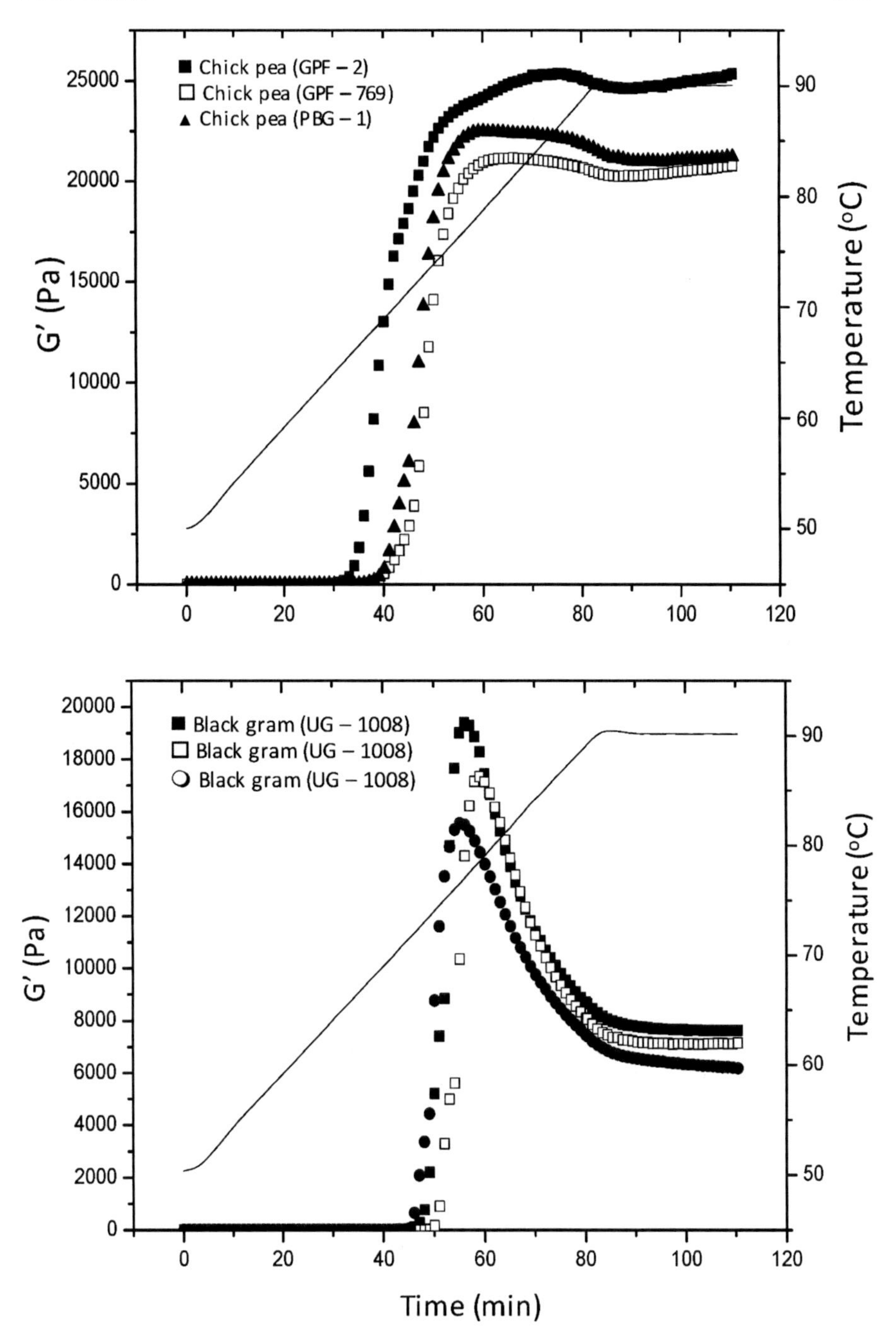

Figure 7.11 Changes in storage modulus of chickpea and black gram starch as function of temperature *Source*: Singh *et al.*[33]

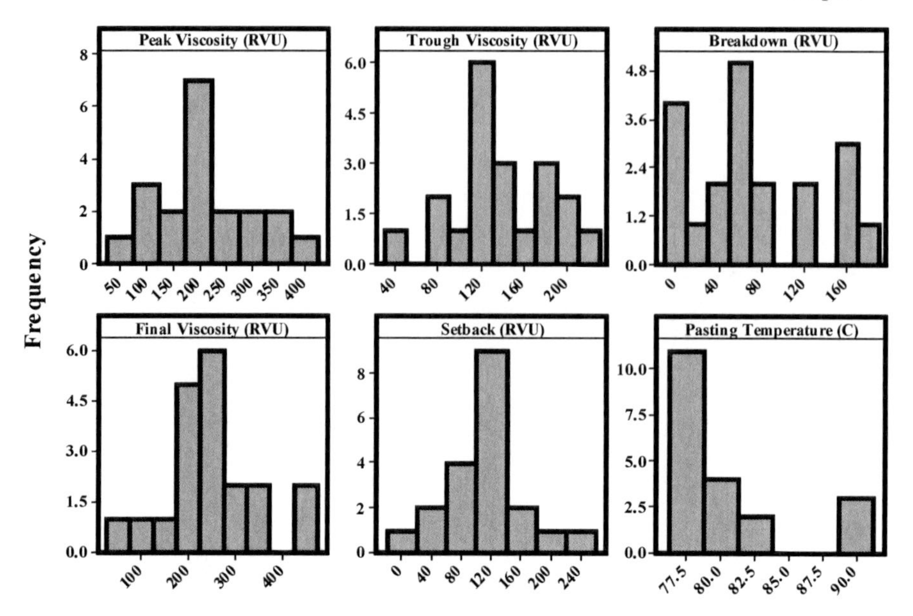

Figure 7.12 Diversity in the pasting properties of starches from different kidney bean lines *Source*: Singh *et al.*[34]

of crystallinity and amylose leaching.[81,82] When the amorphous region swells, it imparts a stress on the crystalline regions and this effect strips polymer chains from the surface of starch crystallites.[83] Differential scanning calorimetry (DSC) is the technique most commonly used to measure the

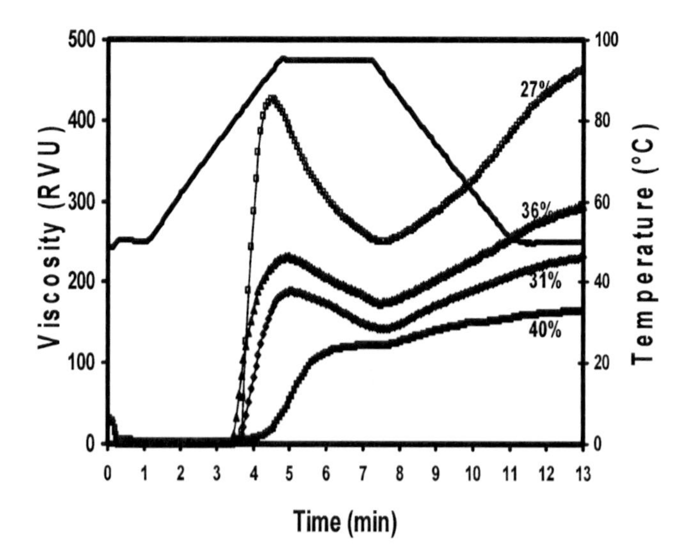

Figure 7.13 Pasting curves of kidney bean starches varying in amylose content (*Source*: Singh *et al.*[34])

gelatinisation transition temperatures (T_o, onset; T_p, peak; T_c, conclusion) and enthalpy of gelatinisation (ΔH_{gel}) of starches. T_o, T_p, T_c and ΔH_{gel} of starches is influenced by the molecular architecture of the crystalline region, which is related to the distribution of amylopectin short chains (DP6–11) and not by the proportion of crystalline region, which corresponds to the amylose to amylopectin ratio.[84] ΔH_{gel} gives an overall measure of crystallinity (quality and quantity) and is an indicator of the loss of molecular order within the granule.[85] T_o, T_p and T_c of chickpea starches range between 59.3 and 60.25 °C, 66.6 and 68.6 °C, and 76.1 and 77.3 °C, respectively, compared to 66.8–79.6 °C, 71.4–74.6 °C and 77–79.6 °C for black gram starches (Table 7.4). Kidney bean and pigeon pea starch showed values of 63.5 and 68.4 °C for T_o, 67.8 and 71.56 °C for T_p, and 73.2 and 76.3 °C for T_c, respectively. Field pea starch had lower transition temperatures while pigeon pea starch showed the highest. The presence of a more amorphous region in field pea has been attributed to a lower transition temperature in these starches.[33] Pulse starch with higher crystallinity has higher transition temperatures as well as ΔH_{gel}.[32] The ΔH_{gel} has been attributed to the disruption of the double helices rather than long-range disruption of crystallinity.[85] The lower transition temperatures and higher ΔH_{gel} of field pea starch was suggested to be due to more pronounced disruption of double helices (in amorphous and crystalline regions) during gelatinisation in this starch than in other starches (chick pea, kidney bean, black gram and pigeon pea) by Singh *et al.*[33] Pigeon pea and black gram starches showed a higher proportion of long side chains of amylopectin and a lower amount of amylose plus intermediate

Table 7.4 Thermal properties of starch separated from different pulses

Pulses	T_o *(°C)*	T_p *(°C)*	T_c *(°C)*	ΔH *(J g⁻¹)*
Kidney beans	61–67,[a] 65–73[j]	67–70,[a] 68–77[j]	76–91,[a] 72–82[j]	11–15[a]
Chickpeas	58–65,[a] 59–60,[b] 61–65[c]	63–72,[a] 63–65,[b] 66–69[c]	70–81,[a] 77–79,[b] 71–74[c]	11–18,[a] 11–13,[b] 7–9[c]
Lentils	58–68,[a] 61–64,[d] 52–56[h]	66–76,[a] 68–71,[d] 61–62[h]	71–82 ,[a] 77–80,[d] 69–73[h]	3–13,[a] 2–3[h]
Mung beans	58–62,[a] 58,[f] 67,[k] 61[m]	67,[a] 67,[f] 72,[k] 69[m]	72–82,[a] 82,[f] 76,[k] 78[m]	8–18,[a] 18.5,[f] 5[k]
Pigeon peas	69–74[a]	75–81[a]	81–87[a]	10–11[a]
Peas	61–64,[a] 61[g]	67–71,[a] 67[g]	73–807[a],5–76[g]	10–14 1[a=],1[g]
Adzuki beans	70,[a] 60,[l] 70[m]	73.4,[a] 66,[l] 73[m]	77,[a] 77[m]	11.7[a]
Black beans	61–67,[a] 61–66,[d] 62–67[e]	70–76,[a] 71–75[d] 70–76[e]	81–87,[a] 81–87,[d] 83–84[e]	11–13,[a] 12–13[e]
Navy beans	61–67,[a] 61–63,[e] 61[m]	69–75 ,[a] 74–75,[e] 70[m]	71–91,[a] 85,[e] 78[m]	9–15,[a] 13[e]
Great northern beans	63–64,[a] 64[m]	66–73,[a] 73[m]	70–80,[a] 80[m]	13–15[a]
White kidney beans	66[i]	74[i]	82[i]	16[i]

[a]Hoover *et al.*;[27] [b]Hughes *et al.*;[107] [c]Singh *et al.*;[10] [d]Zhou *et al.*;[109] [e]Hoover and Ratnayake;[68] [f]Hoover *et al.*;[111] [g]Ratnayake;[8] [h]Hoover and Manuel;[110] [i]Guzel and Sayar;[116] [j]Singh *et al.*;[33] [k]Li *et al.*;[115] [l]Yousif *et al.*;[117] [m]Su *et al.*[118]

fractions as compared to other starches, which is responsible for the difference in transition temperature. High transition temperatures are believed to result from a high degree of crystallinity, which imparts structural stability, making the granules more resistant to gelatinisation.[23,86] High transition temperatures may also reflect more stable amorphous regions or a lower degree of chain branching.[87,88] The longer chains in amylopectin required a higher temperature to dissociate completely than that required for shorter double helices.[89] Kidney bean starches with higher proportion of the longer side chain amylopectin fraction had a higher transition temperature which was consistent with their greater crystallinity and higher ΔH_{gel}.[34] Tester[90] postulated that the gelatinisation and swelling properties are controlled in part by the molecular structure of amylopectin (perfection and ordering of amylopectin crystallites, extent of branching, molecular weight and polydispersity), starch composition (amylose/amylopectin ratio, lipid-complexed amylose chains) and granule architecture (crystalline to amorphous ratio). Higher transition temperatures resulted from a higher degree of crystallinity, which provides structural stability and makes the granules more resistant to gelatinisation.[86] The starches with long-branch chain length amylopectin generally display higher ΔH_{gel}, indicating that more energy is required to gelatinise the crystallites of long chain length in such starches. An inverse relationship of DP6–12 and positive relationship of DP13–30 with Tp and Tc was observed.[29] However, the correlation of DP6–10 with T_p ($r = -0.483$ to -0.650, $P \leqslant 0.05$) and T_c ($r = -0.426$ to -0.585) was significant. DP14–18 showed significant correlation with T_p ($r = 0.580, 0.630, 0.551, 0.506$ and 0.445, respectively, $P \leqslant 0.05$) and DP14–16 showed significant correlation with T_c ($r = 0.456, 0.507$ and 0.465, respectively, $P \leqslant 0.05$). An inverse relationship of short chains DP6–12 with T_o, T_p and T_c among wheat starches has been observed.[91] A negative correlation between the amount of amylopectin short chains of DP6–12 and T_o and T_p has also been observed.[92] Short chains are known to be located on the external part of the crystalline structure;[93] they seem to form a less stable double-helical structure and are consequently get disrupted by heat at lower temperatures. Amylopectin chains with long DP showed significant positive correlations with T_p and T_c. These results reflected a more stable crystalline network formation by these chains in the starch granule.

7.7. RETROGRADATION

The molecular interactions (hydrogen bonding between starch chains) that occur in the gelatinised starch paste during cooling are referred as retrogradation. Starch retrogradation is generally considered undesirable in food products, such as bread, puddings, soups or sauces. During retrogradation, amylose forms double-helical associations of 40–70 glucose units,[94] whereas amylopectin crystallisation occurs by association of the outermost short branches.[95] Starch retrogradation rate and extent is dependent upon a number of factors, such as starch source, amylose/amylopectin ratio, storage temperature, starch concentration and presence of other constituents such as lipids, sugar, salt, *etc.* Rheology and DSC-based methods are used to evaluate the retrogradation behaviour of gelatinised starch. DSC is used to measure the transition temperatures and enthalpy of retrogradation (ΔH_{ret}) of gelatinised starch. Retrograded starch shows lower transition temperatures and ΔH_{ret} as compared to transition temperatures and ΔH_{gel} of native starch. Starch ΔH_{ret} is usually 60–80% smaller than ΔH_{gel} and transition temperatures are 10–26 °C lower than those for gelatinisation of starch granules.[96–98] The crystalline forms of retrograded starch are different in nature from those present in the native starch granules.[99] The extent of decrease in transition temperatures and enthalpy as a function of retrogradation has been attributed to the extent of retrogradation.[100] The change in G' of cooked pastes during cooling and holding at lower temperature monitored by dynamic rheometer was also used to measure the retrogradation tendency of cooked starch pastes. Field pea, chickpea and kidney bean starches showed a greater tendency towards retrogradation, as indicated by increase in G' at 10 °C in 10 h, as compared to black gram and pigeon pea starches (Figure 7.14). The changes in G' during 10 h at 10 °C revealed retrogradation in the order: field pea> kidney bean> chickpea>black gram > pigeon pea starch.[33] Black gram and pigeon pea starches also showed increase in moduli, but the increase was lower than other pulse starches. Chickpea, field pea and kidney bean starches with higher amylose content and intermediate fraction (mixture of short amylose and long side-chains of amylopectin) and lower short side chain amylopectin fractions showed greater change in moduli at 10 °C during a holding period of 10 h. Black gram and pigeon pea starches with lower amounts of amylose and intermediate fraction and higher amounts of short side chains of amylopectin showed less change in moduli. Both the amylose and long-branch chain amylopectin probably contribute more to the increase in moduli during retrogradation. The pulse starches have a high tendency towards retrogradation.[7,101] Pulse starches with high degree of retrogradation show greater syneresis, which makes these starches unsuitable for use in many food products requiring low-temperature storage. A greater amount of amylose has traditionally been linked to a greater retrogradation tendency in starches,[102] but amylopectin and intermediate materials also play an important role in starch retrogradation during refrigerated storage.[89]

7.8. STARCH DIGESTIBILITY

Starch is classified into three groups according to the rate of glucose release and its absorption in the gastrointestinal tract: rapidly digestible starch (RDS), slowly digestible starch (SDS) and resistant starch (RS). RDS is the group of starches that can be rapidly hydrolysed by digestive enzymes, SDS is the group that is digested at a relatively slow rate,[102] and RS is not digested by digestive enzymes and is consequently transferred into the colon. RDS and SDS are measured as the glucose released after 20 and 100 min, respectively, of incubation. The term 'resistant starch' was first used by Englyst and others[103]

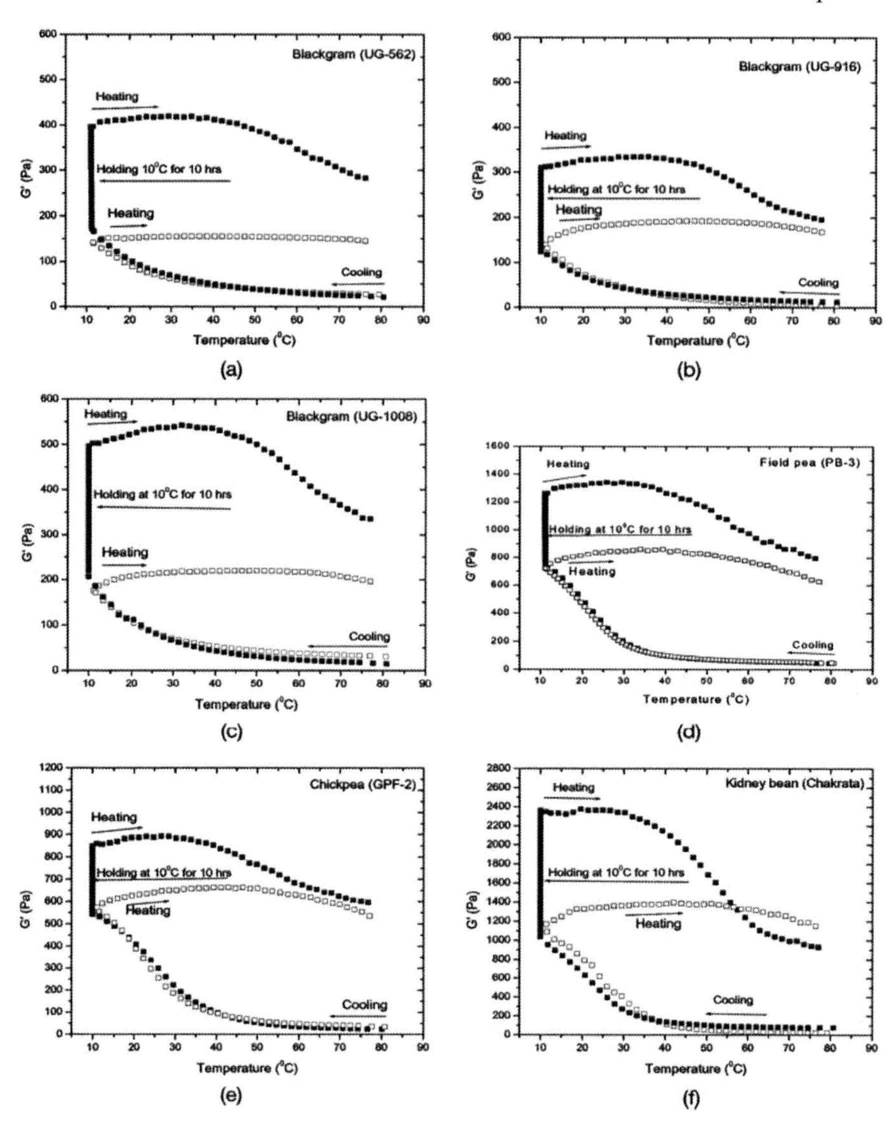

Figure 7.14 Changes in G' during cooling, holding for 10 h and reheating of pastes of different pulse starches. Light boxes (□) showed the cooling and heating cycle without holding whereas dark boxes (■) showed cooling and heating with holding (*Source*: Singh *et al.*[35])

to describe a fraction of starch that was resistant to *in vitro* hydrolysis by exhaustive α-amylase and pullulanase treatment. RS is that portion of starch that not hydrolysed in 120 min.[103] RS is more commonly defined as the sum of starch and products of starch degradation not absorbed in the small intestine of healthy individuals. RS escapes digestion in the small intestine and is digested/fermented by bacteria in the colon.[104] The lower digestibility of pulse starches has been attributed to their higher amylose content, C-type crystalline structure and absence of pores on the granule surface. RS may not be digested because the compact molecular structure limits the accessibility of digestive enzymes,[105] or because the starch granules are structured in a way which prevents the digestive enzymes from breaking these down, *e.g.* raw potatoes, unripe bananas, and high-amylose starches.[106] The starch granules are disrupted during cooking in an excess of water in a process known as gelatinisation, which renders the molecules fully accessible to digestive enzymes. The cooling of gelatinised starch lead to retrogradation and formation of crystals resistant to enzymic digestion. Both the amylose long linear chains and the amylopectin external linear chains retrograde, although amylose retrogrades much faster than amylopectin. Most pulse starches have a higher content of amylose compared with cereal and tuber starches, hence they have a greater tendency to retrogradation. This lower digestion rate renders pulse starches either slowly digestible and/or resistant to digestion. The starch present in most pulses is known for its slow digestibility or resistant character.

REFERENCES

1. S. S. Deshpande, S. K. Sathe, D. K. Salunkhe and D. P. Cornforth, *J. Food Sci.*, 1982, **47**, 1846–1850.
2. D. K. Salunkhe, S. S. Kadam and J. K. Chavan, *Postharvest Biotechnology of Food Legumes*, CRC Press, Boca Raton, FL, 1985.
3. T. J. Schoch and E. C. Maywald, *Cereal Chem.*, 1968, **45**, 564–573.
4. R. D. Reichert and C. G. Youngs, *Cereal Chem.*, 1978, **55**, 469–480.
5. S. Kawamura, Y. Tuboi and T. Huzii, *Tech. Bull. Kagawa Agric. Coll.*, 1955, **7**, 87–90.
6. R. Hoover and F. W. Sosulski, *Can. J. Physiol. Pharmacol.*, 1991, **69**, 79–92.
7. R. Hoover and F. W. Sosulski, *Starch*, 1985, **37**, 181–191.
8. W. S. Ratnayake, *Food Chem.*, 2001, **74**, 189–202.
9. A. H. Schulman and K. Kammiovirta, *Starch/Stärke*, 1991, 43, 387–389.
10. N. Singh, K. S. Sandhu and M. Kaur, *J. Food Eng.*, 2004a63, 441–449.
11. S. Naguleswaran and T. Vasanthan, *Food Chem.*, 2010, **118**, 627–633.
12. F. Meuser, N. Pahne and M. Möller, *Starch/Stärke*, 1995, **47**, 56–61.
13. R. T. Tyler, C. G. Youngs and F. W. Sosulski, *Cereal Chem.*, 1981, **58**, 144–148.
14. I. D. Evans and D. R. Haisman, *J. Texture Stud.*, 1979, **17**, 253–257.
15. V. M.Leloup, P. Colonna and A. Buleon, *J. Cereal Sci.*, 1991, **13**, 1–13.
16. J. Mua, D. Jackson, *J. Agric.Food Chem.*, 1997, 45, 3848–3854.
17. C. G. Biliaderis, in *Polysaccharide Association Structure in Food*, ed. R. H. Walter, Marcel Dekker, New York, 1998, pp. 57–168.
18. A. Buleon, P. Colonna, V. Planchot and S Ball, *Int. J. Biol. Macromol.*, 1998, **23**, 85–112.
19. N. Singh, M. Kaur, K. S. Sandhu and N. S.Sodhi, *J. Sci. Food Agric.*, 2004, **84**, 977–982.

20. S. K. Sathe, S. S. Deshpande and D. K. Salunkhe, *J. Food Sci.*, 1982, **47**, 503–506.
21. D. R. Lineback and C. H. Ke, *Cereal Chem.*, 1975, **52**, 535–545.
22. H. A. El-Faki, H. S. R. Desikachar, S. V. Paramahans and R. N. Thavanathan, *Starch*, 1983, **35**, 118–122.
23. R. Hoover and F. W. Sosulski, *Starch*, 1986, **38**, 149–155.
24. F. R. T. Rosenthal, L. Espindola, M. I. S. Serabiao and S. M. O. Silva, *Starch/Stärke*, 1971, **23**, 18–23.
25. W. Banks and C. T. Greenwood, *Starch and Its Components*, Edinburgh, UK: Edinburgh University Press, 1975.
26. W. R. Morrison and B. Laignelet, *J. Cereal Sci.*, 1983, **1**, 9–20.
27. R. Hoover, T. Hughes, H. J. Chung and Q. Liu, *Food Res. Int.*, 2010, **43**, 399–413.
28. Y. Nakamura, A. Sakurai, Y. Inaba, K. Kimura, N. Iwasawa and T. Nagamine. *Starch/Stärke*, 2002, **54**, 117–131.
29. N. Singh, N. Kaur, J. C. Rana and S. K. Shrama, *Food Chem.*, 2010, **122**, 518–525.
30. H. J. Chung, Q. Liu, R. Hoover, T. D. Warkentin and B. Vandenberg, *Food Chem.*, 2008, **111**, 316–321.
31. N. Singh, S. Kaur, N. Isono, Y. Ichihashi, T. Noda, A. Kaur and J. C. Rana, *Food Res. Int.*, 2012, **46**, 194–200.
32. M. Asaoka, K. Okuno, H. Fuwa, *Agric. Biol. Chem.*, 1985, **49**, 373–379.
33. N. Singh, Y. Nakaura, N. Inouchi, K. Nishinari, *Starch/Stärke*, 2008, **60**, 349–357.
34. N. Singh, S. Kaur, J. C. Rana, Y. Nakaura and N. Inouchi, *Food Res. Int.*, 2011 (In press).
35. N. Singh, Y. Nakaura, N. Inouchi and K. Nishinari, *Starch/ Stärke*, 2007, **59**, 349–357.
36. N. Singh, N. Inouchi and K. Nishinari, *Food Hydrocoll*, 2006, **20**, 923–935.
37. N. Singh, J. Singh, L. Kaur, N. S. Sodhi and B. S. Gill, *Food Chem.*, 2003, **81**, 219–231.
38. A. Blennow, A. M. Bay–Smidt, B. Wischmann, C. E. Olsen and B. L. Møller, *Carbohydr. Res.*, 1998, **307**, 45–54.
39. A. Blennow, S. B. Engelsen, L. Munck and B. L. Møller, *Carbohydr. Polym.*, 2000, **41**, 163–174.
40. A. Blennow, A. M. Bay-Smidt, C. E. Olsen and B. L. Møller, *Int. J. Biol. Macromol.*, 2000, **27**, 211–218.
41. A. Blennow, S. B. Engelsen., T. H. Nielsen, L. Baunsgaard and M. René, *Trends Plant Sci.*, 2002, **7**, 445–450.
42. T. J. Schoch, *J. Am. Chem. Soc.*, 1942, **64**, 2954–2956.
43. T. J. Schoch, *J. Am. Chem. Soc.*, 1942, **64**, 2957–2961.
44. S. A. S. Craig, C. C. Maningat, P. A. Seib and R. C. Hoseney, *Cereal Chem.*, 1989, **66**, 173–182.
45. J. J. M. Swinkles, *Starch/Stärke*, 1985, **37**, 1–5.

46. W. R. Morrison, T. P. Milligan and M. N. Azudin, *J. Cereal Sci.*, 1984, **2**, 257–271.
47. W. R. Morrison, R. F. Tester, C. E. Snape, R. Law and M. J. Gidley, *Cereal Chem.*, 1993, **70**, 385–389.
48. J. L. Jane, T. Kasemsuwan and J. F. Chen, *Cereals Foods World*, 1996, **41**, 827–838.
49. O. Smith, in *Potato Processing*, ed. W. F. Tallburt and O. A. Smith, 4th ed., Van Nostrand Reinhold, New York, 1987, pp. 203–286.
50. H. F. Zobel, *Starch/Stärke*, 1988, **40**, 1–7.
51. H. F. Zobel, *Starch/Stärke*, 1988, **40**, 44–50.
52. P. J. Jenkins, R. E. Cameron and A. M. Donald, *Starch*, 1993, **45**, 417–420.
53. D. French, in *Starch: Chemistry and Technology*, ed. R. L. Whistler, J. N. BeMiller and E. F. Paschall, Academic Press, London, 1984, pp. 183–247.
54. R. E. Cameron and A. M. Donald, *Polymer*, 1992, **33**, 2628–2635.
55. S. Hizukuri, in *Carbohydrates in Food*, ed. A. C. Elliason, Marcel Dekker, New York, 1996, pp. 347–429.
56. C. Sterling, *J. Polymer Sci.*, 1962, **56**, S10–S12.
57. A. H. Muhr, J. M. V. Blanshard and D. R. Bates, *Carbohydr. Polym.*, 1984, **4**, 399–425.
58. G. T. Oostergetel and E. F. J. Van Bruggen, *Starch/Stärke*, 1989, **41**, 331–335.
59. H. C. H. Wu and A. Sarko, *Carbohydr. Res.*, 1978, 61, 27–40.
60. R. F. Tester, J. Karkalas and X. Qi, *World's Poultry Sci. J.*, 2004, **60**, 186–195.
61. Gernat, S. Radosta, G. Damaschun and F. Schierbaum, *Starch*, 1990, **42**, 175–178.
62. H. J. Lee, RMIT University, PhD Thesis, 2007.
63. T. V. Bogracheva, V. J. Morris, S. G. Ring and C. L. Hedley, *Biopolymers*, 1998, **45**, 323–332.
64. S. Wang, P. Sharp and L. Copeland, *Food Chem.*, 2011, **126**, 1546–1552.
65. S. Nara and T. Komiy, *Starch/Stärke*, 1983, **35**, 407–410.
66. P. H. Hermans and A. Weidinger, *J. Polym. Sci.*, 1949, **4**, 135–144.
67. J. H. Wakelin, H. S. Virgin and E. Crystal, *Appl. Phys.*, 1959, **30**, 1954.
68. J. V. M. Blanshard, in *Starch: Properties and Potential; Critical Reports on Applied Chemistry*, ed. T. Galliard, Wiley, New York, 1987, pp. 16–54.
69. R. Hoover and W. S. Ratnayake, *Food Chem.*, 2002, **78**, 489–498.
70. P. Cairns, V. J. Morris, N. Singh and A. C. Smith, *J. Cereal Sci.*, 1997, **26**, 223–227.
71. A. Lopez-Rubio, B. M. Flanagan, E. P. Gilbert and M. J. Gidley, *Biopolymers*, 2008, **89**, 761–768.
72. N. P. Badenhuizen, *Cereal Sci. Today*, 1969, **14**, 280–286.
73. P. Colonna, A. Buleon, M. Lemaguer and C. Mercier, *Carbohydr. Polym.*, 1982, **2**, 43–59.
74. V. Aggrawal, N. Singh, S. S Kamboj and P. S. Brar, *Food Chem.*, 2004, **85**, 585–590.

75. P. Colonna, D. Gallant and C. Mercier, *J. Food Sci.*, 1980, **45**, 1629–1636.
76. P. Ambigaipalan, R. Hoover, E. Donner, Q. Liu, S. Jaiswal, R. Chibbar, K. K. M. Nantanga and K. Seetharaman, *Food Res.*, 2011, **44**, 2962–2974.
77. W. A. Atwell, L. F. Hood, D. R. Lineback, E. Varriano-Marston and H. F. Zobel, *Cereal Foods World*, 1988, **33**, 306–311.
78. R. F. Tester and W. R. Morrison, *Cereal Chem.*, 1990, **67**, 551–557.
79. M. J. Miles, V. J. Morris, P. D. Orford and S. G. Ring, *Carbohydr. Res.*, 1985, **135**, 271–278.
80. N. Singh and L. Kaur, *J Sci Food Agri.*, 2004, **84**, 1241–1252.
81. D. Stevens and G. Elton, *Starch/Stärke*, 1971, **23**, 8–11.
82. J. W. Donovan, *Biopolymers*, 1979, **18**, 263–275.
83. K. O. Adebowale and O. S. Lawal, *Food Chem.*, 2003, **83**, 237–246.
84. T. Noda, Y. Takahata, T. Sato, H. Ikoma and H. Mochida. *Starch/Stärke*, 1996, **48**, 395–399.
85. D. Cooke and M. J. Gidley, *Carbohydr. Res.*, 1992, **227**, 103–112.
86. V. Barichello, Y. Yada and R. H. Coffin, *J. Food Sci. Agric.*, 1990, **56**, 385–397.
87. C. G. Biliaderis, D. R. Grant and J. R. Vose, *Cereal Chem.*, 1979, **56**, 475–480.
88. M. J. Leszkowiat, R. Y. Yada, R. H. Coffin and D. W. Stanley, *J. Food Sci.*, 1990, **55**, 1338–1340.
89. F. F. Yamin, M. Lee, L. M. Pollak and P. J. White, *Cereal Chem.*, 1999, **76**, 175–181.
90. R. F. Tester, *Int. J. Biol. Macromol.*, 1997, **21**, 37–45.
91. H. Singh, N. S.Sodhi and N. Singh, *Int. J. Food Prop.*, 2009, 12, 713–725.
92. T. Noda, Y. Nishiba, T. Sato and I. Suda, *Cereal Chem.*, 2003, 80, 193–197.
93. S. Hizukuri, in *Carbohydrates in Food*, ed. A. C. Elliason, Marcel Dekker, New York, 1996, pp. 347–429.
94. J. L. Jane and J. F. Robyt, *Carbohydr. Res.*, 1984, **132**, 105–118.
95. S. G. Ring, P. Colonna, K. J. Ianson, M. T. Kalichevsky, M. J. Miles, V. J. Morris and P. D. Orford, *Carbohydr. Res.*, 1987, **162**, 277–293.
96. P. White, I. Abbas and L. Johnson, *Starch/Stärke*, 1989, **41**, 176–180.
97. L. A. Baker and P. Rayas-Duarte, *Cereal Chem.*, 1998, **75**, 301–307.
98. R. Yuan, D. Thompson and C. Boyer, *Cereal Chem.*, 1993, **70**, 81–81.
99. A. A. Karim, M. H. Norziah and C. C. Seow, *Food Chem.*, 2000, **71**, 9–36.
100. N. Singh, J. Singh, and S. K. Saxena, *J. Food Eng.*, 2002, **52**, 9–16.
101. Thahjadi and W. M. Breene, *J. Food Sci.*, 1984, **49**, 558–565.
102. R. L. Whistler and J. N. BeMiller, in *Carbohydrate Chemistry for Food Scientists*, Eagan Press, St. Paul, MN, 1997, pp. 203–210.
103. H. N. Englyst, S. M. Kingman and J. H. Cummings, *Eur. J. Clin. Nutr.*, 1992, **46**, S33–S50.
104. N. G. Asp, *Eur. J. Clin. Nutr.* 1992, **46** (Suppl. 2), S1–148.
105. S. G. Haralampu, *Carbohydr. Polym.*, 2000, **41**, 285–292.
106. A. P. Nugent, *BNF Nutr. Bull.*, 2005, **30**, 27–54.

107. T. Hughes, R. Hoover, Q. Liu, E. Donner, R. Chibbar and S. Jaiswal, *Food Res. Int.*, 2009, **42**, 627–635.
108. N. Singh, K. S. Sandhu and M. Kaur, *J. Food Eng.*, 2004, **63**, 441–449.
109. Y. Zhou, R. Hoover and Q. Liu, *Carbohydr. Polym.*, 2004, **57**, 299–317.
110. R. Hoover and H. Manuel, *Food Res. Int.*, 1996, **29**, 731–750.
111. R. Hoover, Y. X. Li, G. Hynes and N. Senanayake, *Food Hydrocolloids*, 1997, **11**, 401–408.
112. K. S. Sandhu and S. T. Lim, *Carbohydr. Polym.*, 2008, **71**, 245–252.
113. H. J. Chung and Q. Liu, *Carbohydr. Polym.*, 2009, **77**, 807–815.
114. R. Hoover, *Carbohydr. Res.*, 1993, **246**, 185–203.
115. S. Li, R. Ward and Q. Gao, *Food Hydrocolloids*, 2011, **25**, 1702–1709.
116. D. Güzel and S. Sayar, *Food Res. Int.*, 2010, **43**, 2132–2137.
117. A. M. Yousif, I. L. Bateyb, O. R. Larroqueb, B. Curtinb, F. Bekesb and H. C. Deetha, *LWT – Food Sci. Technol.*, 2003, **36**, 601–607.
118. H. S. Su, W. Lu and K. C. Chang, *LWT – Food Sci. Technol.*, 1997, **31**, 2165–2173.

CHAPTER 8

Properties of Pulses

8.1. INTRODUCTION

Knowledge of grain properties is a key to understanding the changes in the grain structure in the course of different post-harvest operations. This understanding will facilitate the design of equipment for the processing and utilisation of grains. A property of a biological material can be defined as any observable attribute or characteristic of the material or system. The relevant properties of pulses can be classified as follows:

- **Physical properties:** grain dimensions, 1000 grain weight, sphericity, roundness, size, volume, shape, surface area, bulk density, fractional porosity, static coefficient of friction against different materials and angle of repose.
- **Thermal properties:** specific heat, thermal conductivity, and thermal diffusivity.
- **Optical properties:** colour, gloss of grain.
- **Mechanical properties:** grain hardness, fracturability, rupture strength.
- **Transport properties:** density, angle of repose, coefficient of friction, terminal velocity.

The physical properties of any biological material including food grains can be defined as "properties that lend themselves to description and quantification by physical rather than chemical means".[1] Hence, physical properties of grains are measurable entities that can be observed or measured without changing the chemical properties, *e.g.* grain shape, size, colour, hardness, density. Thermal properties are those which are related to heat transfer control in grains and can be further classified as thermodynamic properties (enthalpy and entropy) and heat transport properties (thermal conductivity and thermal diffusivity). The heat transport properties of grains play a

Pulse Chemistry and Technology
Brijesh K. Tiwari and Narpinder Singh

Published by the Royal Society of Chemistry, www.rsc.org

significant role in the designing of dryers. The optical properties of foods are determined by the mode in which they interact with electromagnetic radiation in the visible region of the spectrum, *e.g.* colour coordinates (*L*, lightness; *a*, red/green; *b*, yellow/blue) of grain. The mechanical properties are those which are mainly related to the physical structure and rheology which lead to grain deformation or resistance to stress, *e.g.* grain hardness. The mechanical properties of food materials can be defined as its behaviour under applied stress.[2] Transport properties of grains can be defined as the ability of the grain to transport matter (*e.g.* water) or energy (*e.g.* heat transfer during drying) along a gradient.

As for any biological material, pulse grain properties are extensively studied and are classified into different categories including a few interchangeable terms. For example, grain bulk density or true density can be classified as both physical and transport properties of grain. Electrical conductivity and diffusivity can be considered as both thermal and transport properties. Broadly, these properties can be grouped together as engineering properties of grain. The available literature on grain properties mainly describes the physical properties or physicochemical properties of the grain. However, no clear distinction between physical and physicochemical properties of grain has been made in the literature. For instance, physical properties may also be otherwise known for the measures of the bulk behaviour of grain and its interactions with energy, while physicochemical properties are the measures of grain behaviour and energy interactions, which depend on the chemical constituents of grain (*e.g.* protein, starch).

8.2. IMPORTANCE OF GRAIN PROPERTIES

The methodical study of grains includes physical, mechanical and thermal properties, which are of paramount importance to grain processors particularly in technological and engineering design, control of grain processes and processing equipment. Grain properties allow plant breeders to differentiate the variations within cultivars and within species. Many of the market trade quality attributes are linked with physical properties; for example, 1000-seed weight is often used to determine conformity to standards during quality control of raw materials and often determines the classification of grains. Grain properties of pulses provides essential data required for:

- Grain handling and design of aeration and storage structure, *e.g.* bulk density, porosity
- Design and control of key post-harvest unit operations such as mechanical separation, drying, and processing, *e.g.* bulk density, coefficient of friction, angle of repose
- Analysis and determination of the efficiency of a machine or an operation, *e.g.* grain dimensions during dehulling
- Developing a new consumer product, *e.g.* hydration characteristics for ready-to-eat cooked beans

- Evaluation and retention of the quality of final product
- Estimation of time and temperatures of preheating and resting stages during grain conditioning and drying period
- Cleaning and grading of pulses, as grain properties differ in respect of amount of other impurities, which helps in removal of chaffy materials, shrivelled grains, straw, insects and other foreign materials by either manual sieving or mechanical cleaners
- Predicting the pressures and loads exerted by the grain masses on storage structures
- Predicting the drying rates of grains, as moisture diffusion coefficients vary with grain geometry.

Mechanical properties help in determining the behaviour of grains when subjected to external forces such as flow ability of grain during suction or movement during conveying. The evaluation of pulse grain properties has a practical utility in machine and structural design and in control engineering operations.[3] Various types of cleaning, grading, separation and drying systems are designed on the basis of physical properties of seeds.[4] Grain dimensions are important for their electrostatic separation from undesirable materials,[2] while grain shape and size are important for the prediction of drying behaviour and kinetics.[5] Knowledge of grain density (bulk, true or apparent) is useful in designing grain hoppers for dehulling or grinding and storage facilities as these influence heat and moisture transfer during drying and aeration processes.[6] A grain bed with low porosity, *i.e.* higher bulk density, will impede the air flow during aeration and pose resistance to water vapour escape during the drying process, which demands higher power to drive the aeration fans. Angle of repose and coefficient of friction provides essential information to determine the angle of chute or channel in order to achieve consistent grain flow and energy required for transportation during grain handling[7,8] and prediction of pressures and loads on storage structures.[2,9,10] Knowledge of the seed dimension and grain weight can be useful in particular end-use applications. For example, large-size cowpea cultivars are preferred for canning, which means fewer beans are required to attain a specific cooked bean weight.[11]

8.3. VARIATIONS IN GRAIN PROPERTIES

Grain properties vary within species or genotypes, mainly depending on moisture content and temperature. These properties of grains are influenced by climatic, genetic and agronomic conditions. Table 8.1 shows some physical properties of pulses at different moisture content, whereas Table 8.2 presents some of the physical properties of pulses at similar moisture content. The moisture play an important role during various post-harvest processing processes such as size reduction, grain handling and conveying, and product development. Estimation of grain properties at different moisture content is also one of the most important factors because pulses are traded at different

Table 8.1 Physical properties of some pulses at varying moisture content

Pulse	*Moisture content (% wb)*	*Length (mm)*	*Width (mm)*	*Thick-ness (mm)*	*Geometric mean diam. (mm)*	*Sphericity (%)*	*1000-grain weight (g)*	*Bulk density (kg m^{-3})*	*Kernel density (kg m^{-3})*	*Porosity (%)*	*Seed volume (cm3)*	*Angle of repose (°)*
Kidney beans	8.21	16.66	8.86	7.17	10.17	61.03	694.53	467.21	1182.78	60.50	0.62	11.66
	11.83	16.66	8.94	7.20	10.21	61.28	709.80	455.54	1163.51	60.84	0.64	12.79
	18.01	16.77	8.99	7.24	10.28	61.31	746.60	446.45	1150.02	**61.11**	0.66	13.43
Field peas	8.20	7.46	6.02	4.49	5.85	78.51	154.43	503.72	1263.14	60.10	0.13	14.08
	12.12	7.50	6.09	4.54	5.90	78.76	169.67	500.95	1248.22	60.13	0.14	15.62
	14.56	7.52	6.17	4.58	5.95	79.12	170.13	482.01	1235.51	60.98	0.16	16.41
Blackeye peas	5.66	9.19	6.96	6.26	7.32	79.72	253.53	431.58	1155.01	62.67	0.24	10.30
	9.80	9.22	6.97	6.30	7.38	80.11	271.30	429.18	1150.02	62.66	0.24	11.50
	13.25	9.47	7.19	6.47	7.59	80.21	273.97	426.26	1144.23	62.63	0.24	12.63

Table 8.2 Selected properties of pulses at a moisture content of 12.7 % db

Pulses	*Bulk density (kg m^{-3})*	*True density (kg m^{-3})*	*Thousand grain mass (g)*	*Porosity (%)*	*Terminal velocity (m s^{-1})*	*Angle of repose (°)*
Moth gram	825	1330	35·16	37·97	9·9	27·91
Green gram	793·7	1360	30·15	41·63	10·5	27·6
Chickpeas	736·6	1306	172·03	43·58	NA	29·8
Black gram	NA	1260	52·10	39·68	10·2	28·9
Pigeon peas	770·2	1321	80·99	41·68	NA	24·9
Faba beans	826	NA	422·69	NA	NA	22·3
Lentils	1095	NA	NA	29·30	11·3	NA
Soybeans	732	1195	111	38·80	NA	27·1

moisture contents. The effect of moisture on various grain properties is discussed in relevant sections within this chapter. Processing treatments also influence the physical properties of pulse grains. For example, roasting of chickpea induces changes in its structural and physical properties; chickpeas have a high density, but upon roasting a number of air pockets are incorporated within the cotyledon. The following section discusses the important properties of grain that have to be considered before processing.

8.4. PHYSICAL PROPERTIES OF GRAIN

8.4.1. Thousand Grain Weight

The thousand grain weight (TGW) is an indirect measure of seed size and indicates grain quality. It is defined as the weight of 1000 grains represented in grams. In pulses with large grains the 100-grain weight (HGW) is used. Wide variations in grain weight amongst pulses, mainly due to genotypic (cultivar), environmental (growing conditions), agronomic practices (growing conditions) and grain maturity are well known. TGW or HGW can help pulse producers to account for grain size variations for grain handling during post-harvest operations, also for the calculation of seeding rate and calibrating seed drills. It can be measured by counting 100 grains of large-sized pulses (*e.g.* chickpea, kidney bean, cowpea) or 1000 grains of small grains (*e.g.* lentils, green gram, black gram) either manually or by using an electronic grain counter and subsequently weighing.

The moisture content of the grain also plays a major role in increasing the TGW. The TGW increases linearly, with an increase in moisture content from 10.9% to 28.4% for chickpea[12] and 8.0% to 20.0% for soybean.[13] Nimkar *et al.*[14] also observed a similar increase in TGW from 33.26 to 40.52 g with increase in the moisture content of moth gram from 7.33% to 33.57%. Comparing different pulses at constant moisture of 12.7%, TGW of moth gram is 35.16 g, which is greater than green gram (30.15 g) lower than chickpea (172.03 g), pigeon pea (80.99 g), black gram (52.10 g), soya bean (111 g) and faba bean (422.69 g).[14]

8.4.2. Grain Dimensions

Grain principal dimensions (length, width and thickness), sphericity/roundness, size, volume, shape and surface area are considered as important properties of grains. The length, width and thickness of pulse grains can be measured using a standard vernier caliper. However, in some cases effective diameter [D_e, mm], geometric [D_g, mm] or arithmetic [D_a, mm] average of three principal dimensions for the diameter of an equivalent sphere are calculated using Equations 8.1–8.3.[2]

$$D_g = [LWT]^{\frac{1}{3}} \tag{8.1}$$

$$D_e = \left[L \frac{[W+T]^2}{4} \right]^{1/3} \tag{8.2}$$

$$D_a = \frac{[L+W+T]}{3} \tag{8.3}$$

where L is length, W is width and T is thickness of grains in mm;

The effective diameter [D_e, mm] of pulse grain can also be expressed in terms of TGW and grain density[53] as shown in Equation 8.4:

$$D_e = \left[\frac{6 \times TGW}{1000 \times \rho_\tau \times \pi} \right]^{1/3} \tag{8.4}$$

where TGW is the 100-grain weight (kg) and ρ_t is true density (kg m^{-3}).

Sphericity (S_p) can be defined as the ratio of the surface area of a sphere having the same volume as that of the pulse grain, to the surface area of the pulse grain. Sphericity can be calculated by using Equation 8.5.

$$S_p = \frac{[LWT]^{\frac{1}{3}}}{L} \tag{8.5}$$

Roundness is a measure of the sharpness of a grain's edges and corners, which can be expressed as a measure of the sharpness of its corners and can be defined as the ratio of the largest projected area of an object in its natural rest position to the area of the smallest circumscribing circle.[15,16] There are several available methods to estimate both the sphericity and roundness but the most widely accepted method is that proposed by Curray.[16] According to

Curry[16] sphericity is a ratio of diameter of the largest inscribed circle (D_i, mm) and the diameter of the smallest circumscribed circle (D_c, mm) whereas roundness (R) is the ratio of the largest projected area of grain (A_p, mm^2) and area of smallest circumscribing circle (A_c, mm^2) as given in Equations [8.6–8.7]:

$$S_p = \frac{D_i}{D_c} \tag{8.6}$$

$$R = \frac{A_p}{A_c} \tag{8.7}$$

Dutta *et al.*[12] employed shadowgraphs of chickpea to determine the sphericity and roundness of gram in three mutually perpendicular positions schematically depicted in Figure 8.1, which shows the projected views of a grain normal to (left) short, (middle) medium and (right) major axes. They observed that the sphericity and roundness of the chickpea grain were 0.735 and 0.697, respectively. In addition, they also noted that if roots and a small protruding portion of chick pea grain are ignored, which occupy about 4.5 and 7.5% of the total projected area (Figure 8.1), the sphericity and roundness increases to over 0.810. The sphericity of pulses increases with an increase in moisture content, such as the example of moth gram cited earlier.[14]

The grain volume and surface area of pulse grains can be calculated by using following equation as proposed by Jain and Bal,[17] who considered grain volume, V (Equation 8.8) and surface area, S (Equation 8.9):

$$V = 0.25\left[\left(\frac{\pi}{6}\right)L(W+T)^2\right] \tag{8.8}$$

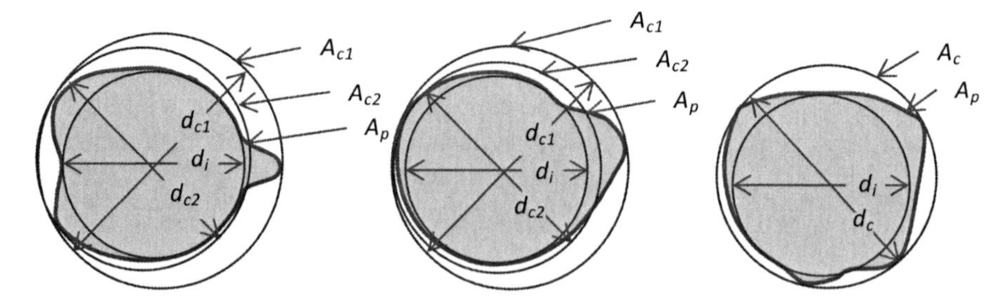

Figure 8.1 Shadowgraphs of a grain at mutually perpendicular positions of (a) normal to short axis, (b) normal to medium axis and (c) normal to major axis (Dutta *et al.*[12])

$$S = \frac{\pi L (WT)^{1/2}}{\left[2L - (WT)^{1/2}\right]} \tag{8.9}$$

However, McCabe *et al.*[18] proposed that surface area can be determined by using Equation 8.10:

$$S = \pi {D_g}^2 \tag{8.10}$$

where D_g is geometric diameter

Bhattacharya *et al.*[19] proposed a method to calculate the radius of curvature and the surface area of dehulled splits (dhals) and whole lentils. Dehulled splits of pigeon pea or lentils are hemispherical in shape (Figure 8.2) where the centre of curvature of the top surface of dehulled splits is O and its radius of curvature is R, and then according to Pythagoras' theorem the height (h) can be determined from Equation 8.11.

$$R^2 = h^2 + r^2 \tag{8.11}$$

Assuming T is the thickness of the splits,

$$T = R - h \tag{8.12}$$

The radius (r) and thickness (T) of the split can be obtained by using a vernier caliper. Thus, radius of curvature (R) can be obtained from Equations 8.11 and 8.12:

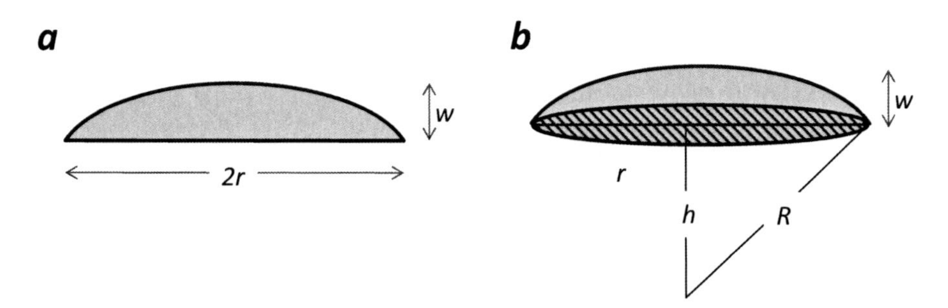

Figure 8.2 Schematic representation of lentil dhal as shown by (a) two-dimensional and (b) three-dimensional views. Here, $2r$ is the diameter of dhal, w is thickness, the hatched portion is the cross-sectional area of dhal parallel to horizontal axis, O is the centre of curvature of the top surface, R is the radius of curvature, h is the perpendicular distance between the centre of curvature and the cross-sectional area (Bhattacharya *et al.*[19])

$$h = R - T$$
$$R^2 = (R - T)^2 + r^2$$
$$R^2 = R^2 + T^2 - 2RT + r^2$$
$$2RT = T^2 + r^2 \tag{8.13}$$

or

$$R = \frac{T^2 + r^2}{2T}$$

Hence, the total surface area of splits can be calculated by using integration (Equation 8.14) as shown below with an assumption that the curved surface may be divided into several cylindrical slices having height dT:

$$A_T = \pi \int_0^T dT \int_0^{2r} dr + \pi r^2 \tag{8.14}$$

where it is assumed that the curved surface may be divided into several cylindrical slices having a height of ΔT and whose diameter decreases from $2r$ at the bottom to a value of zero at the top.[19]

8.4.3. Grain Density

Grain density is one of the most important transport properties and is widely used in various processing operations. In simple terms it is the unit mass (g) per unit volume (mL). Grain density can be expressed in different ways, but bulk density and true density are commonly used for grains in processing. Bulk density is defined as the ratio of the mass of a sample of a food grain to its total volume, sometime also referred to as apparent density. True density is the ratio of the mass of a sample of pulse grain to its actual occupied volume or its true volume. Actual volume of grain can be determined by the liquid displacement method in which the amount of water displaced by grains fully immersed in water is measured.[20] Grain volume has also been measured by toluene displacement and gas displacement.[2] Gas or liquid displacement methods work on the Archimedes principle of fluid displacement. To avoid absorption of liquid by grains during immersion, Shepherd and Bhardwaj[20] suggested using a thin water-resistant coating over the grain prior to immersion in water or toluene so that the grain does not absorb moisture during soaking. A gas displacement method using a pycnometer has also been reported for measurement of grain volume.[21]

Bulk density (ρ_d, g mL^{-1}) can be determined by filling a standard container (*e.g.* a measuring cylinder) of known volume with grain by pouring it from a certain height, striking off the top level and then weighing the content.[22] Bulk density can be calculated using Equation 8.15:

$$\rho_b = \frac{Mass}{Volume} \tag{8.15}$$

can be defined as the ratio of the mass of a grain to its solid volume. Porosity (ε, %) can be measured by using Equation 8.16. Grain porosity can be defined as the ratio of void spaces between grains to that of bulk grain. In simple words, it is a property of the grain which depends on its bulk and true densities and can be determined if bulk density and true density are known.

$$\varepsilon = \frac{[\rho_t - \rho_b]}{\rho_t} \times 100 \tag{8.16}$$

The bulk density and true density of pulses vary according to the moisture content and decreases with an increase in moisture content. The inverse linear relationship of bulk density and true density to moisture content is reported for several pulses including pigeon peas,[20] faba beans,[22] chickpeas,[12,23] lentils,[24] soybeans[13] and green gram.[25] The decrease in bulk density with an increase in moisture indicates that the weight gained due to the absorption of moisture is relatively lower than the corresponding volumetric expansion. However, the porosity of pulses increases with an increase in moisture content from 5.2% to 16.5% for chickpeas[26] (Table 8.3) and from 27.5% to 32.2% for lentils.[24] Estimation of bulk density with respect to moisture content is critical from a storage point of view to calculate the storage space required at various stages of grain processing.

Table 8.3 Physical attributes of chickpea grains[26]

Attribute	*Moisture (% db)*	*Value*
Length	5.2	9.342 ± 0.118 mm
Thickness	5.2	7.752 ± 0.087 mm
Width	5.2	7.722 ± 0.103 mm
Geometric mean diameter	5.2	8.358 ± 0.089 mm
Sphericity	5.2	87.589 ± 0.474 %
Mass	5.2	0.324 ± 0.003 g
Volume	5.2	0.238 ± 0.027 cm^3
Rupture strength (thickness)	5.2–16.5	$528.50 \times m^{-0.582}$
Rupture strength (width)	5.2–16.5	$367.65 \times m^{-0.473}$
Rupture strength (length)	5.2–16.5	$340.97 \times m^{-0.477}$
Static friction (μ_s)		
Rubber	5.2–16.5	0.228Ln(m) + 0.140
Plywood	5.2–16.5	0.165Ln(m) + 0.173
Sheet metal	5.2–16.5	$0.257 \times m^{0.238}$
Dynamic friction (μ_s)		
Rubber	5.2–16.5	0.219Ln(m) + 0.095
Plywood	5.2–16.5	0.157Ln(m) + 0.128
Sheet metal	5.2–16.5	$0.203 \times m^{0.276}$

Table 8.4 Thermal properties of chickpea with moisture and temperature range

Pulses	*Moisture content (%)*	*Temperature (K)*	*Thermal conductivity ($W\,m^{-1}\,K$)*	*Thermal diffusivity ($m^2\,s^{-1}$) × 10^{-8}*	*Specific heat (kJ $kg^{-1}\,K$)*	*Reference*
Beans	0–23.1		0.136–0.245	8.54–7.55	1.210–2.466	53
Broad beans	0–23.1		0.140–0.252	7.98–7.43	1.344–2.600	53
Faba beans	13–26	0.23–0.33	0.23–0.33	10–11.4		55
Chickpeas	11.5–27.2	283–3 12	0.114–0.247			52
Chickpeas	12.5–26.5	293–307		9.46–16.39		52
Chickpeas	12.4–32.4	283–323			1.464–2.904	52
Chickpeas	9.86 65.24				1.375–2.480	49
Chickpeas	7.0–25.1	298–371	0.1535–0.3257			49
Peas	–	10–30	0.129–0.232	7.93–6.82		53
Pigeon peas	10–30		0.132–0.171	8.13–9.7	1.85–2.31	54

8.4.4. Angle of Repose

When grains are poured on to a horizontal surface a conical heap is formed and the angle of repose of grain (θ) is calculated as the steepest angle of fall relative to the horizontal plane. The value of θ is in the range 0–90°. This property is important in the filling of a flat storage facility when grain is not piled at a uniform bed depth but rather is peaked.[2] It also assists in designing the storage structures and the grain hopper during milling. Grains with lower θ form flatter piles than those with higher θ. The value of θ is strongly dependent on other physical properties such as grain shape, size, surface area, density and surface characteristics. For example, grains with a smooth seed coat will have lower θ than those with a rough surface because of friction. Researchers have employed several techniques to determine the angle of repose. Waziri and Mittal[27] measured θ by allowing grains to fall from a certain height on to circular discs of 200, 150, and 100 mm diameter until the maximum height was reached. The height of cone formed by the grain heap was noted and θ was determined by using Equation 8.17. Nimkar and Chattopadhyay[25] measured θ of green gram by using an apparatus as shown in Figure 8.3. Apparatus consisted of a box of dimensions 30 cm × 30 cm × 30 cm, a funnel with a circular disc of 10 cm radius (*r*) attached inside and a discharge gate at the bottom of the box. After filling the box with a test sample, the gate is removed.

The height (*h*, cm) of grain pile retained on the circular disc is measured by using a scale and used to determine the angle of repose by using Equation 8.17.

$$\theta = \tan^{-1}\left[\frac{h}{r}\right] \tag{8.17}$$

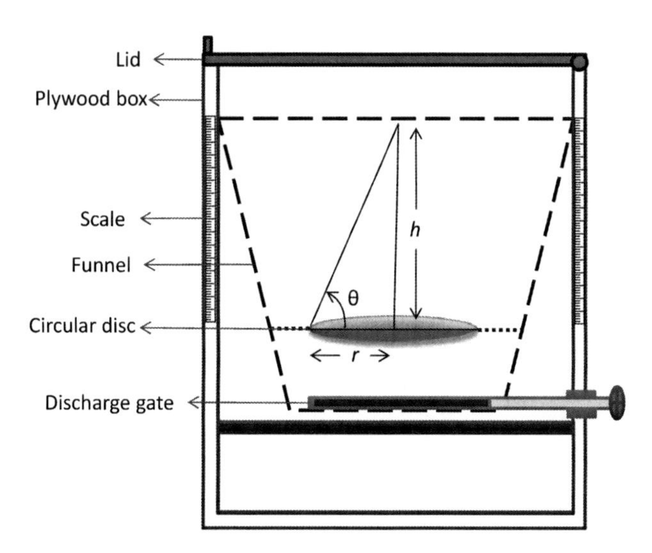

Figure 8.3 Apparatus to measure the angle of repose (Nimkar and Chattopadhyay[25])

Kaleemullah[28] suggested a technique to measure the angle of repose by placing a hollow cylinder of known dimensions on a horizontal surface, filling the cylinder with the test sample and removing the cylinder slowly in a vertical direction allowing the formation of a cone. The height (h) and diameter (D) of the cone can be noted to determine θ by using Equation 8.18.

$$\theta = \tan^{-1}\left[\frac{2h}{D}\right] \tag{8.18}$$

Amin *et al.*[29] employed this method to measure the angle of repose of lentil seeds. The angle of repose of chickpeas increases from 24.5° to 27.9° with an increase in moisture content from 5.2% to 16.5%[26] (Table 8.3). The angle of repose is negatively correlated with the sphericity of the grain, *i.e.* the higher the sphericity, the lower will be the angle of repose. This is because the higher sphericity of the grains allows them to slide and roll on each other.[26]

8.4.5. Coefficient of Friction

Friction is defined as the force resisting the relative motion of grain on a solid surface, whereas the coefficient of friction is defined as the ratio of the force causing a grain to slide along a plane to the normal force pressing the two surfaces together. Both static and dynamic coefficients are important for grains. Static friction (μ_s) is the friction between grains and surface with no relative motion. For example, the static friction resists the flow of grain, while kinetic or dynamic friction (μ_k) is the friction between grains and surface with a relative motion between grain and the surface. The static coefficient of friction is used to determine the angle at which chutes must be positioned in order to achieve consistent flow of materials through the chute. Both the coefficients of friction are dependent on grain characteristics, other physical properties and characteristics of the surface on which the grain slides. The coefficient of friction is generally determined with respect to different surfaces such as steel, wood, concrete, glass, rubber sheet, plywood and galvanised iron.

The static coefficient of friction (μ_s) of pulse grains has been determined on several materials by using a tilting platform of known dimension (35.0 cm × 12.0 cm) with an open-ended plastic box (6.5 cm in diameter and 4.0 cm in height). The cylinder is filled with the grain and placed on the adjustable tilting surface (Figure 8.4). The box is raised slightly to avoid contact with the surface, and the structural surface with the box resting on it is inclined gradually with a screw device until the cylinder just starts to slide down. The angle of tilt is recorded from a graduated scale and the coefficient of friction is calculated by using Equation 8.19:

$$\mu_s = \tan\alpha \tag{8.19}$$

This experimental setup has been used to determine the coefficient of friction for pigeon peas,[20] chickpeas,[12] moth gram (*Vigna aconitifolia*)[14] and peas.[30]

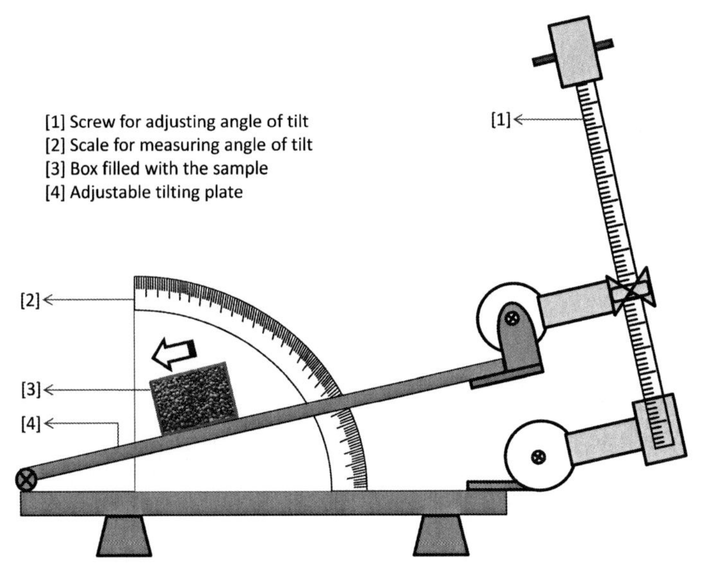

Figure 8.4 Apparatus to measure static coefficient of friction (Singh and Goswami[42])

The coefficient of friction of pulse grain is also measured by employing a friction device[31,32] as shown schematically in Figure 8.5. This device consists of a metal box, a friction surface and an electronic unit consisting of a mechanical force unit, electronic variator, load cell, electronic analogue–digital converter and personal computer. Frictional forces are measured by the load cell and are converted electronically by the analogue–digital converter and the data is recorded by the personal computer.[31] Both the static and dynamic coefficients of friction are calculated using Equations 8.20 and 8.21:

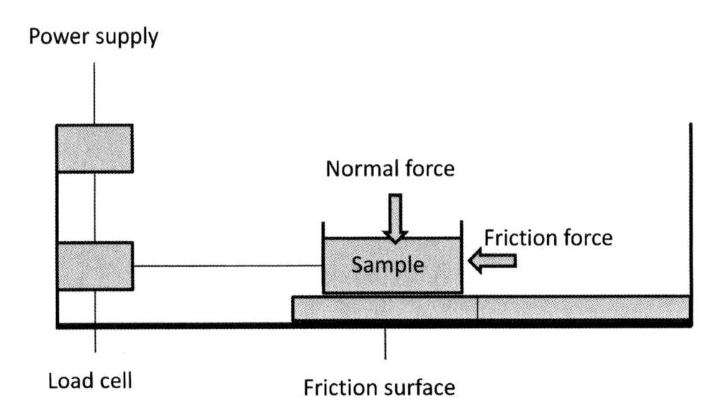

Figure 8.5 Schematic of the measuring device for friction force (Altuntas and Demirtola[31])

$$\mu_s = \frac{F_o}{N_f} \tag{8.20}$$

$$\mu_d = \frac{F}{N_f} \tag{8.21}$$

where μ is the coefficient of friction, F_o is the measured friction (in newtons) just to move the box, F is the force required to move the box with a slight push such that the box moves uniformly with a gentle push, and N_f is the normal force in newtons.

The maximum value of the friction force (F) is obtained when the box with sample starts moving to obtain the static coefficient of friction and when the box continues to slide over the frictional surface at 0.02 m s^{-1} velocity; at this point the dynamic coefficient of friction is measured.[31] Altuntas and Demirtola[31] used the friction device for measurement of static and dynamic coefficient of friction of kidney beans (*Phaselous vulgaris*), peas *(Pisum sativum*), and blackeye peas (*Vigna sinensis*) grains at varying moisture contents. The static and dynamic coefficients of friction are reported to increase with an increase in moisture content irrespective of frictional surfaces, which is due to the increased adhesion between the product and the surface at higher moisture values.[30]

8.4.6. Coefficient of Restitution

The coefficient of restitution (COR) of pulse grains can be defined as a ratio of speeds after and before an impact on a horizontal surface. The COR of grains can be determined by dropping a grain from a known height on a horizontal surface and measuring the height of grain on the rebound. Jayan and Kumar[33] measured the COR by dropping pigeon peas from a height of 50, 100, 150, 200, 250 and 300 mm on a 3 mm horizontal surface made from thick mild steel and rubber sheets. A graduated scale of 500 mm at the background is placed and the maximum height of pigeon pea rebound was recorded using a high-speed digital video camera using the video editing unit. This was replicated 10 times for each kind of seed and the COR calculated using Equation 8.22:

$$COR[-] = \sqrt{\frac{h}{H}} \tag{8.22}$$

where h is the height of rebound (mm) and H is the height of drop (mm).

The COR can be useful to predict the grain loss during conveying or grain transfer. The bouncing properties of pigeon peas have been employed to separate whole unhulled and dehulled whole (gota) pigeon pea grains.[34] Up to 61% separation efficiency has been achieved in the laboratory trials.

8.4.7. Shrinkage

Shrinkage can be defined as the reduction in the grain volume or grain dimensions as a result of moisture loss which mainly occurs during drying. Grain shrinkage can be either isotropic (uniform shrinkage in all dimensions) or anisotropic (shrinkage differs with respect to grain dimensions or non-uniform shrinkage). Generally, shrinkage in pulses occurs uniformly. Moisture shrinkage during drying can be calculated as follows:

$$\textit{Moisture shrinkage}\ [\%] = \frac{M_1 - M_2}{100 - M_2} \times 100 \tag{8.23}$$

where $M_1$is the moisture before drying (%) and M_2 is the moisture after drying (%).

For grains, some authors also evaluated the volume shrinkage coefficient (β), which can be defined as the the proportion of initial specific volume that shrinks as moisture is removed and can be estimated by using Equation 8.24:

$$V = V_o\left[1 + \beta\frac{M}{100}\right] \tag{8.24}$$

where M is the moisture content (% dry basis), V is the specific volume (volume per unit of dry matter) and V_o is the specific grain volume at 0% moisture dry basis.

Equation 8.24 was employed by Abalone *et al.*[35] to calculate the shrinkage coefficient of amaranth seeds. Estimation of the volume shrinkage of pulse grains also provides essential data required for the moisture desorption during drying process.

8.5. AERODYNAMIC PROPERTIES

The terminal velocity of grains is the air velocity at which grains remain in suspension. It can be measured using an apparatus as shown in Figure 8.6. The apparatus consists of an air blower to which a vertical air column is connected with a regulator to adjust the air flow rate. The regulator setting can be adjusted manually or can be varied through a computer controller that can regulate the opening of the air entrance into the blower. To measure the terminal velocity, the grains are poured into the air stream from the top of an air column and the airflow rate is gradually increased till the grains are suspended in the air stream. The air velocity near the location of the grain suspension can be measured by hot-wire anemometers, which are precise instruments commonly used in the study of laminar, transitional and turbulent boundary layer flows. This system has been widely employed to estimate the terminal velocity of peas.[36] This method is widely used for the terminal velocity determination of peas,[30] cowpeas,[37] vetches (*Vicia sativa*),[38] green gram,[25] kidney beans[39] and moth gram.[14]

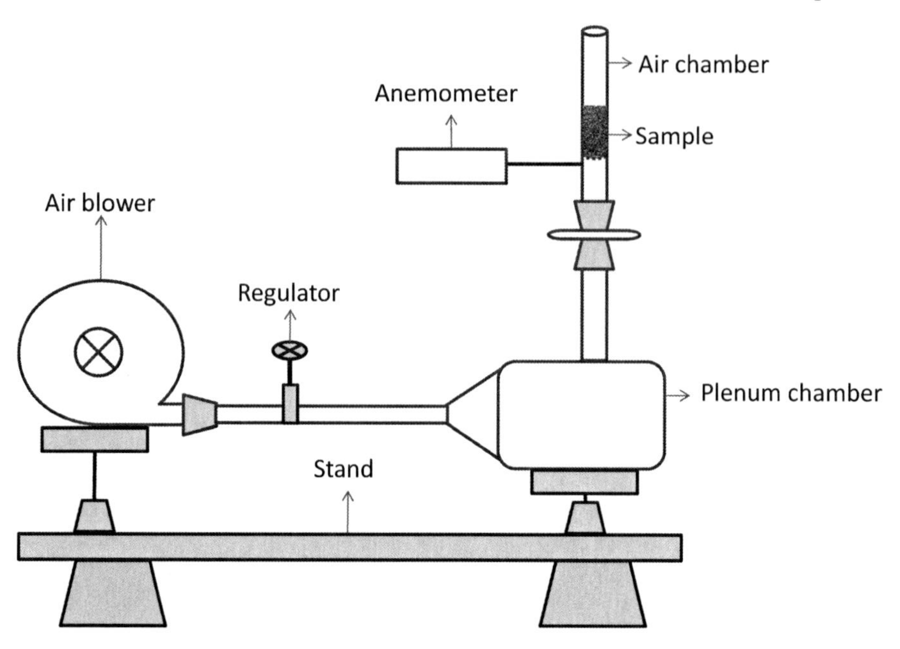

Figure 8.6 Apparatus for measuring terminal velocity (Altuntas and Demirtola[31])

Another method, known as the free fall method, was developed by Keck and Goss[40] to determine the terminal velocity of pulse grains. In this method, the grains are allowed to fall from the top of a dropping tube at varying heights and the duration of the fall is recorded precisely. The linear portion (slope) of the curve obtained from the plot of distance versus time indicates the terminal velocity. This method has been reported for determining the terminal velocities of chickpeas,[26] pigeon peas [41] and lentils.[24]

Terminal velocity also increases with an increase in pulse grain moisture content, which can be attributed to the increase in grain mass of an individual grain per unit frontal area exposed to the airflow and also to the friction of the edges of the grain.[25,42] Generally, terminal velocity increases linearly with an increase in moisture content. The measurement of terminal velocity is important for designing fluidised bed drying systems or when pneumatic or mechanical transport is used during grain conveying. Terminal velocity is also important for the design of cleaning systems because impurities such as chaffy material, husk, shrivelled grains and straw can be removed while conveying grains, due to differences in terminal velocities.

8.6. MECHANICAL PROPERTIES

Rupture strength is one of the mechanical properties commonly used for grain quality evaluation. The rupture strength of grains can be defined in terms of force (in newtons) required to break the grain along its principal axis.

Altuntasa and Yıldız[43] employed a test device to measure the rupture strength of faba beans as shown in Figure 8.7a. This device consisted of three principal components: (1) a moving platform, (2) a driving unit and (3) a data acquisition system which includes a load cell, personal computer card and software to analyse data. The test samples are placed on the moving platform and the grain is pressed with a fixed plate on the load cell until it is ruptured.[43] This test is performed for all three axes of the grain (Figure 8.7b) to calculate the rupture force, specific deformation and rupture energy required from the force versus deformation curve by using the data acquisition system.

For example, the rupture strength along all three major axes is dependent on moisture, which ranges from 5.2% to 16.5% dry basis for chickpeas[26] (Table 8.3), 9.89% to 25.08% dry basis for faba beans[43] and 8% to 18% wet basis for cowpeas.[44] Higher force is required to rupture the grains at low moisture content, whereas at higher moisture levels the grain imbibes water to cause a swelling, leading to easy rupture. However, in some grains lower force is required to rupture at low moisture content as observed by Ige;[44] for example, rupturing cowpeas required lower force mainly due to their brittleness property. Visual observations indicate that cotyledon tends to rupture before seed coat. Altuntasa and Yıldız[43] observed a decrease in the rupture force for faba beans up to a moisture level of 18.32% and later increase with an increase in moisture content from 18.32% to 25.08%. This may be due to the increased absorption of water by the seed coat of faba bean grains when compressed along the z-axis, leading to swelling up of the inner side of coat and filling of the clearance between the inside of grain and the seed coat. The transformed grain became structurally turgid, perhaps resulting in an increase in rupture force again.[43] The rupture strength of grains at given moisture

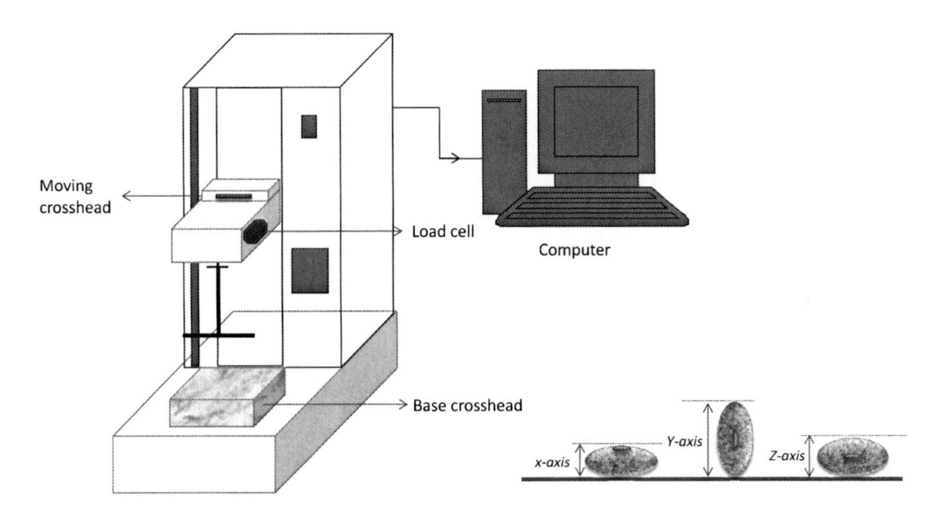

Figure 8.7 (a) Apparatus to measure the mechanical properties of grain (Altuntasa and Yıldız[43]); (b) grain axis

content appears to be dependent on the strength of the seed coat rather than the cotyledons. Rupture strength behaviour is based on moisture content, and also dependent on the chemical composition and axis of rupture. Nowadays, mechanical properties of grains and their products, including rupture strength, can be obtained from a texture analyser as shown in Figure 8.8. The texture analyser is a highly accurate and versatile instrument for the precise measurement of grain strength and the behaviour of grain when compressed. A probe fitted in the desired load cell of the texture analyser moves downwards to rupture the grain sample and measures the force response of the test grain to the deformation imposed.

Essential data relating to mechanical properties can be obtained from the force versus deformation curves. Bhattacharya *et al.*[19] employed the texture analyser to perform a compression study on whole lentils and dehulled lentil splits. The generalised compression curve for whole pulses and dehulled splits are shown in Figure 8.9. The same authors divided the compression curve for whole pulse into several regions as listed below:

- a–a′: An initial unstable zone that arises when the compression surface touches the sample and sample reorientation takes place
- a′–b: An elastic zone when the sample is deformed elastically in a linear fashion
- b–f: A non-linear deformation zone showing a major fracture at point c associated with the creation of a crack on the seed coat

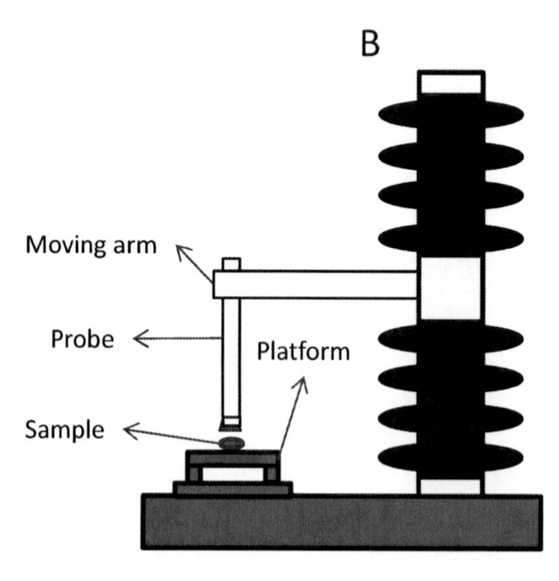

Figure 8.8 Schematic of texture analyser used for determination of mechanical properties of grain

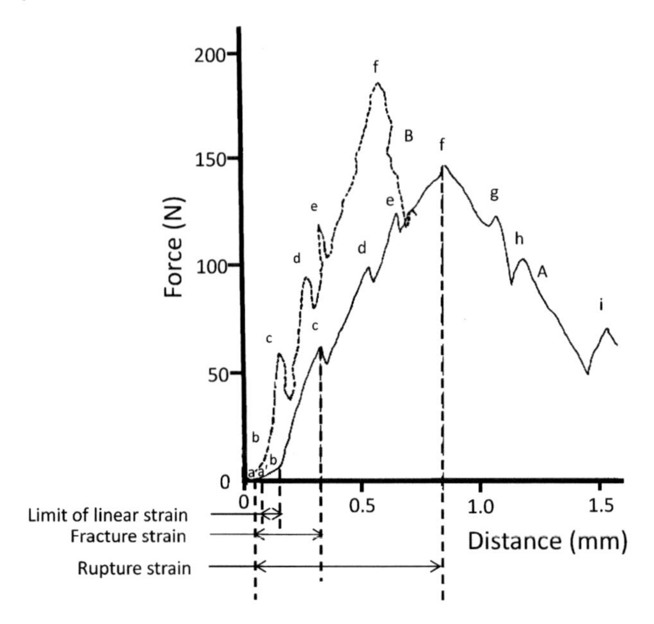

Figure 8.9 Typical compression curve for (a) whole lentil and (b) dehulled splits (Bhattacharya *et al.*[19])

- d and e: A multiple fracture zone when the sample undergoes several fractures; this depends on the grain structure, moisture content of grain and the extent and speed of compression
- f: A rupture point in the cotyledon occurs and the pulse grain splits inside the seed coat
- f–i: A post-rupture zone where more rupture points are present while further splitting or breakage of the splits created at point f occurs.

While comparing the slope of linear zone (a′–b) from the compression curve Bhattacharya *et al.*[19] observed a higher slope for splits (>100 N mm^{-1}) compared to whole lentil grains (<100 N mm^{-1}).

8.7. THERMAL PROPERTIES

The thermal properties of foods can be characterised by heat transfer mechanisms in different post-harvest operations. A knowledge of thermal properties of pulse grains is required in order to understand the heat and moisture transfer phenomena during drying and storage. Chemical composition, physical and structural characteristics of grains and their variation among grains are some of the factors responsible for differences in thermal properties. The thermal properties of grain can vary with moisture content and temperature, especially during storage and grain drying stages. They also vary with moisture content and temperature during drying. Thermal properties such as specific heat capacity, thermal conductivity and other thermal properties are given as functions of temperature and proximate composition. Data on thermal properties are required in product development to ensure whether the correct thermal processing is applied so that products are microbiologically safe.[45] For example, knowledge of the thermal properties of beans at varying moisture content is required in order to design heat and mass transfer for optimum cooking during the bean canning process and ensure that the product is not under-processed or overcooked.

Heat transfer to the surroundings during the drying of grains occurs by the three modes of heat transfer, *i.e.* conduction, convection and radiation. Conduction is the mode of heat transfer in which the transfer of energy takes place at a molecular level, and the efficiency of heat transfer depends on the thermal conductivity of the grain, which is turn depends on its chemical composition as well as the grain structure of the sample, which may vary from grain to grain and cultivar to cultivar. These variations may make it more difficult to measure the thermal conductivity of grains, compared to specific heat measurement.[46] During drying, heat transfer also takes place by natural or forced convection between hot air as a fluid (see Chapter 10) and grains. Heat transfer by radiation occurs between two surfaces by the emission and subsequent absorption of electromagnetic waves.

8.7.1. Specific Heat Capacity

The specific heat capacity of grains determines the quantity of thermal energy that has to be added to or removed from the grain to affect a given temperature change. It is defined as the amount of heat needed to increase or decrease the temperature of the grain by unit temperature.

$$\Delta q = mC_p(\Delta T) \tag{8.25}$$

$$C_p = \frac{\Delta q}{m(\Delta T)} \tag{8.26}$$

where Δq is the change in thermal energy (heat) (J), m is the mass of grains (kg) and ΔT is the change in temperature (°C). Hence, the unit of specific heat (C_p) is J kg^{-1}°C^{-1}.

Information on specific heat is very important for calculating the total amount of heat required to raise the grain temperature to a certain level during drying or storage.

The specific heat of grains can be measured by several methods, which can be grouped into four categories:[47] (1) the mixture method, (2) the comparison method, (3) the adiabatic method and (4) differential scanning calorimetry (DSC). The mixture method is the most widely used technique to measure the specific heat of food materials. In this method, a grain sample of known mass and temperature is placed in a calorimeter of known specific heat containing water or liquid of known temperature and mass.[48] The specific heat of grain sample can be calculated by a heat balance equation between the heat lost by the grain and heat gained by the liquid in the calorimeter (heat lost is equal to heat gained).

$$\Delta q(grain) = m_g C_{pl}(T_{il} - T_f) \tag{8.27}$$

$$\Delta q(liquid) = m_l C_{pl}(T_{il} - T_{fl}) \tag{8.28}$$

$$m_g C_{pg}(T_{il} - T_f) = m_l C_{pl}(T_{il} - T_{fl}) \tag{8.29}$$

Rearranging,

$$C_{pg} = \frac{m_l C_{pl}(T_{il} - T_f)}{m_g(T_{il} - T_f)} \tag{8.30}$$

where C_{pg} is the specific heat of grain; C_{pl} is the specific heat of the liquid in the calorimeter; m_l is the mass of liquid, m_g is the mass of grain, T_{il} is initial temperature of the liquid, T_{ig} is the initial temperature of the grain and T_f is the final equilibrium temperature of the mixture (liquid + grain sample) reached after adding grains to the liquid sample.

The main sources of errors in the mixture method as outlined by Rahman[47] are thermal leakage from the calorimeter, the problem of mixing and the energy added to the system by stirring. The other issue is that the density of the heat transfer medium should be lower than the grain sample so that the grain is completely immersed in the liquid.

The problems encountered when using the mixture method can be overcome by using DSC. This is a thermoanalytic technique developed by Watson and O'Neill in 1962, which was introduced commercially in 1963. DSC was developed to estimate the difference in the amount of heat required to increase the temperature of a test sample while the reference sample is measured as a function of temperature (Figure 8.10). Generally, the temperature program for

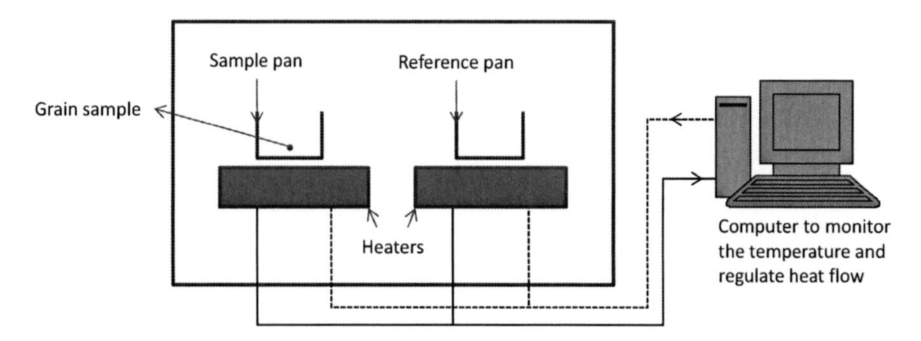

Figure 8.10 Schematic diagram of DSC

a DSC analysis is designed such that the sample holder temperature increases linearly as a function of time. The reference sample should have a well-defined specific heat capacity over the range of temperatures to be scanned. DSC allows precise measurement of specific heat. The DSC is an accurate method well-suited for determining the effect of temperature on specific heat of food samples because it is easy to scan a wide range of temperatures and allows the determination of specific heat as a function of temperature.

8.7.2. Thermal Conductivity

Thermal conductivity is defined as the rate of heat flow through a grain by conduction. Thermal conductivity of food materials depends on its chemical composition as well as the structure of the sample, and because of this the measurement of thermal conductivity of food materials is difficult and more challenging than specific heat measurement.[46] For grains, the heat transfer mechanisms of solid, liquids and gases are not easy to establish because of changes that take place in physical and chemical properties with the change in moisture content and temperature. Heat transfer by conduction is the thermal energy transmission by molecular vibrations. The thermal conductivity of grains can be measured by either steady-state techniques, quasi-steady state or transient techniques. A steady-state method is commonly used to determine the thermal conductivity of grains. This is based on Fourier's law of heat conduction, as follows:

$$q = kA\left[\frac{\Delta T}{\Delta t}\right] \tag{8.31}$$

where q is the heat flow rate (J s^{-1}), A is the area of heat transfer (m^2), $\Delta T/\Delta t$ is the temperature gradient through the grain and k is the thermal conductivity (W m^{-1} K).

Two methods are available to measure the thermal conductivity of grains but each has its own advantages and disadvantages. A disadvantage of the one-dimensional steady-state heat flow method is that a long time is required to attain the steady state. Moreover, there is a possible migration of moisture due to temperature differences maintained across the grain for the time required to attain steady state. However, the transient heat flow method makes it possible to overcome these difficulties.[20,49] Sabapathy[49] employed two methods to measure the thermal conductivity of chickpea grains: an assembled thermal conductivity probe and a commercially available thermal properties analyser. The thermal conductivity probe consisted of brass needle tubing with a T-type thermocouple fitted inside the tube. The assembly for the determination of thermal conductivity consisted of a power supply, the thermal conductivity probe and a data logger to record time–temperature data. To determine the thermal conductivity, the probe needle is inserted into the grains and the time–temperature data are recorded. Figure 8.11 shows a schematic of the apparatus used to measure the thermal conductivity of

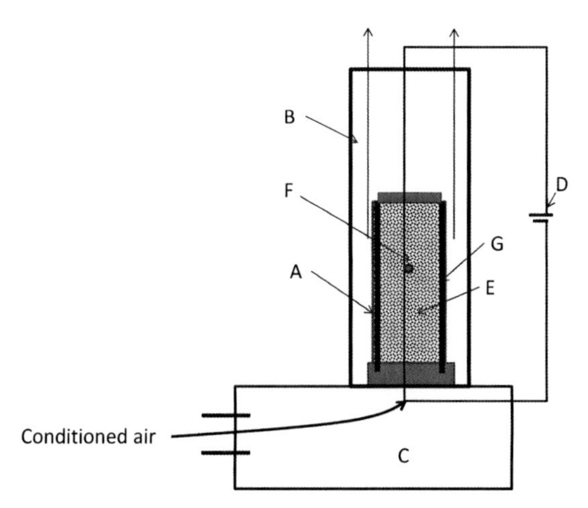

Figure 8.11 Schematic of the apparatus for thermal conductivity measurement; A, bare-wire apparatus; B, air duct; C, air plenum; D, DC power supply; E, heater wire; F, thermocouple for core temperature measurement; G, thermocouple for surface temperature measurement (Yang *et al.*[50])

grains.[50] It consists of a bare-wire thermal conductivity apparatus, a cylindrical air duct, an air-conditioning system, a circulating system and a data acquisition system made up of a data logger and a computer. The bare-wire thermal conductivity apparatus consists of a brass cylindrical sample tube, with a removable top cover and a fixed base. A T-type thermocouple connected to the data acquisition system is installed to measure the core temperature and the outer surface temperature of the sample tube. Nowadays thermal conductivity measuring instruments are available to measure thermal properties; these instruments are fairly accurate and precise.

8.7.3. Thermal Diffusivity

The thermal diffusivity of grains is defined as the rate at which heat flows through a grain. Thermal diffusivity of grains depends on the density, specific heat and thermal conductivity of the grain (Equation 8.32). The rate at which heat diffuses out of or into the grain and the movement of heat depends on the thermal diffusivity of grain. Thermal diffusivity can be measured either by direct measurement or by indirect prediction from thermal conductivity, specific heat, and density data (Equation 8.32):

$$\alpha = \frac{k}{\rho C_p} \tag{8.32}$$

Both direct and indirect measurement methods have been employed for the determination of thermal diffusivity of grains. Direct measurement of thermal diffusivity is based on transient heat transfer as proposed by Dickerson.[51] The direct measurement can be done only if the apparatus consists of a metallic cylinder in which grain sample is placed. The metallic cylinder along with the sample is placed in an agitated water bath and transient temperatures are recorded at the surface and in the centre of the grain sample. A major disadvantage of using a water bath is the necessity to seal the sample holding tube, which may be time consuming and cumbersome, and there may still be a possibility of water leakage into sample during the tests.[50] In order to overcome this problem Yang *et al.*[50] employed high-velocity air over the outer surface of the sample tube to maintain a constant surface temperature of the sample holder during determination of thermal conductivity of seeds and predicted thermal diffusivity by using the indirect method. The thermal diffusivity of grains can also be determined indirectly from the data of specific heat, thermal conductivity and density. Dutta *et al.*[52] used the indirect method to predict the thermal diffusivity of chickpea. They obtained the following equation (Equation 8.33):

$$\begin{aligned}\alpha\left[m^2/h\right] = {} & 4.788\times10^{-8} - 3.04\times10^{-5}T + 5.04\times10^{-8}T^2 \\ & + 1.3644\times10^{-2}M^2 - 4.068\times10^{-5}MT\end{aligned} \tag{8.33}$$

This equation is valid for the boundary conditions $M \geqslant 0.125$ or $\leqslant 0.265$ and $T \geqslant 293$ or $\leqslant 307$ K.

Table 8.3 shows the specific heat, thermal conductivity and thermal diffusivity of some pulses. Specific heat and thermal conductivity increase with an increase in moisture content and temperature. However, thermal diffusivity values of chickpea have been reported to increase with an increase in moisture content, but decrease with increase in temperature from 30 to 80 °C.[49] Computed values of thermal diffusivity are usually sensitive to temperature rise and often decrease as the temperature increases. The magnitude of thermal diffusivity depends on the combined effects of thermal conductivity, specific heat and density. In grain the value of thermal conductivity increases faster than that of density and specific heat at the same temperature and moisture ranges, whereas the thermal diffusivity increases with moisture content.

REFERENCES

1. A. S. Szczesniak, in *Physical Properties of Foods*, ed. M. Peleg and E. B. Bagley, AVI, Westport, CT, 1983, pp. 7–9.
2. N. N. Mohsenin, Structure, *Physical Characteristics and Mechanical Properties*, 2nd ed.: Gordon and Breach, New York, 1986.
3. M. N. Amin, M. A. Hossain and K. C. Roy, *J. Food Eng.*, 2004, **65**, 83–87.
4. K. M. Sahay, K. K. Singh, *Unit Operation in Agricultural Processing*, Vikas Publishing House, New Delhi, 1994.
5. I. Esref and U. Halil, *J. Food Eng.*, 2007, **82**, 209–216.
6. K. Kheiralipour, M. Karimi, A. Tabataeefar, M. Naderi, G. Khoubakht and K. Heidarbeigi, *J. Agric. Technol.*, 2008, 53–64.

7. S. R. Parde, A. Johal, D. S. Jayas and N. D. G. White, *Can. Biosyst. Eng.*, 2003, **45**, 3.19–3.22.
8. M. Ghasemi Varnamkhasti, H. Mobli, A. Jafari, S. Rafiee and M. Heidary Soltanabadi, *Int. J. Agric. Biol.*, 2007, **5**, 763–766.
9. H. A. Janssen, *Z. Vereines Deutsch. Ing.*, 1895, **39**, 1045–1049.
10. D. Singh, E. B. Moysey, *Can. Agric. Eng.*, 1985, 27, 43–48.
11. F. O. Henshaw, *World J. Agric. Sci.*, 2008, **4**, 302–306.
12. S. K. Dutta, V. K. Nema and R. K. Bhardwaj, *J. Agric. Eng. Res.*, 1988, **39**, 259–268.
13. S. D. Deshpande, S. Bal and T. P. Ojha, *J. Agric. Eng. Res.*, 1993, **56**, 89–98.
14. P. M. Nimkar, D. S. Mandwe, R. Dudhe, *Biosyst. Eng.*, 2005, **91**, 2, 183–189.
15. S. K. Dutta, V. K. Nema, R. K. Bhardwaj, *J. Agric. Eng. Res.*, 1988, **39**, 259–268.
16. J. K. Curray, MS dissertation, Pennsylvania State University, University Park, PA, 1951.
17. R. K. Jain, S. Bal, *J. Agric. Eng. Res.*, 1997, **66**, 85–91.
18. W. L. McCabe, J. C. Smith, P. Harriott, *Unit Operations in Chemical Engineering*, McGraw-Hill, New York, 1986.
19. S. Bhattacharya, H. V. Narasimha and S. Bhattacharya, *Int. J. Food Sci. Technol.*, 2005, **40**, 213–221.
20. H. Sheperd and R. K. Bhardwaj, *J. Agric. Eng. Res.*, 1986, **35**, 227–234.
21. R. A. Thompson and G. W. Isaacs, *Trans. ASAE*, 1967, **10**, 693–696.
22. B.M.Fraser, S. S. Verma and W. E. Muir, *J. Agric. Eng. Res.*, 1978, **23**, 53–57.
23. R. K. Gupta and S. PrakashPaper presented at XXVI Annual Convention of Indian Society of Agricultural Engineers, Hissar, 1990.
24. K. Carman, *J. Agric. Eng. Res.*, 1996, **63**, 87–92.
25. P. M. Nimkar and P. K. Chattopadhyay, *J. Agric. Eng. Res.*, 2001, **80**, 183–189.
26. M. Konak, K. Çarman and C. Aydin, *Biosyst. Eng.*, 2002, **82**, 73–78.
27. A. N. Waziri, J. P. Mittal, *AMA Agricultural Mechanization in Asia, Africa and Latin America*, 1983, **14**, 59–62.
28. S. Kaleemullah, *Trop. Sci.*, 1992, **32**, 129–136.
29. M. N. Amin, M. A. Hossain and K. C. Roy. *J. Food Eng*, 2004, **65**, 83–87.
30. I. Yalçın, C. Özarslan and T. Akbas, *J. Food Eng.*, 2007, **79**, 731–735.
31. E. Altuntas, H. Demirtola, *N. Z. J. Crop Hort. Sci.*, 2007, **35**, 423–433.
32. A. Kasap and E. Altuntas, *N. Z. J. Crop Hort. Sci*, 2006, **34**, 311–318.
33. P. R. Jayan and V. J. F. Kumar, *J. Trop. Agric.*, 2004, **42**, 69–71.
34. P. P. Kurien, N. Ramakrishnaial and V. M. Pratap. *Res. Ind.* 1993, **38**, 77–82.
35. R. Abalone, A. Cassinera, A. Gaston and M. A. Lara, *Biosyst. Eng.*, 2004, **89**, 109–117.
36. I. Yalçın, C. Özarslan, T. Akbas, *J. Food Eng.*, 2007, **79**, 731–735.
37. I. Yalçın, *J. Food Eng.*, 2007, **79**, 57–62.
38. İ. Yalçın and C. Özarslan, *Biosyst. Eng.*, 2004, **88**, 507–512.
39. E. Işik and H. Ünal, *J. Food Eng.*, 2007, **82**, 209–216.

40. H. Keck and J. R. Goss, *Trans. ASAE*, 1965, **12**, 553–557.
41. E. A. Baryeh and B. K. Mangope, *J. Food Eng.*, 2002, **56**, 59–65.
42. K. K. Singh and T. K. Goswami, *J. Agric. Eng. Res.*, 1996, **64**, 93–98.
43. E. Altuntas and M. Yildiz, *J. Food Eng.*, 2007, **78**, 174–183.
44. M. T. Ige, *J. Agric. Eng. Res.*, 1977, **22**, 127–133.
45. P. Nesvadba, M. Houška, W. Wolf, V. Gekas, D. Jarvis, P. A. Sadd and A. I. John, *J. Food Eng.*, 2004, **61**, 497–503.
46. V. E. Sweat, in M. A. Rao and S. S. H. Rizvi, eds. Engineering Properties of Foods, 2nd ed. Marcel Dekker, New York, 1995.
47. S. Rahman, *Food Properties Handbook*. CRC Press, Boca Raton, FL, 1995.
48. N. N. Mohsenin, *Physical Properties of Plant and Animal Materials*, Gordon and Breach, New York, 1970.
49. N. D. SabapathyMaster Thesis, University of Saskatchewan, Saskatoon, Saskachewan, 2005.
50. W. Yang, S. Sokhansanj, J. Zang and P. Winter, *Biosyst. Eng.*, 2002, **82**, 169–176.
51. R. W. Dickerson, *Food Technol.*, 1965, **19**, 198–204.
52. S. K. Dutta, V. K. Nema and R. K. Bhardwaj, *J. Agric. Eng. Res.*, 1988, **39**, 269–275.
53. S. Pabis, E. Bilovitska and S. P. Gadai, *J. Eng. Phys.*, 1970, **19**, 1150.
54. K. K. Singh and N. Kotwaliwale, *J. Food Process. Preserv.*, 2010, **34**, 845–857.
55. C. H. Vinod and M. B. Bera, *J Food Sci. Technol.*, 1995, **32**, 94–97.

CHAPTER 9

Post-Harvest Handling

9.1. INTRODUCTION

The term harvest originated from two words, *hær* and *fest* together *hærfest* which indicates the joyful celebration of gathering together as the mature crop is ready to be reaped from the field. Like cereals, pulses are also harvested at or after attainment of maximum grain yield, known as physiological maturity. The physiological maturity of grain refers to a particular stage in plant development. Generally the growth stage of the plant determines the readiness of the crop for harvest at its physiological maturity. Pulses are harvested either by manual plucking of mature pods or by cutting pulse stalks using hand tools or mechanically. The manual plucking of pods incurs post-harvest losses because most cultivated pulses have a tendency towards non-synchronous maturity and pod shattering. For this reason, several plant breeders are in the process of developing new pulse cultivars with uniform maturity. Synchronous maturity is one of the primary breeding objectives to ensure a single cost-effective harvesting method that contributes to farm productivity.[1] In most Asian and African countries, pulses are harvested manually by using various hand tools such as sickles, scythes or cradles (Figure 9.1). However, manual harvesting is a most labour-intensive activity, used on the small farms where farm mechanisation is not feasible.

Mechanical harvesting is the process of using agricultural machinery such as a combine harvester or land roller. Mechanical harvesting improves farm output and the productivity of farm workers. Modern harvesting techniques such as powered machinery have replaced humans and farm animals, while minimising harvesting losses For example, using a land roller can improve the ease of harvesting low, lodged and tangled pulse crops such as lentils, which also reduces the risk of "earth-tag" on the seed by compressing soil ridges.[2] This chapter discusses various pulse post-harvest handling operations and

Pulse Chemistry and Technology
Brijesh K. Tiwari and Narpinder Singh

Published by the Royal Society of Chemistry, www.rsc.org

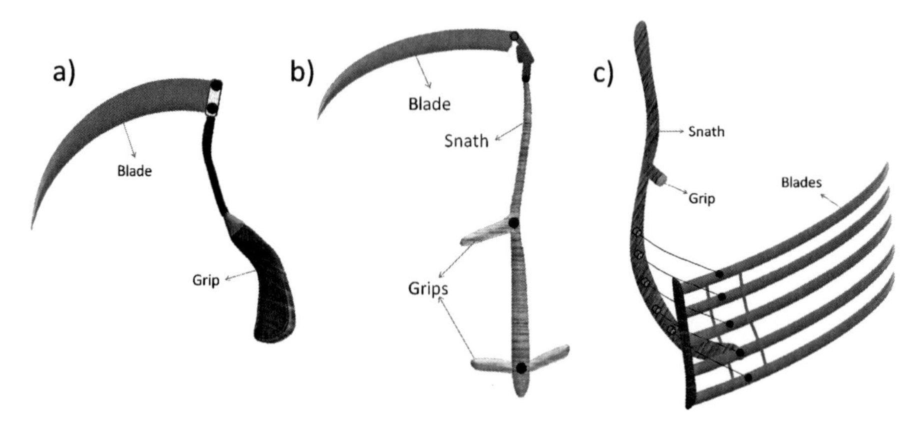

Figure 9.1 Harvesting implements: (a) sickle; (b) scythe; (c) cradle

losses which might occur, and also discusses ways to reduce post-harvest losses during handling of pulses.

9.2. POST-HARVEST HANDLING TECHNIQUES

Post-harvest handling and technologies, *i.e.* techniques that are commonly applied to handle agricultural commodities such as cereals, pulses, fruits and vegetables, constitute an interdisciplinary science. The post-harvest handling concept also extends to cover other after-harvest processing operations for the purposes of preservation, quality maintenance, processing, packaging, storage and subsequent distribution and marketing. The study of post-harvest handling and technology increases agricultural production, prevents post-harvest losses, improves nutritional quality and creates economic opportunities through job creation and adding value to agricultural commodities. Study of post-harvest technology requires an interdisciplinary scientific approach from plant biologist, agronomists, food technologists, food and agriculture stakeholders and policy-makers.

Pulse production systems can be broadly classified into two stages, *i.e.* pre-harvest and post-harvest. Figure 9.2 shows the important components of pulse

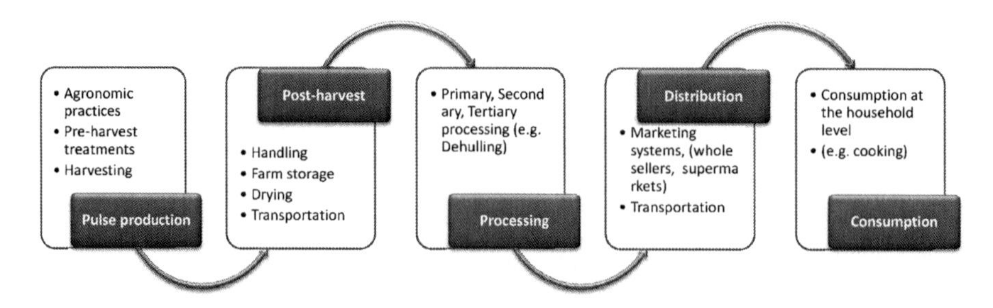

Figure 9.2 Components of pulse production and processing systems

production and processing systems. Harvesting of the pulse crop is the process of collecting mature crops from the fields. The post-harvest sector for pulses is similar to that for cereals and includes all operations taking place from last stage of production in the field to the food being placed on a plate for consumption (Figure 9.2). These operations include harvesting, on-farm handling, on-farm or bulk storage, processing at various levels, packaging, transportation, and subsequent marketing and consumption. The post-harvest handling of pulse crops differs from post-harvest handling of fruit and vegetables because these are more perishable, leading to qualitative and quantitative losses primarily due to the high respiration rate and high moisture content of fruits and vegetables. Pulses are harvested at much lower moisture content of about 20%, compared to fresh fruit and vegetables which have a moisture content of about 80–90% depending on the type. Table 9.1 shows some differences in post-harvest handling of perishable and non-perishable foods.

Generally, harvesting of pulse crops is begun as soon as the crop attains maturity or reaches the stage of physiological maturity. However, a pulse crop can also be harvested any time after it reaches physiological maturity, but it is necessary to dry the grains in order to reduce the moisture content to about 20% during on-field or on-farm level storage. In most circumstances, however, the pulse crop harvest does not begin as soon as the crop is ready to harvest. The harvesting time normally depends on many factors: infrastructure

Table 9.1 Comparison of perishable and non-perishable food commodities in relation to post-harvest handling

Parameters	*Non-perishable agricultural commodities (pulses, cereals)*	*Perishable agricultural commodities (fruit and vegetables)*
Size	Small size and low volume	Relatively large volume and weight
Harvest moisture level	10–25%	>50%
On-farm storage facilities	Long-term storage, may not require on-farm storage	Need for short-term storage facilities
Respiration losses	Low respiratory activity	High respiratory activity
Harvesting damage	Minimal, due to smaller size, hard tissues, good protection against injuries	Bruises may significantly affect the quality due to its soft tissues
Pre-treatment immediately after harvesting	Not required	Requires pre-cooling to remove field heat
Product life	Can be stored for long duration (>1 year) at ambient conditions (room temperature)	Short shelf life. Maximum shelf life (<6 months) depending on the commodity
Storage losses	Mainly due to external factors such as moisture, insects or rodents	Both external (*e.g.* microorganisms and insects) and internal (*e.g.* respiration) factors

availability to deal with high-moisture grains, climatic conditions during the harvesting period and availability of workers in case of manual harvesting. Furthermore, the non-synchronous maturity of pulse pods may also delay the harvest time for pulse breeders or growers. The harvest time greatly influences the quality of the grain and is a crucial factor for maximising yield and grain quality. If the grains are harvested early, *i.e.* before physiological maturity, the yield is greatly affected; while late harvesting, *i.e.* after physiological maturity, may lead to the loss of grain quality and yield. However, the pulse breeders or growers need to remember that there may be a trade-off between delaying harvest date and grain quality and yield. Thus, harvest time is crucial and appropriate timing of harvest ensures the maximum number of grains (grain yield increase) which have attained maximum weight and no grains shattering due to pod splitting. Unfavourable environmental conditions during harvesting also play a crucial role in deciding harvesting time. On-field germination of mature crop is a sporadic but serious problem in most pulses. Rainfall during crop maturation, or during the later stages of crop ripening, increases the potential for on-field germination of the mature pulse crop.[3] Additionally, if the mature crop is left in the field after maturity, then any rainfall is likely to induce germination. This may result in both quantitative and qualitative losses of grains, highlighting the importance of harvesting time.

9.3. POST-HARVEST LOSSES

Post-harvest losses represent loss not only of food grains but also of all the resources utilised in production of that loss. For example, the total post-harvest loss in the chickpea production system is about 15%, which includes the loss of food grains and loss due to agricultural inputs (seeds, fertilisers, irrigation, labour), and land utilised to grow and cultivate chickpeas. Hence, post-harvest loss translates not just into human hunger (lack of available food) and financial loss to farmers but also leads to tremendous environmental waste.[4]

Harvesting losses can be minimised by adopting the following measures:

- Harvesting should be done at physiological maturity to ensure optimum grain quality and quantity for trade and market needs.
- Regular pre-harvest inspection and removal of weeds is necessary in order to avoid growing foreign grains.
- Harvesting before the maturity of crop usually result in lower yields and a higher proportion of immature grains (shrivelled grains) with poor quality and high probability for infestation during storage.
- Delay in harvesting should be avoided to reduce pod shattering and other losses caused by birds, rodents, mammals, insects and other external factors.
- The optimum time to harvest a crop is when the pods are large, *i.e.* 80 % of the pods are fully matured.
- Harvesting should be avoided during adverse weather conditions, *i.e.* high precipitation level and/or overcast weather.

- Using the right kind of harvest implements (*e.g.* sickle, scythe, cradle) depending on the pulse crop will facilitate reduced harvest loss.
- The harvested bundles should be aligned in one direction for efficiency of threshing.
- The harvested product should be stacked in a dry and clean place, in rectangular piles in order to facilitate the circulation of air.

9.3.1. Types of Losses

Quantitative and qualitative losses in pulse production system occur at pre-harvest stages, during harvesting and at post-harvest stages. Pre-harvest losses may be due to insects, birds, rodents, or the presence of weeds and diseased plants in the field during harvesting; the pre-harvest shattering of over-matured pods of standing crops also leads to losses. Losses during harvesting may occur due to shattering or windrowing (heaping), and also during mechanical harvesting, *i.e.* grain loss occurs when using a combine harvester or land roller. Generally the post-harvest losses occur between harvest and the moment before the grains are processed for human consumption. Apart from some direct causes of post-harvest losses as shown in Figure 9.3, the post-harvest losses also depend on indirect causes such as marketing practices, food policies, inadequate capital, lack of trained manpower and other socio-economic and political aspects.

Post-harvest loss denotes loss of food grains and this can be measured in quantitative, qualitative and economic terms. Quantitative post-harvest losses are mainly due to the reduction in weight, which can be easily defined and estimated. However, qualitative losses of pulses are difficult to assess in terms of locally or

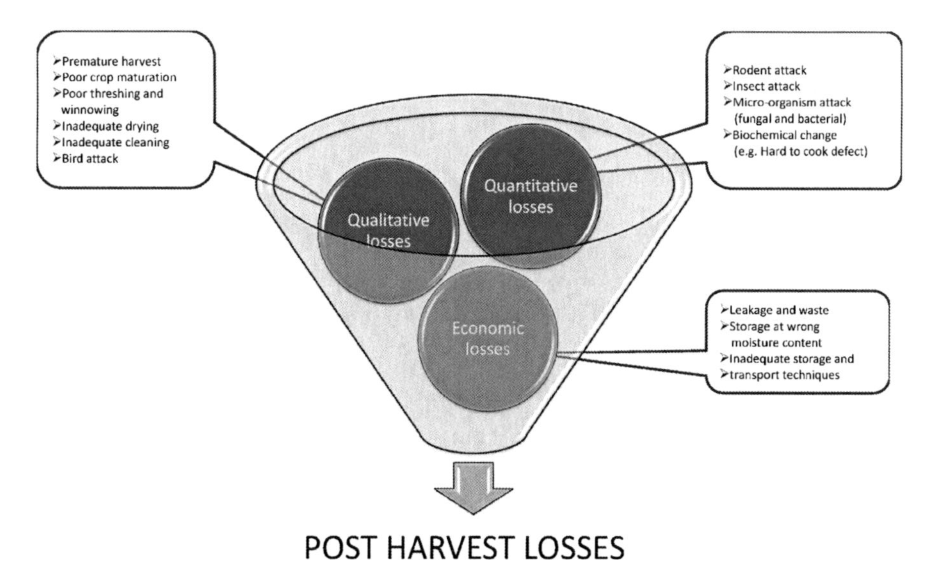

Figure 9.3 Causes for post-harvest losses of pulses

internationally acceptable pulse quality standards because they are frequently based on subjective judgment. Apart from loss of physical or visual quality, *e.g.* discolouration of grains and presence of foreign materials, nutritional losses occur as a result of reduction in either quality or quantity or may be a combination of both. Nutritional quality losses mainly due to biochemical changes occur during improper storage conditions, such as being hard to cook, poor protein digestibility, development of rancidity (chickpea and other pulses with a high fat content) due to insect and fungal attack, and the pulses become inedible. Thus, physical and nutritional quality losses are difficult to establish. Economic losses are considered as a reduction in monetary value of pulse grains because of physical losses during transportation, theft or improper storage.

More than 40% of the food losses occur during post-harvest and processing levels in developing countries. Similar amounts of losses also occur in industrialised countries due to variance in retail and consumer levels.[5] The food waste at consumer level in industrialised countries is about 222 million tons which is almost as high as the total net food production in sub-Saharan Africa (*i.e.* 230 million tons).[5] Post-harvest losses in any country depend on the technology and infrastructure available to handle agricultural commodities, and the availability of markets for agricultural produce.[6]

There are three interrelated global drivers that provide an overall structure for characterising supply chains and future trends in developing and transitional countries:

- Urbanisation and the contraction of the agricultural sector
- Dietary transition
- Increased globalisation of trade.

Figure 9.4 shows the importance of these three global drivers along with the post-harvest infrastructure along the economic gradient.[6]

Quantitative, qualitative and economic losses of widely varying magnitude occur at all stages in the post-harvest system, *i.e.* from harvesting, through handling, storage, processing and marketing to final delivery to the consumer. Post-harvest losses in pulses from mechanised, semi-mechanised and manual techniques adopted during various pulse production stages are up to 20–35%. Many factors influencing the post-harvest losses of pulses may vary with location and farm to farm management practices, which mainly results in more complex systems of marketing. For instance, if the losses are reduced by on-farm grain threshing, winnowing and drying, transportation, storage and processing, this will enhance pulse productivity and improve the pulse grower's income by a small and marginal amount.

Threshing losses occur as a result of grain spillage, incomplete removal of the grains from pods or chaff, and in some cases damage to grains during threshing. Harvesting of immature pulse crops leads to poor separation of grains from the pod during threshing and losses occur during manual winnowing or mechanical cleaning. Cleaning is a critical step prior to on-farm storage to avoid storage losses. On-farm cleaning is usually a combination of

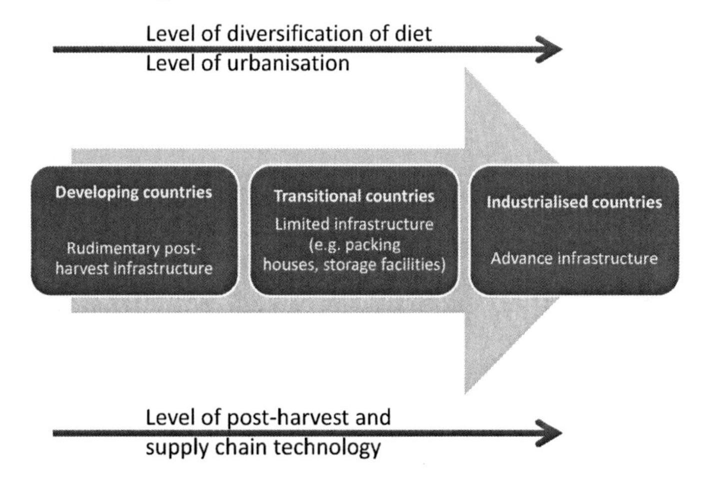

Figure 9.4 Schematic developments of food supply chains in relation to post-harvest infrastructure (Parfitt *et al.*[6])

winnowing to remove chaffy materials and other low-density impurities or foreign materials such as earth and manual removal of stones and gravels by hand. In many Asian and African countries on-farm drying of pulse crops is carried out prior to threshing and cleaning. This may also cause additional on-farm losses due to partial consumption by birds, rodents and in some cases theft. Such on-farm drying utilises solar energy and natural wind, which depend on prevailing climatic conditions; unfavourable climatic conditions may result in the entire pulse harvest becoming inedible.

These losses are of high magnitude in developing countries compared to countries where agriculture is generally mechanised. That these losses mainly occur in developing countries is probably due to the lack of post-harvest management information or prior knowledge of farm management for growing pulses. Moreover, most pulse growers are reluctant to use modern technologies for harvesting and for post-harvest management practices, either because of the higher capital cost or because they are not convinced of the benefits of using modern, energy-efficient post-harvest management practices.

To conclude, post-harvest losses are affected by several factors such as cultivar variations, seasonal variations, and agronomic practices, method of harvesting, threshing, handling of harvested products such as drying, storage, processing, transportation, and distribution (Table 9.2). Such variability may be make it hard to predict the level of losses during post-harvest stages.

9.4. POST-HARVEST TECHNOLOGY

9.4.1. Harvesting

The time of harvesting is determined by the degree of maturity. With pulses, optimum harvesting stage and grain moisture percentage during harvesting are

Table 9.2 Important stages of post-harvest system for pulses and associated losses

STAGES	*CRITICAL FACTORS*	*ESTIMATED LOSS*
HARVESTING	Harvesting stage Harvesting time (before or after physiological maturity) Method of harvesting (manual or mechanical) Climatic conditions at harvesting	5–8%
THRESHING and WINNOWING	Traditional method Mechanised	5–15%
DRYING	Mechanical drying Types of dryer Capacity of dryers Drying performance Sun drying Drying surface (cemented or uncemented) Location Drying conditions	5–15%
STORAGE	On-farm storage conditions Temperature Humidity Grain moisture Insect, Rodent attacks Microorganisms	5–10%
PROCESSING	Primary processing Secondary processing Tertiary processing	5–20%
PACKAGING and TRANSPORTATION	Weighing of product Method of packaging and mode of transportation	10–12%

the most important factors that affect subequent operations, particularly storage and preservation. Drying before harvesting, mainly pre-harvest field drying, ensures good preservation but also increases the risk of attack by birds, rodents, insects, moulds and theft. On the other hand, harvesting before maturity entails the risk of loss through moulds and the decay of some of the seeds due to high moisture content in immature grains. Harvesting is an important operation in the pulse production system, but it is a challenge for the pulse producer to maintain an optimum balance between yield and grain loss due to non-synchronous maturity of pulses. Generally, pulse growers rely on visual observations of pod and plant leaf colour. Harvesting delay, with overambitious attempts to obtain maximum yields, often leads to on-field losses.

If the grains are harvested at low moisture content (<11%) especially for large-sized pulses like chickpea, this may cause breakage during threshing and other post-harvest operations. Thus, optimum harvesting time may be crucial for reducing harvesting losses and minimising quality deterioration during post-harvest handling. The harvest time of most pulse crops is conventionally identified from the colour of the leaf and the pod, which is often judged as golden yellow or brown, and the moisture content of the grain, assessed by pressing between the teeth. Generally, grains should be firm but not brittle when squeezed between the teeth. Monitoring of seed colour is an important criterion for determining optimum harvesting time for chickpeas.[7] Yet another index for harvesting time is when the leaves start senescence and shedding. Chickpeas can be harvested at 16–18% seed moisture, thus making a delicate balance between yield and quality of grains. Figure 9.5 shows that seed yield (g/plant) is lower for a chickpea crop harvested early or late compared to the seed yield at optimal harvest in two different sites. The late harvest of the chickpea crop was reported to be due to the pods dropping and shattering, along with a significant reduction in protein content when compared to optimal harvest. Similarly, the seed protein content was significantly lower in the early harvest compared to those harvested at optimal harvest time. As a result, both yield and nutritional value of chickpea are significantly affected by time of harvest.[7]

Cassells and Armstrong,[8] in a study conducted in Wagga Wagga, New South Wales, Australia on the impact of harvest timing on yield and quality of field pea cultivars (*e.g.* Jupiter and Bohatyr), demonstrated the importance of harvesting the crops by machine at an optimum moisture content of 14–15 % wet basis in order to avoid yield and shattering losses. Early harvesting at physiological maturity with moisture content of about 20 % wet basis compared to late harvest (*i.e.* two weeks after the on-time harvest) results in maximum harvesting losses, as shown in Figure 9.6. Harvesting losses at late harvest occur mainly due to shattering of pods before and after harvest, whereas early harvest may pose difficulty in removing grains from the pod. If rainfall occurs during harvesting, it may cause increased moisture in the grain and in that case, if the pulse grower decides to wait for on-field drying of standing crops, shattering losses of high magnitude will be encountered. In case of late harvesting at much lower moisture (<14%), defects in visual grain quality occur due to bleaching of seed coat colour and also affect germination capacity. Breakage during dehulling as a result of repeated on-field wetting and drying of grains before harvest occurs due to changes in fracturability of pulses and grains, which cannot resist frictional forces during dehulling. The late harvesting may result in lodging of the mature plants because, as plants grow beyond physiological maturity, seeds gain weight and stems become weaker due to senescence. Natural winds also cause lodging. Lodged crops are difficult to harvest, and pods pick up dirt due to contact with the soil surface. Late harvesting may also cause carryover of weed seeds, particularly as a result of the growth of late weeds and plant diseases occurring at later stages of plant growth.

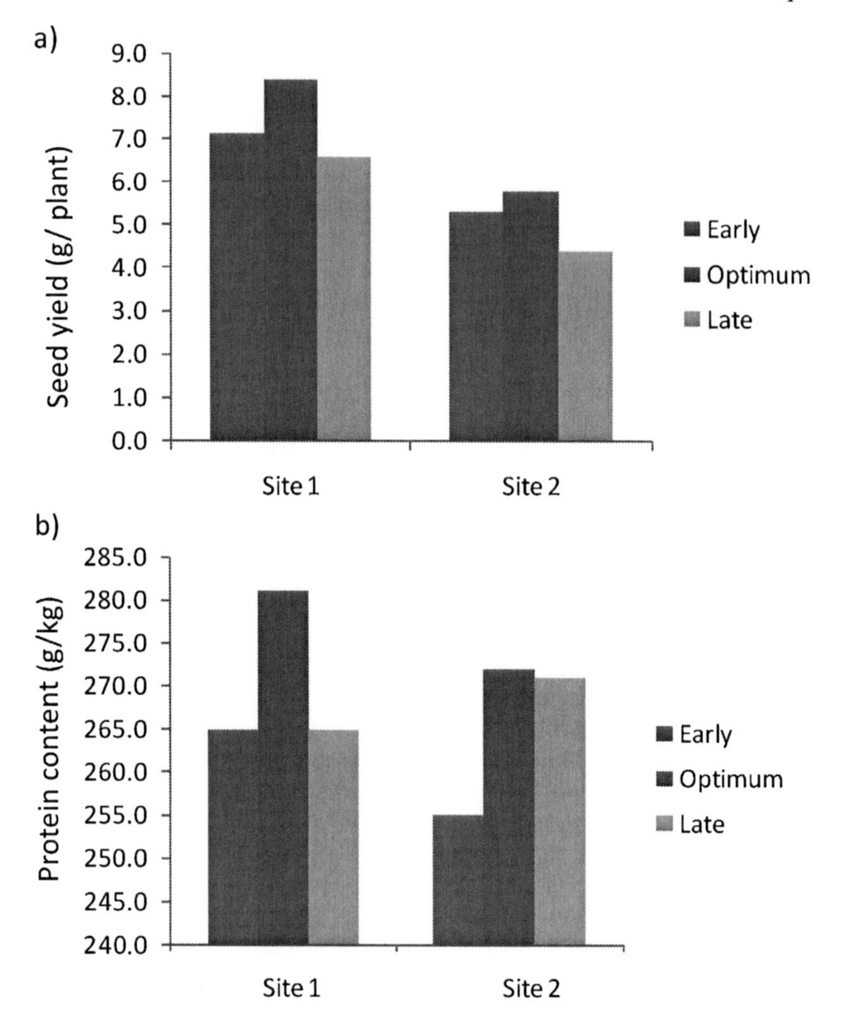

Figure 9.5 Effect of harvest timing on (a) seed yield and (b) protein content of chickpeas (kabuli type) grown under rain-fed conditions at two sites in Ankara, Turkey (adapted from Adak *et al.*[7])

Pulse crops are harvested at the base either manually by using a sickle or cradle or mechanically by using a combine harvester or roller. The manual harvesting of pulses often involves uprooting plants and pre-drying of plants along with grains and pods intact. Uprooting of plants is generally done early in the morning while the night dampness minimises the risk of pod shattering.

In some countries, defoliating chemicals are sprayed on the plants few days before harvesting. This hastens plant drying and reduces the plant biomass going through threshing. Defoliated leaves may act as a source of fertiliser for next growing season and the chemical treatment to terminate the plant growth at physiological maturity is known as desiccation. Desiccation advances pulse grain maturity and avoids several harvesting problems of pod shattering and

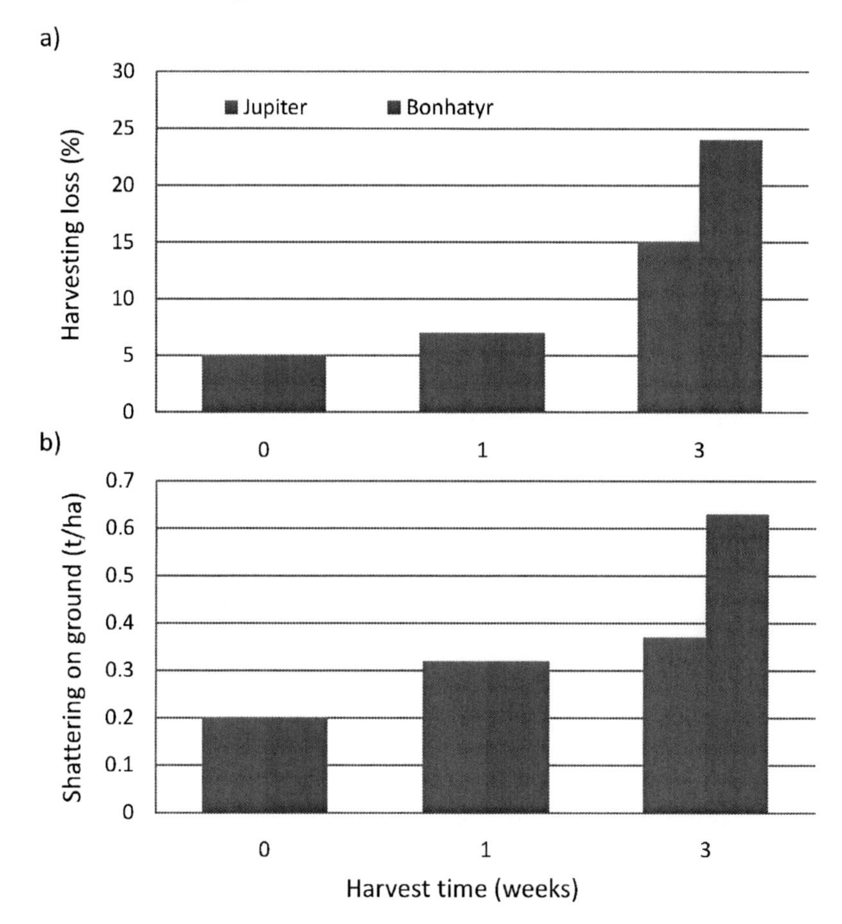

Figure 9.6 (a) Machine harvest loss (%) and (b) shattering loss on ground after harvest (t h^{-1}) for two field pea cultivars (Jupiter and Bohatyr) harvested at different harvesting time at grain moisture content of 20% (week 0), 15.2 (week 1) and 14.1 (week 3) % wet basis for Jupiter and 20.1% (week 0), 15.5 (week 1) and 13.8 (week 3) for Bohatyr respectively

non-uniform maturity of crops, enhancing harvesting efficiency. The desiccation process also reduces quality defects such as bleaching of seed pod colour due to late harvesting, restricts late disease development and reduces the carryover effect of weed seeds to subsequent harvesting and storage. Figure 9.7 shows the drying pattern of field pea seeds, which dry rapidly up to 30% moisture level and moisture content, reducing at a slower rate after physiological maturity. However, the application of chemical defoliants enhances seed drying in the pod and thereby reduces the harvesting time, quality deterioration rate and pod shattering.

Table 9.3 lists some advantages and disadvantages of pulse crop harvesting systems. The choice of a fully manual, semi-mechanised or fully mechanised harvesting system depends on several factors:

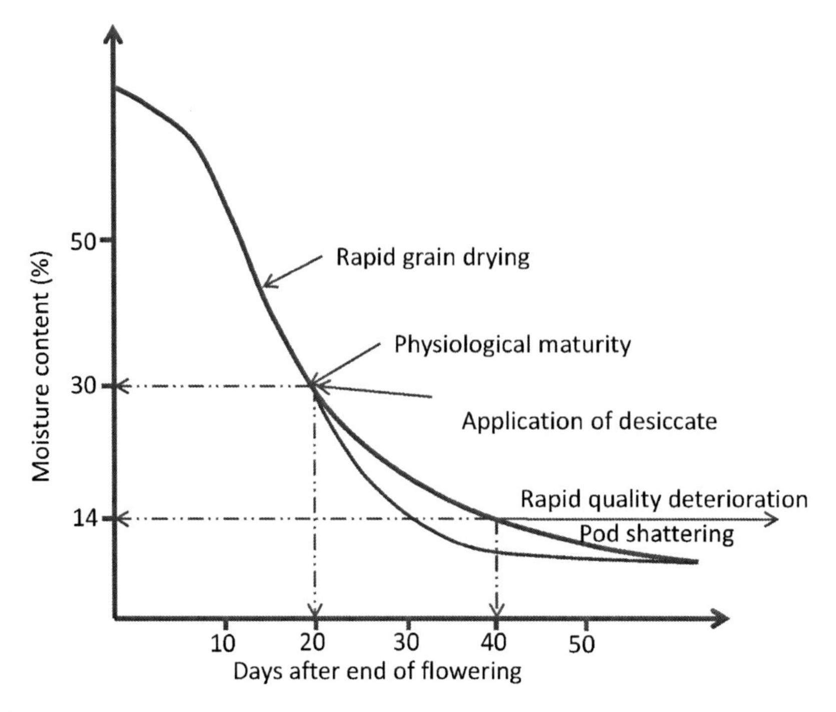

Figure 9.7 Drying pattern of field pea seeds with and without desiccation

- Availability of trained labour or workers
- Size and accessibility of the agriculture holding
- Availability of time to complete harvesting process
- Field crop status; *e.g.* lodging may cause more losses if combine harvester is used for harvesting system
- Requirement for stalks; *e.g.* combine harvester chops the straw and makes it less marketable.

After harvest, the crops are normally allowed to dry in the sun for few days, in the field in case of manual harvesting or in a cemented drying yard, depending on the method of harvesting and availability of infrastructure. The dried grains are threshed manually by beating the plants with sticks to remove grains from the pods. Pulse growers in some Asian and African countries thresh their crop by crushing the dried crop with the help of bullocks, camels or horses, followed by manual winnowing. The trampling of pulse crops by farm animals is performed by spreading dried crops on the horizontal cemented or uncemented drying yard or area and the animals are then allowed to walk on the crops, depending on the amount of material to be threshed. However, for efficient threshing this requires continuous stirring of material or crops at regular intervals.[9] This will avoid physical damage such as cracking or breakage during manual threshing. Table 9.4 shows some common harvesting techniques for pulses adopted by small or marginal pulse growers.

Table 9.3 Advantages and disadvantages of pulse crop harvesting methods

Harvesting method	*Type of operation*	*Advantages*	*Disadvantages*
	Fully manual	Effective for lodged pulse crop or for small farm holding	Labour intensive, requires skilled labour for cutting and winnowing
Manual harvesting, threshing and winnowing	Cutting by farm implement, manual gathering, threshing by beating or trampling and manual winnowing by using fan or natural wind	Less weather dependent	Susceptible to grain damage during threshing
	Semi-mechanised	Less weather dependent. Effective during unfavourable weather conditions	Requires high capital cost for mechanical threshers, maintenance and labour cost
Manual harvesting and mechanical threshing/ winnowing	Cutting by farm implement, manual gathering, threshing by using power thresher or and winnowing by machine or manual winnowing	More efficient than fully manual operation. Less labour requirement	
		Higher capacity than manual	
		Low labour requirements	
	Fully mechanised	Low labour requirement	High capital cost and maintenance cost
Mechanical harvesting and threshing	Crops are harvested, threshed and subsequently cleaned in a single operation	High throughput	Less effective if crop is lodged
		Produces clean grain and separates grains from stalks efficiently	Not suited for small farms
		Less power required for subsequent threshing and cleaning operations	Requires trained operators
		Easy to cut and chop the stalks	

Table 9.4 Harvesting of some common pulses

Pulse crop	*Harvesting time and method*
Lentils	Harvesting is done in 15–20 days after the end of flowering and when the leaves and pods begin to turn yellow and/or brown
	The plants are either pulled out along with the roots or cut at the ground level depending on the soil condition
	The crops are left behind to dry in the field or in the threshing floor for about 6–7 days
	Dried plants are beaten with sticks or trampled with farm animals to remove the pods and break the pod wall when pods are brittle followed by winnowing to remove lighter impurities such as chaff
Black gram or green gram (mung bean)	Harvesting is done when most pods have turned black
	Synchronous maturity cultivars: The plants are pulled out or cut at the ground level either manually by using a farm implement (*e.g.* sickle) and stacked for 6–8 days in small bundles for drying in the field or on the drying yard. Grains are removed from the pods by beating with a stick or trampling with farm animals, or threshed in a mechanical thresher.
	Non-synchronous maturity cultivars: When the pod maturity extends over long periods of plant growth, to avoid shattering of grains from mature pods or to avoid sprouting of mature grains on the plants itself or due cause of uncertain rain, the pods are manually picked. Manual picking requires 2–5 pickings to gather full yields, which is a labour-intensive and time consuming process. Manually picked pods are dried and threshed to remove grains from the pod. During drying of pods, the grains are removed automatically due to shattering but sometimes the crops may require some beating with sticks to remove remaining grains from the pod
Chickpeas	Chickpeas are usually harvested when most of the pods are yellow and leaves turn yellow in colour. Harvesting is generally done by pulling out the whole plants or by cutting with sickles at ground level. The harvested mature plants are bundled and transported to threshing floor or drying yard. Grains are removed from the plants as for lentils
Pigeon peas	Pigeon pea grains are harvested when pods turn yellow or brown, depending on the cultivar. Plants are cut at the base and are left in the field for drying. Sometimes the plants are tied and stacked to facilitate drying of the pods. Dried pods are beaten with sticks or trampled by farm animals to separate grains from pods and subsequently dried after winnowing
	Green pods are also harvested manually and used as a vegetable
Cowpeas	The pods are mature about 20–25 days after flowering and crops are harvested when pod colour turns yellow to brown. The mature pods are handpicked and harvested as for green gram/mung beans
	The immature green pods of cowpea are also harvested for use as a vegetable
Garden peas	Garden pea pods are harvested at several stages for vegetable purposes, including whole green pea pods; half mature pea grains in green pods
	Fully mature peas are harvested when the leaves and pods turn yellow. Harvesting method is similar to that for black or green gram

Large pulse growers normally employ mechanical harvesters and threshers to remove grains from the plant. The use of mechanical techniques is more efficient and reduces post-harvest losses.

Harvesting of pulses requires a uniform rate of crop maturation and possible delays in harvesting of the crops usually occurs when it reaches the maximum harvest maturity, which results in yield losses due to pod dropping; shattering can be expected in late harvested crops.[10] Harvesting time also influences the protein content of pulses significantly, as indicated in Figure 9.5. Depending on the type of pulse crop, a reduction in yield and nutritional quality is reported when it is harvested either earlier or later than the targeted harvesting period, *e.g.* chickpeas. Hence, an appropriate harvesting time is essential for both yield and nutritional quality.

9.4.2. Threshing

The purpose of threshing is to detach grains from the pod and it is critical step to obtain sound and unbroken grains free from foreign materials. Threshing is the process of loosening the edible part of the pulse grain from stalks or straw, and it is performed after harvesting and before winnowing. Threshing is done either manually by beating with sticks or by farm animals trampling the pods, with or without plant, or by mechanical threshers. When farm animals are employed to trample the plant product to loosen grains, the animals are allowed to walk in circles on a hard horizontal surface with regular turn down so that damage to the grains can be prevented. In some countries, pulse grains are spread on the road so that grains can be threshed by the wheels of passing vehicles. However, the impact of heavier vehicles may lead to breakage of sound grains. Mechanical threshing is done by using power threshers.

Traditional or mechanised threshing is performed only when the harvested crop has dried completely, otherwise there would be threshing loss of grains in the form of by-products and chaffy material. If the threshing is done when the grains are in still damp and followed by subsequent heaping or storing in a granary or bags, these become more susceptible to attack from microorganisms, in turn adding to the qualitative and quantitative loss of grains during preservation.

Grains are removed from the stalk during threshing and the efficiency of removal depends on the method adopted for threshing. Mechanised pulse growers employ a combine harvester for harvesting, threshing and winnowing pulses in a single operation. The combine harvester combines these three separate operations and separates grains from stalks and waste, *i.e.* stalks are left behind on the field and either chopped and spread on the field or used for animal feed or bedding. The combine harvester was invented in 1838 by Hiram Moore in Michigan, USA. The first combine harvester, propelled by 16 or more horses, could cut, thresh, clean, and bag 61 ha of wheat in a season. Later models were powered by steam engines using the straw as a fuel. Modern combine harvesters are diesel-powered and are more fuel efficient. Use of the combine harvester system requires high periodic maintenance costs and technically trained labour to operate the machine, which could be a limitation for small pulse growers. The combine harvester consists of (1) the cutting section to cut the crop from the field and to convey it to the threshing unit; (2) the thresher which separates the grains from the straw; (3) a device to separate the straw, (4) a cleaner to remove chaff and other foreign materials from the harvested grain and (5) a grain collection system. Self-propelling combine harvesters are common, consisting mainly of a cutting apparatus, a threshing chamber composed of a revolving threshing drum (with teeth) and a stationary counter-thresher, and a device for cleaning the grain.

9.4.3. Drying

The length of time required for full drying of ears of the grain and the grains themselves depends on the weather conditions. In structures designed for lengthy drying such as cribs, or even unroofed threshing floors or terraces, the harvest is exposed to wandering livestock and the depredations of birds, rodents, or small ruminants. Apart from the actual wastage, the droppings left by these marauders often result in greater losses than what they actually eat. On the other hand, if grain is not dry enough, it is vulnerable to mould and can rot during storage. If grain is too dry it becomes brittle and can crack after threshing, during hulling or milling. In case of on-farm open sun drying, pulse grains are spread in a thin layer on the ground and exposed directly to solar radiation, wind and other ambient conditions. The drying of pulse grains is discussed in detail in Chapter 10.

9.4.4. Storage

Dry grains are stored either short-term or long-term depending on the available facilities, hygiene and monitoring system for effective maintenance. Storage also can be divided into closed structure and open structure storage systems. In closed structures (granaries, warehouses, hermetic bins), control of cleanliness, temperature and humidity is particularly important. Damage caused by pests (insects, rodents) and moulds can lead to deterioration of facilities (*e.g.* mites in wooden posts) which result in losses of grain quality and food value as well as quantity. Safe storage of pulses at farm level can be ensured when the following principles are considered:

- Grain storage structures at the farm level should be located on a raised, well-drained site.
- The storage site should be easily accessible to ensure regular inspection.
- The storage site and structure should be protected from moisture, heat, insects, rodents and other animals.
- The storage structure should be constructed on a platform to avoid moisture gain from the ground surface.
- The storage structure should be properly cleaned before storing pulses to avoid cross-contamination and carryover from previously stored grains.

- Cracks, holes and gaps the in storage facilities should be checked regularly and sealed to avoid giving shelter to insects and rodents.
- The storage structure should be fumigated to avoid gain infestation.
- Pulses should be properly cleaned, *e.g.* of chaffy and other foreign materials which may carry microorganisms and insects from the field.
- Grains must be dried to a safe moisture level to avoid quality deterioration and pest attack.
- Maintaining hygienic conditions in storage facilities will ensure a check on infestation.
- Pulses should be stored separately and must not be mixed with year grain stock.
- The mode of transport used to transfer grains must be cleaned prior to the transport of grains from field to storage facilities.
- Bagged pulses should be kept on wooden or plastic crates before stacking bags and dunnage should be used to avoid moisture pick-up from the storage floor. Stacked grain bags should be covered with polyethylene or tarpaulin sheets to provide additional safety to bagged pulses.
- Storage facilities should be properly aerated during the storage period.
- Regular inspection of stored pulses should be carried out to check infestation.

9.4.5. Processing

The processing of grains is of two types (primary and secondary) required to increase the palatability and digestibility of the grain. The primary processing operations involve cleaning; grading, hulling, *etc.*, and the secondary processing can also result in significant losses. Various aspects of pulse processing methods are described in detail in Chapter 11.

9.4.6. Transportation

Transportation involves the passage of grain or produce from one stage of post-harvest system to another. In general, transportation plays a key role in the post-harvest system irrespective of the harvesting system employed in pulse production. An effective transport system is required to transport pulses (1) from harvest fields to the threshing or drying yard; (2) from on-farm storage facilities to collection-centre warehouse; (3) from warehouse to grain processing units; (4) from grain processing units to wholesalers or retailers.

Much care is needed in transporting mature harvested grains from the field to the storage place, from threshing place or delivery point in order to prevent detached grains from falling on the road before reaching the storage or threshing place. Nevertheless, the collection and initial transport of the harvest grains depend on the place and the conditions prevailing during transportation. Especially if the transport system involves a poor road network and/or if no mechanised system of transportation is available, the transport of grains from field to storage facility is often done by head load, bicycle or bullock cart. The choice of mode of transport varies depending on several factors such as the quantity of grain, availability of transport and infrastructure that allows the transportation of grain using heavy duty vehicles (*e.g.* tractors, trucks). However, most pulse growers rent a truck or tractor to transport grain as the simplest and economic method with no maintenance cost. Grain losses during transportation mainly result when grains are exposed to insects, birds and possibly physical theft. The following aspects should be considered when selecting the mode of transport for pulses:

- The mode of transportation should be cost effective among the available alternatives.
- Loading and unloading of pulses should be convenient.
- The mode of transport should provide protection during transportation under adverse weather conditions, *e.g.* rain.
- It should be safe from pilferage and spillage during transportation.
- It should be easily available, particularly during the post-harvest period.
- Long-distance transportation should be given due consideration.

Grain loss during transportation is a major part of the post-harvest losses. The transport loss indicates the weight difference between quantity loaded and the quantity unloaded. Besides measurable quantitative losses, qualitative losses may also occur owing to the changes in grain due to microbial contamination or moisture pick-up in case of rain during transportation, inducing several biochemical changes which eventually pose problems during processing. It is therefore important to reduce the transport time and protect grains in order to preserve them in good condition. Further, loading or unloading may cause damage to bags containing pulse grains that may cause leakage due to rough handling or using hooks during manual loading or unloading.

9.5. IMPORTANCE OF POST-HARVEST TECHNOLOGIES

Improvements in the post-harvest handling, storage and preservation of crops would reduce losses caused by physical and biological agents which make more food available with reduced diet-related diseases. Reduced post-harvest losses can also contribute to food security by increasing the amount of food available for consumption. Reduction in post-harvest losses can improve the farm productivity and farm earning through marketing of grains and enable farmers to sell their produce profitably. This may create a favourable market for farmers to avoid losses which may occur during storage. Post-harvest primary, secondary or tertiary processing and marketing can create employment and thus income.

Inefficient post-harvest handling of legumes in developing countries has affected the grain quality, which increases the necessity to import from other countries. Reduction in post- harvest qualitative losses of grains will ensure meeting national or international grain quality standards. By preventing qualitative losses from fungi or moulds the chances of grain contamination with mycotoxins can be reduced during adverse post-harvest storage, which makes it possible to address one of the important food safety issues. Any food safety issue associated with microbial contamination can be avoided through proper post-harvest handling and technologies. The improved post-harvest handling of pulses will not only reduce losses and improve the quality of pulses but also help farmers to establish improved productivity at their own farm level. In addition, improved post-harvest technologies may create a value-added processing business which allows increased job opportunities and thus improves the living condition or lifestyle of pulse growers.

9.6. STRATEGIES FOR REDUCTION OF POST-HARVEST LOSSES

- Harvesting grains at optimum moisture level.
- Careful transportation of the pulse grain from field to threshing yard or on-farm storage and throughout post-harvest operations.
- Proper handling of machinery employed for harvesting and/or threshing.
- Prophylaxis measures prior to storage of grains.
- Drying of pulses to a safe moisture level prior to storage.
- Maintaining hygienic storage conditions prior to drying including frequent inspection for free of pests (rodents, mould growth, insects, *etc.*).
- Storage structures should have proper storage conditions and be proofed against the entry of insects and rodents (proofing, rat traps, *etc.*)
- For long-term storage, the grains should be cleaned, graded and free from foreign matter, damaged or shrivelled grains.
- Improper grain loading, unloading and stacking also influence the grain losses during subsequent transportation.
- Pulse growers should attend regular training on post-harvest handling from specialist institutions or associations or grain commissions and regulatory agencies.

REFERENCES

1. P. R. Tah and S. Saxena, *Int. J. Agric. Biol.*, 2009, **11**, 321–324.
2. R. McVicar, in *Saskatchewan Pulse Growers, PulsePoint*, 2005, pp. 37–38. Available online www.saskpulse.com/media/pdfs/pp-2005jun-harvesting-quality.pdf (accesses date 20th August, 2011).
3. U. Tiwari and E. Cummins, *J. Sci. Food Agric.*, 2008, **88**, 2277–2287.
4. World Resources, *Disappearing Food: How Big are Postharvest Losses?*, 1998. Available online at http://earthtrends.wri.org/pdf_library/feature/agr_fea_disappear.pdf (Accessed 1 January 2012).

5. J. Gustavsson, C. Cederberg, U. Sonesson, R. Van Otterdijk and A. Meybeck, *Global Food Losses and Food Waste*. FAO, Rome, 2011 (http://www.fao.org/fileadmin/user_upload/ags/publications/GfL_web.pdf).
6. J. Parfitt, M. Barthel and S. Macnaughton, *Phil. Trans. Roy. Soc. B: Biol. Sci.*, 2010, **365**, 3065–3081.
7. M. S. Adak, N. Kayan, A. Gunes, A. Inal, M. Alpaslan, N. Cicek and T. Guzelordu, *J. Plant Nutr.*, 2007, **30**, 1397–1407.
8. J. Cassells and E. Armstrong, *Farming Ahead*, 1998, **81**, 63–64.
9. F. Singh and B. Diwakar, in *Skill Development Series No. 16*, ICRISAT Training and Fellowships Program, Andhra Pradesh, India, 1995.
10. P. Miller, K. McKay, B. Jenks, J. Riesselman, K. Neill, D. Buschena and A. J. Bussan, *Growing Chickpea in the Northern Great Plains*. Extension Services Guide, MT200204 AG 3/2002, Montana State University, 2002.

CHAPTER 10
Drying of Pulses

10.1. INTRODUCTION

Drying is one of the oldest methods of preserving a range of agricultural commodities. It preserves food by removing moisture, thereby hindering the growth of spoilage microorganisms (bacteria, yeasts, moulds and fungi). It also slows down the enzyme activity responsible for food spoilage and retards attack on the grains by storage pests as well. Drying of food grains using solar radiation to prevent spoilage and improve food longevity has been practised ever since prehistoric times and is the oldest method of preservation. Peas, beans, and other pulses are harvested at a higher moisture content to avoid pod shattering (see Chapter 9). At the time of harvesting grains have a moisture content high enough to be easily invaded by spoilage microorganisms or insect pests, thereby causing severe qualitative and quantitative losses during storage and subsequent processing. Therefore, there is a need for post-harvest drying of pulses for safe storage and easy handling, without spoilage. In general, immediately after harvest, pulses contain about 20–25% moisture depending on the pulse crop, and for safe storage for about 1 year they should be dried to about 13% moisture. The term "drying" most commonly refers to the removal of small amounts of moisture from a solid or nearly solid material by evaporation and can be considered as one of the methods of preservation of grains and food crops. This chapter discusses the various concepts in grain drying technology, followed by methods and technologies for drying of pulse grains.

10.2. DRYING BASICS

In theory, grain drying is a mass transfer process consisting of the removal of moisture from grains by evaporation. Accordingly, one can easily visualise it as an operation involving the transfer of both heat and moisture simultaneously,

Pulse Chemistry and Technology
Brijesh K. Tiwari and Narpinder Singh

Published by the Royal Society of Chemistry, www.rsc.org

with the objective of bringing moisture to a safe level ideal for storage. In pulse grains moisture is present both on the surface and in the kernel. Surface moisture evaporates rapidly when pulses are exposed to hot air, whereas internal moisture content evaporates at a much slower rate. Internal moisture has to migrate from the centre to the surface of the grain before it can be removed by forced or natural air. Drying of grains involves removal of free moisture or unbound water by the virtue of a difference in water vapour partial pressure between the surface of the grain and air in the vicinity of the grain to be dried. At any time there could be three possibilities:

- $V_g > V_a$: loss of moisture from the grain will occur (drying)
- $V_g < V_a$: gain in moisture from the surroundings
- $V_g = V_a$: no drying, but hygroscopic equilibrium between grain and surrounding air will exist

where V_g is the partial water vapour pressure on the surface of the grain being subjected to drying and V_a is the partial water vapour pressure in the air.

To remove moisture from the grains, the most efficient approach is to select the drying conditions, and a basic understanding of the drying process, changes occurring within the grain and the mechanism of moisture removal is a prerequisite.

10.2.1. Equilibrium Relative Humidity

Water activity (a_w) of grains is the free moisture content which can be removed during the drying process. Theoretically, water activity is defined as the ratio of partial pressure of water in the grain (p_g) and the vapour pressure of pure water (p_w) at the same temperature, or it is a measure of the energy status of water in the grain:

$$a_w = \frac{p_g}{p_w} \tag{10.1}$$

The relationship between equilibrium relative humidity (ERH) and a_w is

$$a_w = \frac{ERH}{100} \tag{10.2}$$

The water activity of food materials varies from 0 (no moisture) to 1 (100% moisture); hence the water activity of pure water is 1 and for completely a dry product it is 0. Figure 10.1 shows a hypothetical relationship between water activity and moisture content of food and varying with the type of food products, represented by a moisture sorption isotherm. A moisture sorption isotherm describes the relationship between moisture content and water activity for a given product at a given temperature and humidity value. For each value of constant humidity and temperature, a moisture sorption isotherm indicates the corresponding moisture content depending on the grain.

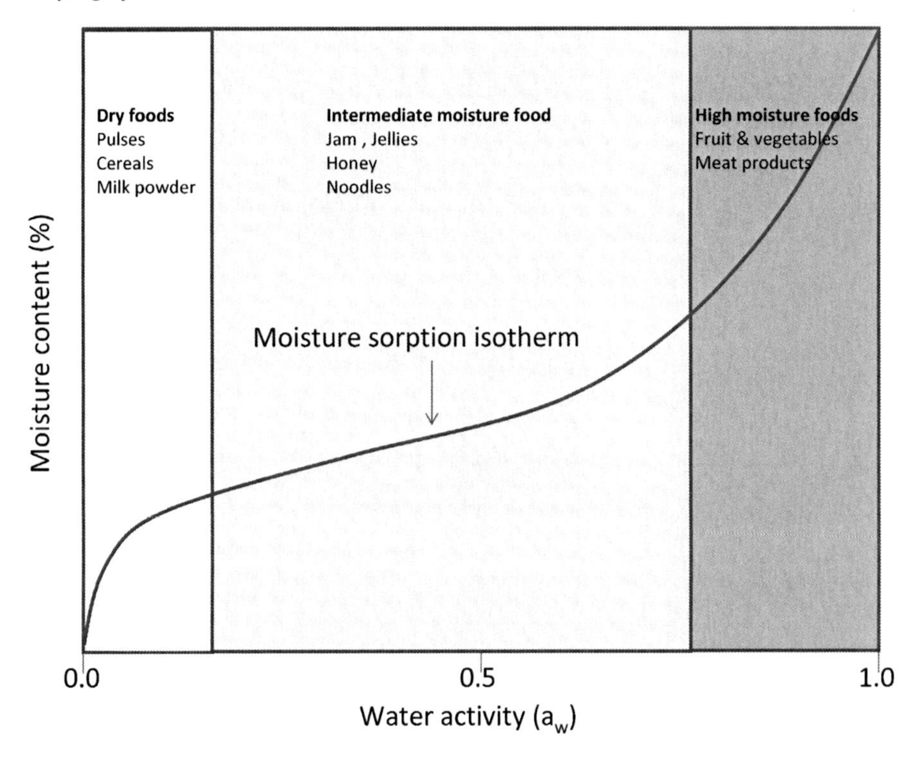

Figure 10.1 Hypothetical moisture isotherm and relationship between water activity and moisture content of food products

The relationship between moisture content and water activity represented by a moisture isotherm is not linear. An increase in water activity is usually accompanied by a non-linear increase in moisture content. The moisture sorption process is complex, thus making it difficult to calculate, but it can be determined experimentally for pulses. Knowledge of the moisture sorption characteristics of pulses is important for predicting drying conditions and stability during storage. Hence, determination of water activity or ERH is important for drying and other food processing operations. Water activity is measured by equilibrating a product with the vapour phase and measuring the relative humidity of the vapour phase. The water activity of grains can be determined by using commercially available hygrometers. In general, a gravimetric method is most commonly used to determine the water activity of pulse grains. This method is known as the isopiestic equilibration method. In this method a grain sample of known mass is stored in a closed chamber and allowed to reach equilibrium with an atmosphere of known ERH or with a reference material of known sorption isotherm (*e.g.* LiCl, NaCl, KCl, KNO_3) for maintaining ERH levels.[1] The change in the moisture content of the test pulse sample is noted till equilibrium is reached, *i.e.* when there is no weight change in a humidity chamber (Figure 10.2). The test is repeated at different

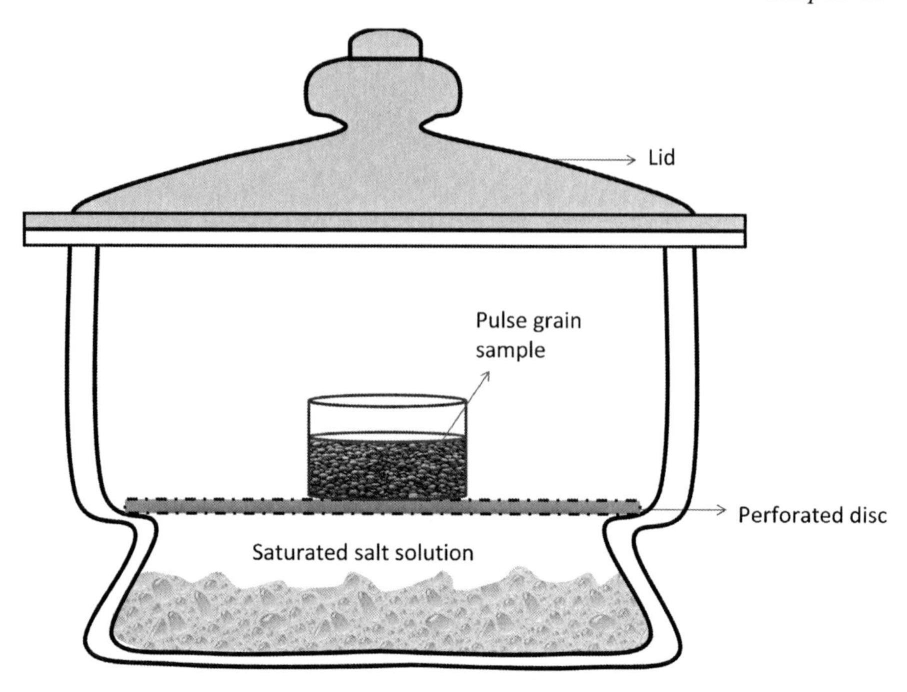

Figure 10.2 Humidity control chamber for the determination of ERH

temperatures and RH conditions obtained from different salts and their combinations.

The data obtained from the isopiestic equilibration method is fitted to theoretical mathematical expressions such as Brunaner, Emmett, Teller (BET) and Guggenheim, Anderson, de Boer (GAB). The practical use of sorption isotherms is limited by the fact that moisture sorption isotherms are valid only for a specific grain at a specific temperature. Moisture absorption is not the same as of desorption. Moisture sorption (moisture pickup) curves are slightly lower in moisture contents than moisture desorption (moisture removal) curves. This behaviour of moisture absorption and desorption is known as hysteresis. This phenomenon occurs due to capillary forces binding the moisture in the grain and volume expansion. The moisture absorption and desorption curve forms sorption hysteresis, as shown in Figure 10.3 for lentils and chickpeas.

10.2.2. Equilibrium Moisture Content

The equilibrium moisture content (EMC) of pulse grains is the moisture content to which grain will dry after it has been exposed to the drying medium (generally hot air) for an infinite period of drying time. The EMC is dependent upon the temperature and related humidity of the environment as well as species, variety and maturity of the grain.[2] A plot of EMC values of a pulse grain at a specific temperature and RH of the air results in a sigmoid curve. It

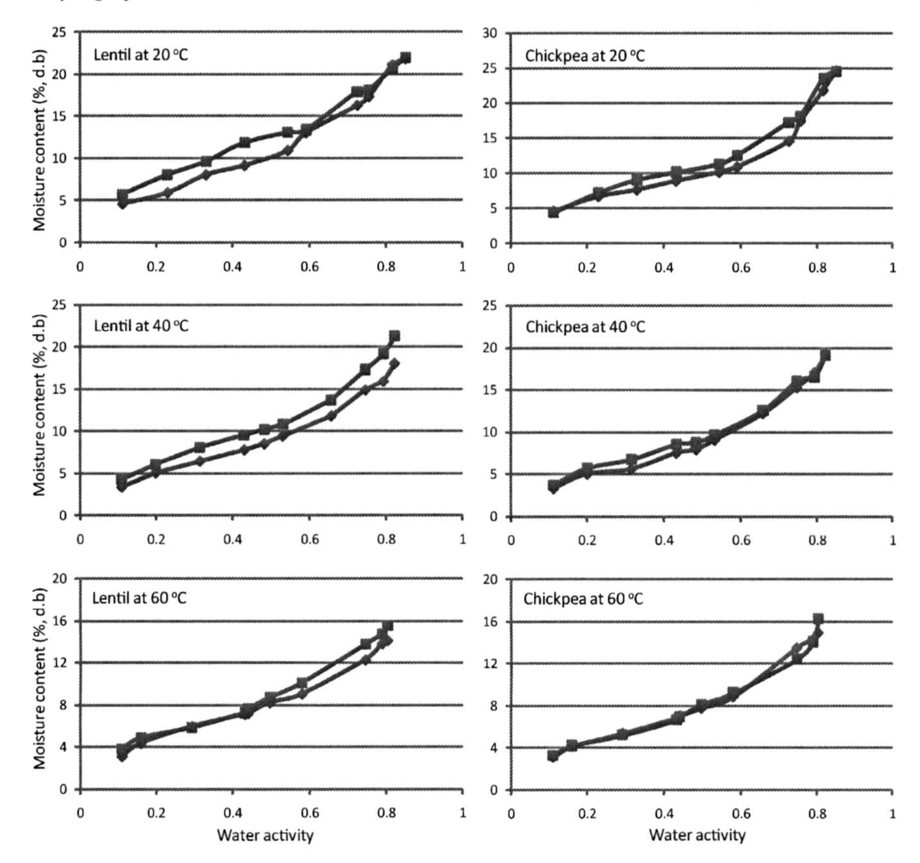

Figure 10.3 Absorption and Desorption moisture isotherms of lentils and chickpea grains at 20, 40 and 60 °C (Adapted from Menkov[12,13])

is very important to understand the practical significance of the EMC. Under no circumstances is it possible to reduce grain moisture content below EMC at a particular temperature and RH of the drying air. Proper optimization of the grain drying process requires knowledge of the relationship between ERH and moisture content at different temperatures. The energy requirements, as well as the state and mode of moisture sorption within the product, should also be known. The methods used for the determination of equilibrium moisture sorption isotherms of food grains can be classified as follows:

- The test material is brought into equilibrium with air at a fixed temperature and RH and the moisture content of the material is measured (EMC or gravimetric method).
- Air is brought to equilibrium with material at a fixed temperature and moisture content and the RH of the air is determined (ERH method).

The EMC is an important concept which helps to determine whether the grains will lose or gain moisture under a given set of drying conditions; more precisely, for known temperature and RH. A wet material placed in a medium having lower water partial pressure (at the same temperature) will dehydrate until equilibrium is reached. Therefore, we can easily appreciate its direct importance to grain drying and storage. When grain or any other wet solid is subjected to air (drying medium) at a constant temperature and humidity and known partial vapour pressure p, then the solid is likely to either lose moisture to the medium or gain moisture from the medium until the partial pressure of the water in the grains equilibrates to reach the value p. Thus, the moisture content of the grains (or wet solid) in equilibrium with air of given humidity and temperature is known as the EMC. It is very important to note that the EMC is a function of the temperature and RH of the environment (or drying medium), the cultivar/variety of the grains and the moisture content of the grains. As a corollary, we can now say that different grains and wet materials have differing EMCs. In order to describe the process of drying, the ambient cooling of grain and the effect on water activity, and to improve physical control in storage, a sound knowledge of the relationship between EMC and ERH is essential.[3] A plot of ERH *vs* EMC for a particular grain or food material at a given temperature (usually 25 °C) is known as the equilibrium moisture curve or isotherm (Figure 10.1). The isotherms for grains are generally S-shaped (sigmoid), owing to the multimolecular layer adsorption of the water molecules (a detailed description of the molecular aspects of water binding is beyond the scope of the present chapter).

The EMC for grains may be determined by two methods, the static method and the dynamic method. The former method involves the equilibration of the grains with the surrounding still air without any agitation, whereas the latter involves the mechanical movement of the air. By common sense, the static method is time consuming and at high RH the chances of mould growth in the grain are much higher, even before equilibrium is reached. It is to be cautioned that the EMC must be determined under constant RH and temperature conditions of air. Quite often a thermostat is employed to control the temperature and aqueous acid/salt solutions of varying concentrations are used to keep the RH of air under control. Literature reveals that many equations have been used over the years to describe moisture isotherms of different biological materials and their relationship to temperature. However, none of these equations can be claimed as the "universal" equation to describe sorption data of cereal grains and seeds.[4] Chirife and Iglesias[5] reviewed 23 different equations for modelling sorption isotherms of different agricultural materials and concluded that each model had some success in predicting the EMC data for a given product, for a given range of RHs and temperatures. Hence, there is a need to select the most appropriate moisture sorption isotherm equation for a specific food grain and range of RHs and temperatures.

10.2.3. Pulse Grain Drying Curve

A basic understanding of the drying curve is essential for selecting drying conditions in a grain dryer. The drying behaviour of different grains under different constant conditions of temperature, relative humidity and moisture content will give a typical drying curve as shown in Figure 10.4. Figure 10.4a shows the change in grain temperature and moisture content over drying time under constant drying conditions. The rate of moisture removal from the grains is known as the grain drying rate. The drying rate is not constant but changes with respect to drying time. A typical drying curve includes two segments: (1) the falling rate period and (2) the constant rate period. Variations in drying curve of pulses mainly occur in the falling rate segment. The process of moisture removal from grains at series of drying rates within the falling rate and constant rate periods, as shown in Figure 10.4b. Practically, there are three distinctive phases during drying of grains. These are (1) the preheating period when grain with a high moisture content is exposed to hot air; there is slow change in the moisture content because the heat energy transferred from the hot air to grain is used up in heating the grain; (2) the constant rate period, where moisture evaporates at a constant rate from the grain surface and heat energy from the hot air is used for removing moisture from the grain, and the drying rate is constant with respect to time; (3) the falling rate period where most of the surface moisture is removed and the drying rate decreases with respect to time because heat energy is used in moving internal moisture to the surface from where moisture evaporates. The shape of the third phase varies with the grain's physical and structural properties. Briefly, initial drying rate is slow, then constant and finally decreases to a critical moisture content (CMC) of the grain which is also known as the EMC. The constant drying rate period continues as long as moisture is available to the surface as fast as it evaporates.

- A→B: Initial removal of moisture occurs at the grain with an increase in grain temperature (preheating period).
- B→C: Indicates constant drying period where rate of moisture removal is constant and grains are at constant temperature.
- C→E: Indicates falling rate after CMC, till it reaches EMC. During this period grain temperature increases; a small amount of heat energy is used

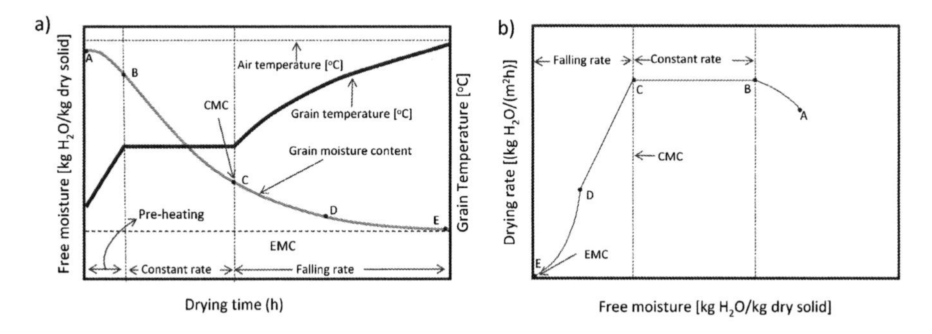

Figure 10.4 Typical drying curves representing (a) free moisture, grain temperature *vs* drying time and (b) drying rate *vs* free moisture content

to remove moisture at a much lower rate, and large amount of heat is used up in heating grains. This phase of drying time, during which migration of internal moisture to the outer surface becomes the limiting factor, may be longest and may reduce the drying rate.

10.2.4. Drying Kinetics

The kinetics of grain drying is important for understanding the drying process as well as the design and simulation of dryers. The kinetic study involves determination of moisture content of the grains as a function of time and this data can provide information about the ease with which a product can be dried under a specific set of drying conditions. The moisture content M of the grains based on dry weight of the grains at any time is given by

$$M = \frac{w - w_s}{w_s} \quad (10.3)$$

where w is the total mass of the grain sample at time t and w_s is the mass of the dry grains.

The initial section A–B of the drying curve (Figure 10.4) represents the period of proper drying and is the constant drying rate period. After long drying times, the grains reach equilibrium with the drying air and no further drying takes place. The moisture content of the grains under consideration, at which the drying rate ceases to be constant is known as the CMC. The free moisture present in the grains, which is available for drying, can be represented by the following equation:

$$M_f = M - M_e \quad (10.4)$$

where M_e is the EMC, which has been discussed earlier.

The drying rate ($\dot{m}$) for the constant rate period may be defined by the following equation:

$$m = -\frac{w_s}{a}\left(\frac{dM}{dt}\right) \quad (10.5)$$

where a is the area of the grain bed available to the drying medium. The solution to this equation can be obtained by separation of variables and integration within proper limits to obtain the total drying time for a given mass of food grains. The best way of obtaining the numerical temporal derivative (dM/dt) is by fitting a polynomial regression for M *vs* t to the experimental data and finding an analytical derivative of the resulting polynomial.

Moisture removal in a constant drying period is very easy and can be considered as if pure water is being vaporised and removed. During the constant rate period the surface of the grain bed is saturated with water and the rate of drying is limited by the rate of heat transfer from air to the grains. However, this is followed by a falling rate period in the drying curve of grains, where the moisture movement may be controlled by combined internal–external resistances or by pure internal resistances.[6]

The complexity of the drying process arises from the fact that drying is also dependent on several other factors such as the rate of flow of drying medium (in convective drying), temperature and RH of the drying medium and several other physical and chemical transformations in the grain itself, as the drying proceeds. These variables affect the quality of the dried product and the mechanism of transport phenomena too (heat and mass transfer). The grains undergo several physical, chemical and biochemical changes simultaneously during the process of drying, which may or may not be desirable, *e.g.* changes in colour, texture, and odour. Several other more advanced concepts associated with drying include shrinkage, puffing, crystallization, and glass transitions in grains.

10.3. PULSE GRAIN DRYING TECHNIQUES

There are several ways of classifying grain drying technologies, but no systematic method of classification can be found in the literature. At a very basic level, drying processes can be classified either as batch, where the material enters into the drying equipment and drying proceeds for a given period of time, or as continuous, where the material is continuously added to the dryer and dried material is continuously removed. Based on the mode of heat transfer, drying technologies may be classified as

- Conduction drying (contact driers or indirect driers)
- Convection drying (direct driers)
- Radiation drying (sun drying or infrared driers).

Some of the drying techniques and their salient features are listed in Table 10.1. Besides these, there are other notable technologies such as dielectric drying, which includes radiofrequency and microwave drying, and chemical drying. Sack drying is also another method which is suitable for small quantities of seed to prevent mixing of varieties and conserve strain purity and viability. However, among these convective drying is most commonly employed for grain drying (accounting for over 85% of industrial driers) with hot air or direct combustion gases as the drying medium. Conduction drying is usually employed for drying of parboiled grains. At present, grain processors have a wide choice for selection of the drying technology to be employed. Ideally the drying technology selected must be optimal in terms of energy consumption, quality of dried product, safety in operation, ability to control the dryer in the event of process upsets, ability to perform optimally even with large changes in throughput, ease of control, and minimal environmental impact due to emissions or combustion of fossil fuels used to provide energy for drying. Many different drying methods are used for pulse grain drying.

Table 10.1 Drying techniques for pulses

Method of drying	*Special features*
Sun drying	No technique involved Low cost Labour-intensive Relatively higher qualitative and quantitative losses compared to mechanical dryers
Low-temperature drying	Energy-efficient Drying can be done in storage structures Requires bulk handling system and skills Maintains grain quality Relatively long drying time
Batch hot-air fixed drying	Possible for small-scale drying operations Relatively less expensive Operation does not require skilled labour Fast drying rate Uniform drying is difficult to achieve Moisture gradient may develop, leading to stress cracks Excess heating of bottom grain layer may reduce grain quality
Continuous hot-air fixed drying	Suitable for large-scale drying operation Requires skilled labour and high capital investment Grain quality can be maintained by involving tempering steps in between
Recirculating batch dryer • Mixed flow • Cross flow	Mixing of grain is a possibility Can operate at large scale Uniform drying can be achieved Requires trained skilled labour Good quality of grains Chances of wear and tear of dryer components
In-store drying	Requires pre-drying of pulses if harvested at higher moisture content Maintains excellent grain quality Can fit well with large-scale operations Requires bulk grain handling systems Energy-intensive

Drying methods are quite similar to those used for cereals. Figure 10.5 shows some of the drying techniques used in pulse grain drying. The methods employed for drying of pulses vary from country to country and depend largely on the availability of infrastructure and level of mechanisation. However, there is no ideal drying method available for drying pulses since they all have their own advantages and disadvantages. The drying techniques and methods used for pulses, like those for cereals, can be classified depending on the operating drying temperature employed:

- Low-temperature drying
- Medium-temperature drying

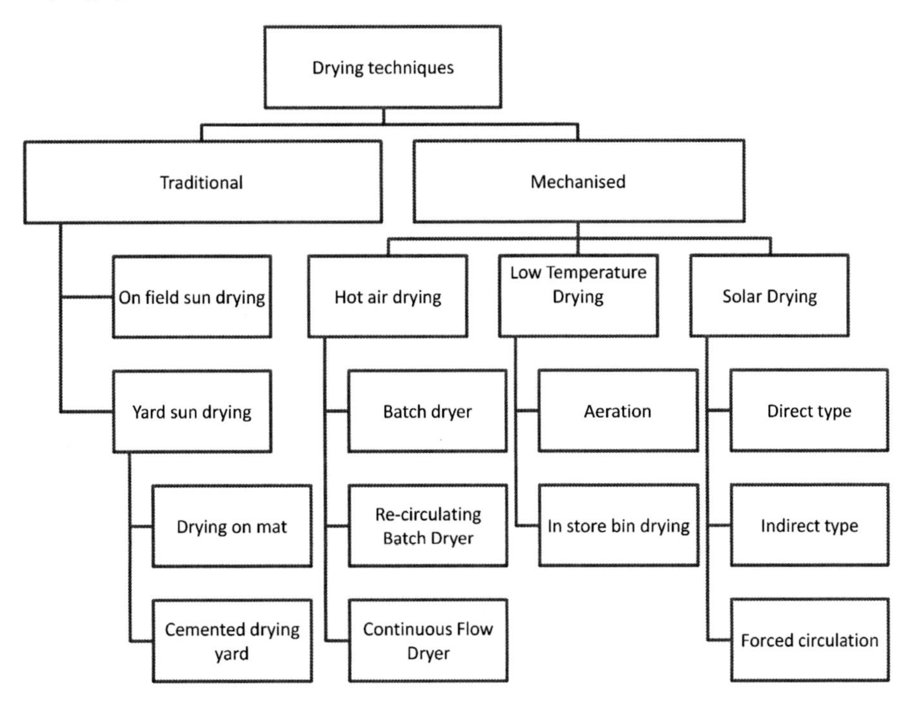

Figure 10.5 Different types of drying techniques

- High-temperature drying
- Combination drying.

Low-temperature drying involves the use of ambient air (unheated, or heated up to 8 °C in temperate countries). Low-temperature dryers use a fan to blow ambient or near-ambient air through a fixed bed of grain. Medium-temperature drying involves application of hot air where grain temperature is in the of 40–60 °C range. The drying temperature is kept below 40–43 °C for grains that are to be used for seed purposes and below 60 °C for grains that are subsequently processed. High-temperature drying involves hot air application, where grain temperature is below 82 °C. High-temperature dryers are used to reduce the moisture content of pulse grain within a short time. Generally, recirculating batch dryers, continuous-flow dryers, rotary dryers and fluidized-bed dryers employ high drying temperatures. High-temperature drying is usually employed for grains intended for animal feed purpose, whereas combination drying uses both high and low air temperature to reduce the grain moisture to a safe storage level.[7]

The drying temperature is not only important for safe storage of pulses, but also governs end product quality. For example, in India black gram is extensively used for a range of fermented food products such as *idli*, *dosa* and *papad*.[8] Most of the pulse processors dry grains to a higher temperature for faster drying. A higher drying temperature reduces the natural fermentability

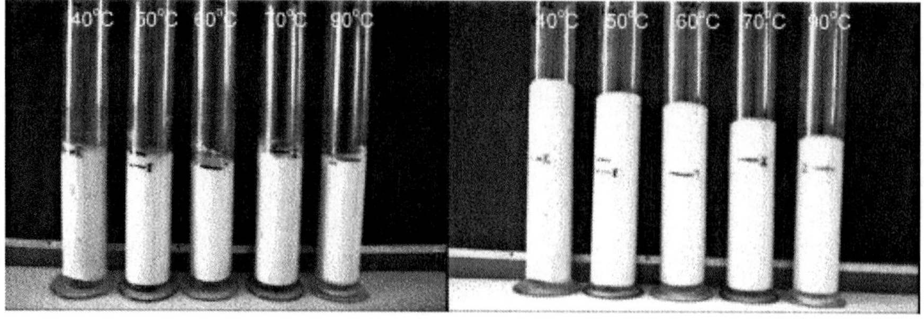

Batter immediately after grinding (0 h) Batter after 24 h fermentation time

Figure 10.6 Effect of drying temperature on fermented batter volume (Tiwari *et al.*[9])

of black gram and reduces product quality.[9] Figure 10.6 shows the effect of drying temperature on fermented batter volume of black gram. Batter volume significantly decreases with an increase in drying temperature beyond 60 °C. The decrease in fermented volume may be due to structural changes occurring in proteins and minor carbohydrates.[9]

10.3.1. Field Drying

Field drying is generally done by leaving pulses along with their stalks before threshing, or as a pre-drying treatment to remove moisture. This method is adopted in many developing countries where harvested produce is bundled or heaped for drying. The field drying of pulses is labour intensive and cheap, but causes severe qualitative and quantitative losses due to rain, birds and rodents.

10.3.2. Sun Drying

Sun drying is the traditional method of drying where grains are evenly spread out in a drying yard and stirred regularly for uniform drying. In many developing tropical and subtropical regions sun drying is the preferred method for drying of pulses. Unlike wheat and rice processing, pulse processing in many countries is not organised and most of the processing units operate at small-scale or cottage level. Sun drying is the preferred method of drying for economic reasons. Traditional sun drying has changed little over the years. It is an unreliable method of drying because of its dependence on the weather, and is also a labour-intensive, cost-effective method. Unfavourable climatic conditions such as rain and wind, as well as birds or rodents, may cause severe losses. It is appropriate to mention that field drying is actually a sun drying process. Although sun drying is practised in various forms, the most common method is to spread the pulses in thin layers of 5 to 15 cm thickness on a concrete, cemented or uncemented floor or a mat (made from agricultural by-products). In thin-layer drying one assumes that the ratio of drying air volume to grain volume is infinite and thus defines a characteristic drying rate dependent only on type of grain, physical properties of grain (see Chapter 7), initial moisture content of pulses and the temperature of air and its velocity. Thus, drying (rate of change of grain moisture over time) rate may be approximated by the following equation:

$$\frac{dM}{dt} = a\left(V_g - V\right) \tag{10.6}$$

where M is the moisture content of grain (%db), t is time (h), a is a constant (water vapour transfer coefficient of the grain and air in vicinity of grain), V_g is the partial water vapour pressure on the surface of the grain subjected to drying and V is the partial water vapour pressure in the air.

Assuming a linear relationship between water vapour pressure and ERH, and between ERH and moisture content of grain over the range in which drying occurs, the equation [10.4] may be expressed as follows:

$$\frac{dM}{dt} = b(M - M_e) \tag{10.7}$$

where M_e is the EMC of grain (%db) and b is a constant, representing grain properties related to drying. Thus, the solution to the equation [10.5] is

$$\frac{M - M_e}{M_o - M_e} = e^{(-kt)} \tag{10.8}$$

where M_o is the initial moisture content of grain ($t = 0$).

The drying rate of grain in drying systems as shown in equations [10.7–10.8] entirely depends upon the drying rate of individual grains. In general, small grains such as green gram, black gram, lentils or horse gram lose moisture more readily than large grains such as pigeon peas, chickpeas or kidney beans. Naked grains (dehulled grains) dry faster than hulled grains. The drying kinetics of pulses is very similar to that of cereals and dries at a decreasing rate. The principle of sun drying is illustrated in Figure 10.7. The solar radiation falling on the grain surface is partly reflected and partly absorbed. The absorbed radiation and hot air in the vicinity of grain heat up the grain surface. Part of this heat is used to remove moisture from the grain surface to the surrounding air and part is lost through radiation to the atmosphere and conduction to the ground surface.

The processes of heat and mass transfer occurs simultaneously in sun drying. The rate of drying depends on numerous extrinsic and intrinsic factors as listed below:

- Extrinsic parameters:

  Solar radiation
  Ambient temperature

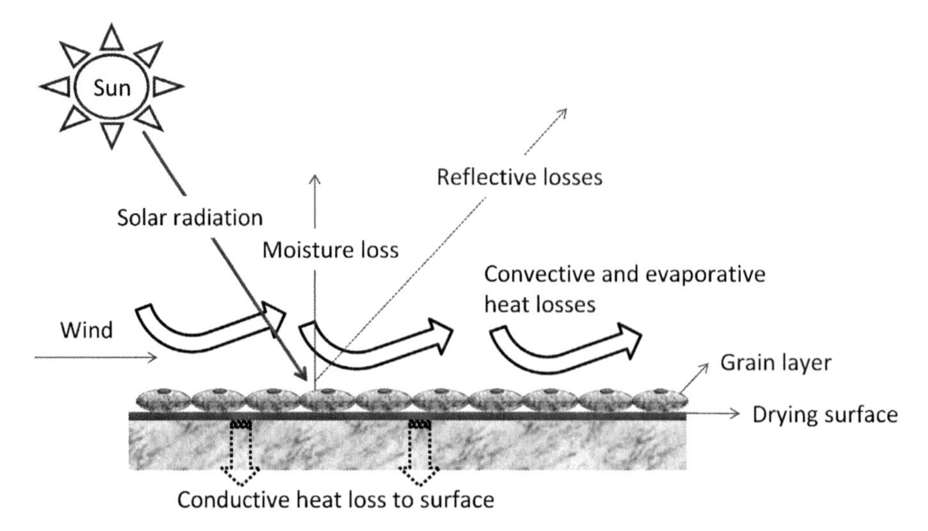

Figure 10.7 Principle of on-farm sun drying

Wind velocity, and
Relative humidity
Drying surface
Thickness of grain layer
Mixing and turning of grain layers

- Intrinsic parameters:

Initial moisture content
Physical properties of grain (*e.g.* size, shape)
Thermal properties of grains

10.3.3. Convective Drying

Convective drying is most popular method for grain drying, which include batch type (*e.g.* tray or compartmental driers) or continuous type driers (*e.g.* pneumatic driers, rotary driers, tunnel driers). In convective drying, the drying medium supplies the heat needed to evaporate water from the grains and also carry away the vaporised moisture. The major mechanism of heat transfer to the grains at a macroscopic scale, therefore, is essentially by convection. Convective drying of gains is characterised by the following key features:

- The drying rate is dependent on the heat transfer from the drying medium, *i.e.* hot air/gases, to the moist grains. The medium itself carries away the vaporised moisture.
- Steam-heated air or direct combustion gases obtained from combustion of agricultural wastes, *etc.* are used as the drying agent.

- The rate of drying is affected by the RH of the air/drying medium, at temperatures below the boiling point of water. Thus, the RH also dictates the final moisture content of the grains.
- A large variation in the drying temperature profiles can be observed.
- If natural atmospheric air is used to effect the grain drying process, the air may require dehumidification.

It is worth noting that the fuel consumption incurred for every unit of moisture evaporated is always higher than that for conduction drying.

10.3.4. Thin-Layer and Deep-Bed Drying

Ideally, in a thin-layer drying process (usually up to 20 cm bed thickness) grains are exposed to the drying medium which flows over it at a constant temperature and RH. Most commercial flow dryers are based on the thin-layer drying principle. On the other hand, in deep-bed drying all grains within the dryer are not subjected to the same conditions of drying air. The dependent psychrometric variables of the drying air change within different depths of the grain bed as well as with time. The most noticeable fact is that the rate of air flow per unit mass of grain is much less than in the thin-layer drying of grain. All on-farm static batch driers are based on deep-bed design principles.

10.4. DRYING PROCESS OF GRAINS

Irrespective of the drying method, the uniform drying of pulses always presents a challenge to achieve good quality grain. During the drying process, there is always variability in grain moisture content., uniform drying can be achieved by intermittent stirring of grain in case of sun drying, or turning of grains in fixed-bed dryers. In some cases tempering is done to equilibrate moisture in grain masses. Tempering during drying involves a process where the heating of grains is temporarily stopped so that the moisture within the grain equalises by diffusion; when the drying process is restarted the drying rate increases compared to continuous drying. Tempering in the drying process involves heating and resting periods. High-temperature drying creates moisture gradients within the grain and may result in the development of stress cracks (see Figure 10.8). Thus, tempering of grains is essential and recommended for efficient drying and improving grain quality. It allows redistribution of moisture in grains, limits stress cracking and subsequently reduces the breakage susceptibility of the grains. The development of cracks on grains lead to breakage during the dehulling process. While tempering, the temperature, moisture and viscoelastic stress gradients in grains (which develop while moisture is evaporating from individual grain kernels) are diminished (or equalise) before drying or cooling of the grain is resumed.[7] Figure 10.9 shows the hypothetical drying curve, indicating the variations in drying rate in a fixed-bed dyer. In the wheat and rice drying process tempering

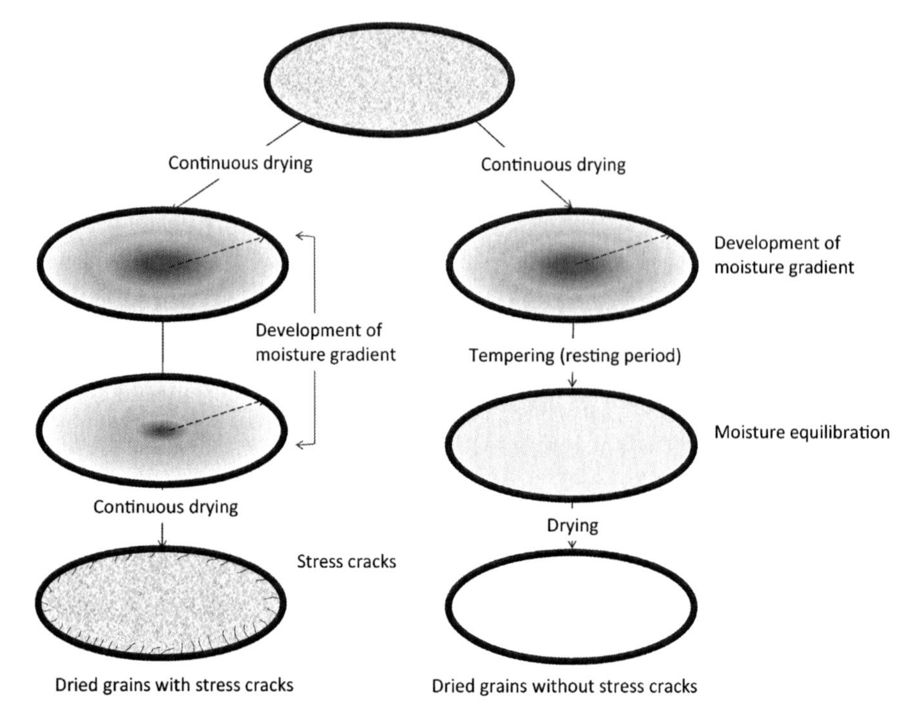

Figure 10.8 Mechanism of stress crack development in pulse grain

is a well-established key step for improving drying efficiency, decreasing microfissures and milling inefficiencies.

10.5. MECHANISM OF GRAIN DRYING

In the convective or conductive drying processes of food grains, heat is usually transferred from the circulating natural or hot air to the surface of the grains and further heat diffusion to the core is established by conduction. A thermal gradient also exists in the grains from the surface to the interior. Transport of moisture within the solid may occur by liquid diffusion if the wet solid is at a

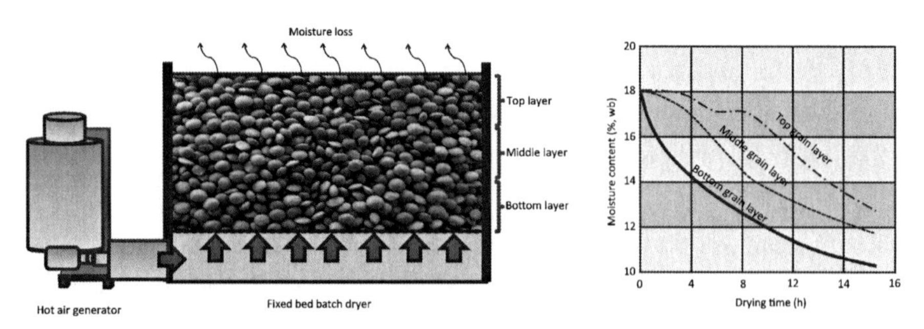

Figure 10.9 Schematic of fixed-bed drying and hypothetical drying curve

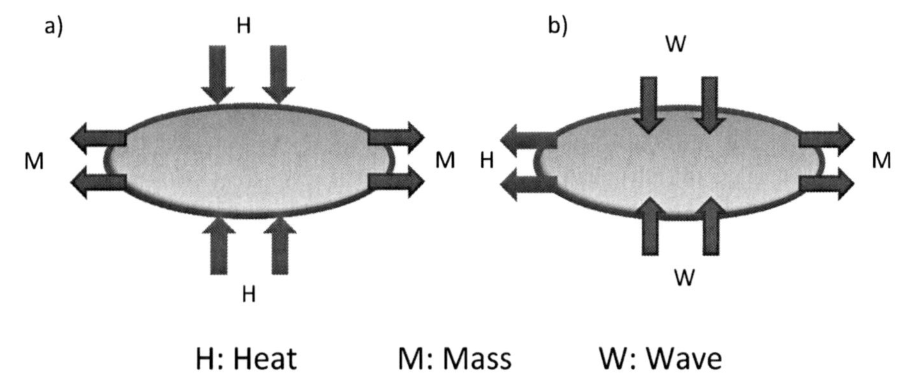

Figure 10.10 Schematic representation of convective and infrared drying (Adapted from Misra.[11])

temperature below the boiling point of the liquid, or by vapour diffusion if the liquid vaporises within the material, or by both processes. Moisture transport can also result from hydrostatic pressure differences, when internal vaporisation rates exceed the rate of vapour transport through the solid to the surroundings (hydrodynamic flow). Further moisture movement also occurs by virtue of surface forces (capillary flow). The structure of the food material being dried plays an important role in the mechanism of water movement within a product. Most importantly, the mode of heat or moisture diffusion changes as the drying proceeds because the physical structure of the grain itself changes when it is subjected to drying.

However, infrared (IR) drying offers many advantages over conventional drying under similar drying conditions. When IR radiation is used to heat or dry moist materials, the radiation impinges on the exposed material, penetrates and the energy of radiation converts into heat.[10] The depth of penetration of radiation depends upon the property of the material and the wavelength of radiation.

A conceptual representation of the difference in heat and mass transfer during convective and infrared drying is shown in Figure 10.10. When a material is exposed to IR radiation, which also includes sun drying, it is heated intensely and the temperature gradient in the material decreases within a short period. Further, by application of intermittent radiation, where heating the material for a period is followed by cooling, an intense displacement of moisture from the core towards the surface can be achieved. Consequently, this results in a greater rate of heat transfer than conventional drying and the product is more uniformly heated, rendering better quality characteristics.

REFERENCES

1. V. R. Sinija and H. N. Mishra, *J. Food Eng.*, 2008, **86**, 494–500.
2. A. Chakraverty, ed, *Post Harvest Technology of Cereals, Pulses and Oilseeds*. Oxford and IBH Publishing Company, Calcutta, 1981.

3. X. Li, Z. Cao, Z. Wei, Q. Feng and J. Wang, *J. Stored Prod. Res.*, 2011, **47**, 39–47.
4. C.-C. Chen and R. V. Morey, *Trans. Am. Soc. Agricultural Eng.*, 1989, **32**, 983–990.
5. J. Chirife and H. A. Iglesias, *J. Food Technol.*, 1978, **13**, 159–174.
6. A. S. Mujumdar, in *Industrial Drying of Foods*, ed. C. G. J. Baker, Blackie, London, 1997, pp. 7–30.
7. D. E. Maier and F. W. Bakker-Arkema,in *Proc. Facility Design Conf.*, St. Charles, IL, July 28–31, 2002.
8. N. Ali, ed, *Asian Productivity Organization (APO)*, Tokyo, Japan, 2003.
9. B. K. Tiwari, R. Jagan Mohan and B. S. Vasan, *J. Food Eng.*, 2007, **78**, 356–360.
10. A. S. Ginzburg and A. Grochowski, *Application of Infra-Red Radiation in Food Processing*, Leonard Hill, London, 1969.
11. N. N. Misra, M.Sc. Food Technology Masters thesis, CFTRI, University of Mysore, 2010.
12. N. D. Menkov, *J. Food Eng.*, 2000, **44**, 205–211.
13. N. D. Menkov, *J. Food Eng.*, 2000, **45**, 189–194.

CHAPTER 11

Storage of Pulses

11.1. INTRODUCTION

Storage is a process by which agricultural products or produce are kept for future use; it is an interim and repeated phase during transit of agricultural produce from producers to processors and its products from processors to consumers.[1] Grain needs to be stored from one harvest to the next in order to maintain its constant supply all year round and to preserve its quality until required for use. Food grain needs to be stored under conditions that will maintain or even enhance the quality of the grain as it was at harvest. Unfortunately, significant amounts of the food grain produced in the developing countries are lost after harvest, thereby aggravating hunger. In Latin America, it has been estimated that there is a loss of 25–50% of harvested cereals and pulses during storage. To further exemplify this, at harvest, 1–5% of cowpea seeds are infested by the most prominent insect pest species, *Callosobruchus maculatus*. Even this relatively low initial infestation rate can lead to 80–100% damage of unprotected seeds after 6 months of storage.[2] Prolonging the time cowpeas can be stored, preferably without the use of pesticides, can help to alleviate this problem.[3] To cope with the current and future food demands, a vital measure is reducing the loss of food grains during and after harvest, which has not received as much attention as it deserves.[4] Even in places where a significant proportion of the crop is retained for the farmer's own use, it has to be stored for up to a year until the next harvest is ready. Food grain losses contribute to high food prices by removing part of the supply from the market. If these losses are minimised, the shortage of food grains in many countries can probably be eliminated.

11.2. IMPORTANCE OF STORAGE

Storage of pulses is vital for any economy, be it a developed nation or developing. Almost every nation producing pulse crops is investing huge

Pulse Chemistry and Technology
Brijesh K. Tiwari and Narpinder Singh

Published by the Royal Society of Chemistry, www.rsc.org

amounts to raise production levels of grains, needless to say, for meeting the pulse needs of its mushrooming population. But these efforts will be in vain if the grains are not protected from the myriad of threats by establishing proper storage. On looking more closely, the problem of deterioration during storage can be seen to contribute to global food insecurity. Once the grain has been harvested, it must be held in storage and care should be exercised to reduce, if not completely eliminate, quantitative and qualitative post-harvest losses caused by various physical, biological and mechanical factors. Due to the inherent seasonal nature of pulse production, storage of pulse grains is necessary to meet the year-round demand. Safe storage, therefore, helps to maintain continuity of supply. Although pulse production is seasonal, processing/milling continues, for which grain legumes in commercial quantities are stored in modern storage structures such as flat godowns and silos. This makes it possible to minimise the losses incurred during storage. Domestic-level grain legume storage containers are traditional. The traditional containers have improved over time and have been replaced by modern airtight metal bins in recent times, that are easily available in a multitude of sizes. Metal bins have the added advantage of easy fumigation.

About 80% of storage loss is due to insects, rodents and microorganisms. Pulses are specifically attacked by bruchids (a type of beetle) at the time of maturity itself. At high humidity and temperature (rainy season) conditions bruchids are difficult to control, whereby farmers cannot store grains even for seed purposes. The loss caused by bruchids may range from 10% to 30%.[5] Most of the storage loss takes place at farm level. Grain handling and storage conditions for pulses are important for insect control and grain quality. Improper storage, generally at higher temperatures ($\geqslant$25 °C) and humidity ($\geqslant$65%) results in hardening of the seeds. This hardening can reduce nutritional and sensorial characteristics of the beans, ultimately reducing their commercial value. The prime objective of storage is to maintain the quality of the grain during the storage period, either short-term storage ($<$1 month) of processed pulses or products such as flour at processor level or long-term storage ($>$1 month) depending upon the end use of the grains. Safe storage of pulses is essential to maintain the physical, chemical and processing quality of pulses. Apart from nutritional, physical and biological safety of grains, storage is also necessary for services such as initial processing including drying, cleaning, aeration, milling; fumigation to control pest infestation; maintaining grain supply and demand; and help in merchandising. Thus, storage of pulse grains is one of the key steps in attaining sustainability of pulse grain production and processing, as well as consumption.

11.3. FACTORS AFFECTING STORAGE LOSSES

Pulses pose problems during storage, often related to harvesting time and variations in maturation due to irregular ripening of the pods. Improper storage conditions coupled with stored pulses harvested under wet conditions may lead to microbial spoilage. All organisms responsible for losses in stored grain and seed are affected by the temperature and moisture of the grains. Such organisms include bacteria, insects, moulds and mites. Stored-product insects are primarily thermophilic in nature, which implies that their growth and survivability is greatly influenced by temperature. The lower developmental threshold for most stored-product pests is approximately 18 °C and the optimum developmental range of many stored-grain insects is approximately 25–35 °C.[6] Therefore, cool and dry grains keep longer if these deteriorating conditions are prevented or retarded.[7] In dry and sound grain, the spores of the microflora on the grain are in a dormant state and remain inactive until conditions become favourable for their growth.

Field peas at about 15% moisture content may develop a surface crust during the winter as a result of moisture migration, particularly when stored warm without aeration, or as a result of snow seepage and melting.[8] Similarly, in beans, improper storage problems result in crusting and discoloration of the top layer caused by mould growth. Moulds causing spoilage, discoloration and heating can develop in the crevices of machine-damaged beans, and in beans that have been improperly dried. Seed coat damage during mechanical handling can also pose a problem which becomes more severe at low temperature and moisture levels. Storage of pulses with other foreign material including immature seeds, chaffy materials, other pulse grains or weed seeds having higher moisture can harbour microorganisms and cause spoilage and heating in the centre of bins, originating from aggregated immature weed seeds, which have a high moisture content.

The quality of pulses deteriorates significantly if they are stored at higher temperature. Field peas and lupin seeds stored at temperature of 35 °C deteriorate faster than grains with high moisture content stored at 27 °C. Storage at 20 °C at 13% and 14% moisture content is ideal for lupins and field peas, with no loss in grain viability and mould growths during 10 months' storage.[9] Other visible qualities such as seed coat colour change with respect to storage temperature, moisture content of grain and storage period. Higher storage temperature and moisture content during prolonged storage periods result in bleaching of seed coat colour. Maximum safe storage for brown beans as a function of seed moisture content and storage temperature is shown in Figure 11.1. Other commonly consumed pulses also show similar trends with slight variations. Figure 11.1 indicates that brown beans can be stored safely as long as 370 weeks at low moisture content (11%) and low storage temperature (5 °C), whereas grains stored at high moisture (23%) and storage temperature (25 °C) may last less than a week.[10] Low moisture content and storage temperatures restrict the growth of microbes and other storage pests, while high moisture content and temperature provide favourable conditions for storage pests. Additionally, storage of high-moisture pulses may cause physical and cooking quality deterioration whereas low moisture may cause breakage during the dehulling process. To conclude, optimum moisture is required for safe storage.

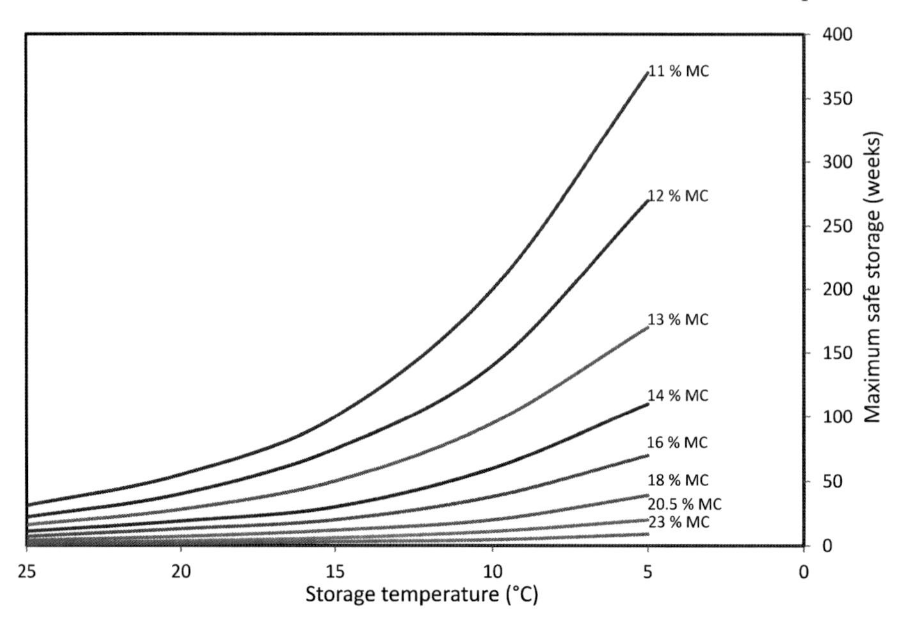

Figure 11.1 Maximum safe storage period for field beans at varying storage temperature and moisture content

11.4. SAFE STORAGE OF PULSES

Safe storage of pulses depends principally on grain moisture content or relative humidity in the grain storage atmosphere, grain temperature and length of storage. Other factors such as harvesting method and presence of impurities are also considered as important factors for safe storage of pulses. Pulses are classified based on moisture content for safe storage and trade. The safe moisture level for storage of pulses varies from one pulse crop to another and country to country depending on the prevailing climatic conditions. For example, the maximum recommended moisture content for storing faba beans is 16% in Canada and 15% in the UK. Faba beans with a moisture content of 14.2% that had not undergone frost damage can be safely stored for up to 2 years in Canada, whereas inferior-quality frost-damaged beans overwintered with moisture content above 15% often heat up during the following summer,[10] leading to mould growth and severely affecting marketability. Storage studies conducted in New South Wales, Australia on field peas and lupins indicate a range of safe storage limits for both crops in short and long-term situations.[9] Dry pea and lupin seeds can be stored up to 3 months at 14% moisture content with storage temperature of 20 °C. Storability of dry pea and lupin seeds can be increased to 10 months at the grain moisture content of 13.5% for peas and 13% for lupin seeds and storage temperature of 20 °C.[9] This indicates that moisture plays a significant role in the safe storage of pulses.

In Canada, peas and beans are classified into three categories:[10]

- Straight grades: those grains that are within accepted limits of moisture (<16.0% for peas)
- Tough: a grain is classified as tough if moisture content exceeds the straight range (*e.g.* 16.1–18.0% for peas)
- Damp: a grain is classified as damp if moisture content exceeds the tough range (*e.g.* >18.0% for peas).

Soybeans are classified in five categories:

- Dry: moisture <14.0%
- Tough: moisture 14.1–16.0%
- Damp: moisture 16.1–18.0%
- Moist: moisture 18.1–20.0
- Wet: moisture >20%.

Key factors influencing the quality of pulses during storage are the initial condition of the seed, moisture content at the time of harvest, grain temperature and storage period.

Figure 11.2 shows the deterioration of good-quality peas (Figure 11.2a) and beans (Figure 11.2b) incubated at six moisture levels (12–18%) under various temperature regimens.[8] Detection of off-odour in stored pulses is an indication of quality deterioration of pulses because these off-odours indicate microbiological growth (fungi or moulds) and/or biochemical changes during

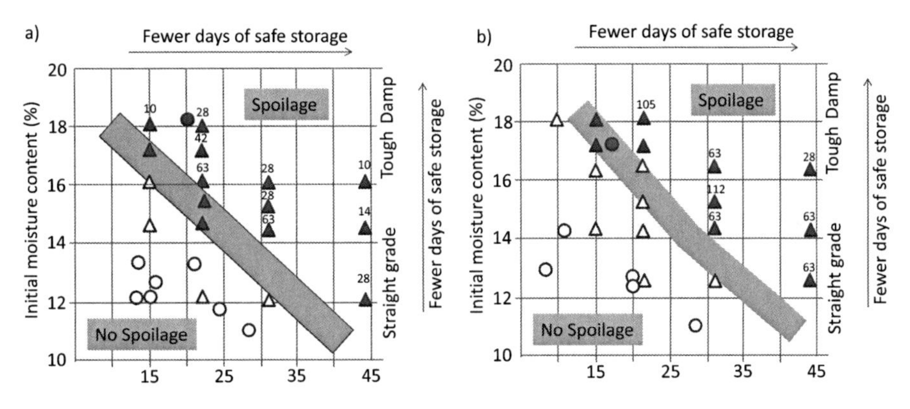

Figure 11.2 Maximum periods (up to 147 days) for storage of (a) field peas and (b) white beans binned at various temperatures and moisture contents at a depth of 1–2 m in the centre of a bin 5 m in diameter, as determined from laboratory and commercial bin data. Laboratory tests: a solid triangle with 28 denotes sign of off-odours within 28 days; a solid triangle with no numbers denotes >25% *Eurorium* spp. present; and an open triangle denotes no indication of spoilage within 147 days. Commercial bins: a solid circle denotes probable spoilage; an open circle denotes no spoilage detected, and the shaded area denotes 1% moisture content safety margin. Dry, tough and damp are moisture grading categories. (Adapted from Mills and Woods[8])

storage. The diagonal lines on Figure 11.2 correspond to the boundary between off-odour (spoilage) and normal odour (no spoilage). Several moulds and fungi are associated with quality deterioration, namely *Eurotium* spp., *Penicillium* spp., and *Aspergillus* spp.

Some of the requirements for safe storage are as follows:

- Low temperatures and seed moisture contents provide safe conditions for the long-term storage of pulses.
- Temperature and moisture interact to determine the speed of quality deterioration.
- Storage experiments on field peas and lupin seeds suggested a range of safe storage limits for both crops in short and long-term situations (see Figure 11.2).
- Dry seed ($\leqslant$10% moisture content) stores well at storage temperatures up to30 °C, but delaying the harvest to achieve this dryness can result in substantial yield and quality losses.
- Pulses harvested at a moisture content of 14% or more will require careful management during storage. The use of aeration is recommended to reduce it and provide a uniform storage temperature.
- The susceptibility of pulses to breakage during handling means there needs to be minimal disturbance during storage.
- Conditions are ideal for moisture movement in undisturbed bulks and potential problems may arise with localised moisture damage and moulding.
- Regular checks of the bulk surface are therefore essential during prolonged storage.

11.5. STORAGE STRUCTURES

Storage facilities are a collection of structures and mechanical devices used for grains after harvesting and as they move through various post-harvest grain handling operations. The grains need to be stored safely until consumed. Storage of pulse grains is done at farmer, trader and industrial levels. Appropriate technology for handling and storage of pulses has been developed in all parts of the globe. All mills have storage facilities. Some pulse processors have storage space within their facilities, but most have separate storage facilities, depending on the volume handled, either owned or rented. Traditional storage practices in developing countries do not guarantee protection against major storage pests of staple food crops, leading to a higher percentage of grain losses, particularly due to post-harvest insect pests and grain pathogens.[11] Apart from causing quantitative losses, pests in stored grain are also linked to mycotoxin contamination and poisoning. Development of efficient and effective storage systems for food grains will inevitably allow substantially cutting down the post-harvest losses. Food grains are usually stored in storage structures ranging from rural underground pits to large metallic silos with high variability in their capacity, material of construction, position with respect to ground level and several other aspects. With demand for pulses starting to outstrip supply, improved storage conditions will provide an opportunity to allow better utilisation of the pulse grains that are grown. There is no well-defined systematic classification of grain storage structures in literature. A classification of the storage structures based on their capacity and structural rigidity is presented in Table 11.1.

An ideal storage structure should provide protection against all possible threats to food grain damage. A large-scale grain storage structure should have the following desirable features:

- Provision of maximum possible protection from insects, rodents, birds and other animals
- Sufficient moisture-proofing to protect from ground moisture and rain
- Protection of the grains against environmental variables like temperature and humidity
- Ease of cleaning
- Provision for periodical inspection
- Provision for application of pesticides through spraying or fumigation.

Storage facilities are generally located far away from possible sources of infection such as kilns, flour mills, bone crushing mills, garbage rumps,

Table 11.1 Classification of storage structures based on material of construction and storage capacity

Storage structure	*Ideal storage capacity*	*Features*
Rigid structures	*May be constructed from metal or concrete or a combination of both*	
Small-size structures	Can range from a few cubic metres (m^3) to 500 m^3	Used for treatment of stacked commodities
Medium-size warehouses and silos	500–2000 m^3	Vertical or horizontal (longer than tall); commodities in store may be stacked or in bulk
Large warehouses and silos	2000–10 000 m^3	Commonly used as central stores
Flexible structures	*Include plastic sheeted structures, bagged stacks sealed in plastic enclosures and bunker type storages*	
Small-size structures	Indoor structures; can range up to 500 m^3 volume	Mostly employed for indoor gaseous treatment
Medium-size flexible structures	Capacities of up to 1500 m^3	Flexible silo linings are contained within a circular wall consisting of metal weld mesh
Large-size structures	1500 m^3 or higher	Mostly bunker type; used for bulk storage

Source: Compiled from Canadian Grain Commission.[10]

slaughterhouses and chemical industries. The location is usually convenient for the reception and transport of food grains, *i.e.* generally near harbours, railway stations or highways. The selection of the most appropriate storage system depends on the time period of harvest (which decides the humidity conditions) and the moisture content of the grains. If the harvesting period lies in a humid or wet season, it is very likely that the dry grains will pick up moisture from the surrounding environment and become susceptible to pest growth and moulds. Therefore, as a rule of thumb under such circumstances the storage structure should ensure completely moisture-proof and airtight conditions.

11.5.1. Hermetic Storage

If the storage ecosystem is sealed to prevent air from entering or leaving it, the respiratory metabolism of insects, moulds and the grain itself lowers the oxygen content and raises the carbon dioxide content of the intergranular atmosphere to a level where aerobic respiration is no longer possible. This constitutes the principle behind hermetic storage (also known as "airtight storage"). The "hermetic" approach to the design of a storage structure covers all the advantages of enabling grain to be protected from pests without employing pesticides. It also takes into account the fact that when grain is taken into storage it is nearly always already infested with stored-product insects that are the main source of losses to grain in warm climates. Hermetic storage technology has emerged as a significant alternative to other methods of storage that protect commodities from insects and moulds. This technology, also termed sealed storage, airtight storage or assisted hermetic storage, is a form of biogenerated modified atmosphere.

Hermetic storage of cowpeas in metal drum containers and triple bagging has been found to be successful in controlling *C. maculatus*.[3] In the past, hermetic storage was difficult to apply in practice because the smaller the storage structure, the more critical is the importance of the hermetic seal, and the more expensive and difficult it is per unit of storage capacity to obtain the required gas tightness.[12] This in turn is because the smaller the structure, the larger the ratio of surface area to volume, and consequently the greater the danger of air infiltrating into the structure due to partial permeability of the structure, leaks, and incomplete sealing.

The use of modern and safer acceptable technologies such as aeration, refrigerated aeration and modified atmospheres are still expensive and require adequate infrastructure. In sharp contrast to the use of chemicals, hermetic storage is an environmentally friendly technology, involving no hazard to the storage operators, consumers or non-target organisms, and as such, its application is beginning to enjoy a high level of consumer acceptance. However, it is also worthwhile noting that the grains to be kept in hermetic storage should be well dried as there is no possibility of further drying during storage. Also, regular inspection is difficult as air entry cannot be avoided at the time of inspection or opening of the airtight structure.

The principle of hermetic storage is used in several airtight structures for both on-farm and commercial-scale grain storage structures. On-farm, conventional and commercial-scale grain silo structures are discussed in the following sections of this chapter.

11.5.2. On-Farm Storage

Farm-level grain storage provides an opportunity to reduce food losses and increase farm family income and security at the same time. Landless labourers may also benefit from good storage, as grain prices flatten out and in-kind wages can be protected from losses in their homes. In agriculturally advanced countries, on-farm storage is used primarily at peak harvest times as a "buffer" in the transport of grain to the central handling systems.

11.5.2.1. Underground Pits

From prehistoric times (over 4000 years ago) until the present day, the traditional method of underground storage in pits has been common among small-scale farmers. The main advantage is that soil temperatures are below that required for the development of insects. Ideal temperatures for growth, reproduction and movement for most stored-product insects are 25–35 °C. This storage method is quite popular in the rural areas of several countries including India, parts of Africa and Latin America. Underground pits are frequently sufficiently airtight not to allow the growth of insects and other aerobic organisms. The oxygen concentration in the grain mass sufficiently reduces below a level permitting insect development, although the grains on top and around the sides are susceptible to moulds. The pits keep grain cool, and some are relatively airtight. These pits in general are prone to termite attack and therefore care is taken to store the grain in a location free from termites. Storage pits are excavated into the soil or rock and are often lined with supporting walls of brick or cement or plastered with cow dung. However, none of these approaches can completely exclude moisture migration. In recent times, inner lining with thick plastic sheets has also become popular.

11.5.2.2. Metallic/Plastic Drums

Metallic or PVC drums, which are originally oil or water drums, are often used for the storage of food grains after thorough cleaning and painting of the inner walls of the drum. In humid and warm places the use of paint checks rusting of the drums. The capacity of these drums can be anywhere between 50 and 200 kg, and they are usually employed for storage periods of up to 1–1.5 year. Often the drums are covered with straw, hay or other locally available materials to avoid direct exposure to sunlight, which can cause heating of the grains, when left on the farm. The drums may also be kept in sheds or under a roof for protection.

11.5.2.3. Basket Storage

Basket granaries made of bamboo strips or other branches have been used traditionally to store pulses in rural areas in dry tropical countries (Figure 11.3). Their capacity can range from 25 to 2000 kg and the storage period can be from 6 to 9 months. The baskets are most often provided with an inner coating of clay, cow dung or mud, which is sometimes even used on the exterior too. Use of a polyethylene sheet inside the basket is also common. This allows better protection against insects than a basket alone, though it excludes the possibility of further drying during storage. To protect against rodent attack, the improved baskets are usually placed on a raised platform with anti-rodent baffles attached to each leg of the platform. In general, the baskets are housed in a shed to protect against rain and enough spacing is provided for ventilation, in case of non-plastered baskets.

11.5.2.4. Bunker Storage

The term bunker storage is used to describe large bulks of grain stored on the ground – in the open – surrounded by low retaining walls, and protected from the weather by plastic sheeting. The storage facility consists of a bunker

Figure 11.3 Bamboo basket commonly employed for storage of cereals and pulses in rural India

bordered on three sides by ramps of earth which should be excavated from both inside and outside the site to form the structural wall of the silo (Figure 11.4). The bunker must be located in a well-drained area where there is no danger of standing water accumulating during the rainy season. The capacities of bunkers can range from small on-farm units of a few hundred tonnes capacity, to units containing up to 20 000–30 000 tonnes each.[13]

This type of storage has been a great success in Israel and Cyprus. Typical dimensions of Israeli bunkers are 150 m long and 50 m wide. Before loading, the floor and ramps are lined with overlapping strips of 0.25 mm polyethylene sheeting laid transversely to form a continuous under-liner. The over-liner is usually 0.83 mm thick PVC formulated sheeting. The oxygen concentration fall to 6% and the carbon dioxide concentration increases to 9% within 3 months in this type of storage. Analysis of the PVC liner at the end of a storage period of up to 15 months has shown that its original elasticity and resistance to tearing are also preserved.[14] The advantage of the Israeli version of bunker storage is that it is based on a high level of sealing using plastic liners that enclose the grain bulk and enable the grain to be stored under hermetic conditions with no need for chemical control treatments. Situations where bunkers can provide rapid, convenient and inexpensive solutions are:[13]

- At terminal silos when bins are filled to capacity.
- At collection sites and regional centres where incoming harvested grain is bottle-necked due to transportation problems.
- When grain is to be held on a temporary basis in anticipation of higher prices.
- For national grain reserves that require long-term storage.
- As buffer stocks for food aid under emergency situations.

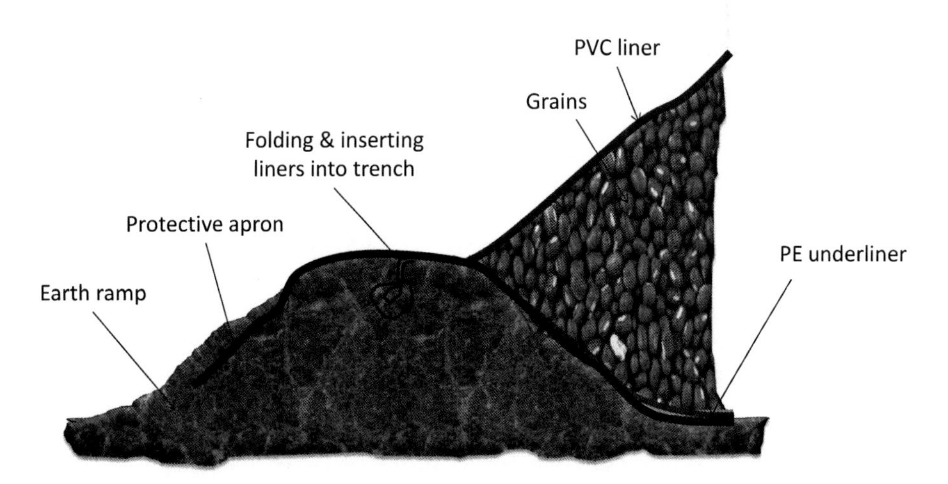

Figure 11.4 A typical bunker storage with a trapezoidal ramp cross-section (FTIC[13])

11.5.2.5. Conventional Grain Storage Structures

The traditional grain storage structures used by farmers in rural areas include gunny bags, mud bins, baked earthen containers, bamboo structures and low-capacity underground pit storage facilities. Some of these have been discussed in the previous section. Most of the traditional methods protect the product reasonably well and need at most slight improvements. On the other hand, it is possible that some traditional methods are unsatisfactory and lead to high losses. Food grains may be stored directly in the open, stored in bulk or packed in bags of different capacities before storage.

11.6. STORAGE FACILITIES

The storage facilities are usually constructed on a raised platform and in a clean surrounding area with low moisture/humidity. The aim is always to keep this zone rodent and termite proof, with easy loading and unloading. Generally, pulses stored in bags are piled under any shelter. Bag storage has the advantage that the grains can be moved easily. Modern warehouses for large-scale, long-term storage have generally proved superior compared to the traditional storage systems. Ideally grain bags should be stacked at least 15 cm above the floor to avoid contact and to allow aeration (see Figure 11.5). This is achieved by using dunnage, *i.e.* material that can be placed between the floor of a warehouse and the sacks of produce to prevent moisture moving from the floor into the grain; this can range from a mat to the commonly employed timber/wood pallets. Stacking is done by placing one row of bags lengthwise and the next widthwise to avoid slippage of bags. The stacking of bagged pulses allows optimal use of space, ease of sweeping the floors, ease in inspection of produce for the presence of rodents and insects, ease of counting sacks and proper ventilation or aeration of bags.

11.6.1. Bag Storage

Sacks made of jute, cotton, hemp or plastic fibres are commonly employed for storage of food grains including pulses. Gunny (fabric) bags may or may not

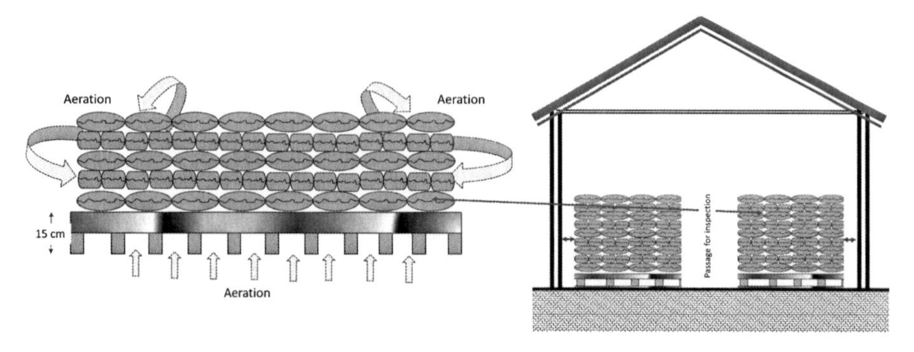

Figure 11.5 Bag storage stacking pattern along with pallet for proper aeration

have a plastic lining. Sack storage works well in dry tropical conditions and is extensively used in India. Bags and sacks can be easily handled for marketing purposes and in transit. Where drying is a problem, bag storage has the advantage that it allows a higher moisture content than bulk storage. The filled bags are usually so stacked that there is provision for the free movement of air to allow cooling and the removal of moisture. This allows the grains to remain safe even if they have a slightly higher moisture content. A raised free-standing platform equipped with rat-guards usually serves the purpose of avoiding rodent attack. The bags are susceptible to moisture uptake and need to be placed on moisture-proof surfaces such as thick plastic sheets or pallets. The development of insect-proof bags and sacks is difficult because tiny holes of about 0.1 mm are sufficient to allow the young moth larvae to invade;[15] in particular, damaged and leaky bags become infested by moths and beetles. Further, storage in bags requires considerable labour and the bags do not offer protection against rodents. Special attention is required when using plastic fibre woven bags as they eventually get destroyed with exposure to sunlight or UV radiation. Depending on the end user, processed dehulled pulses are stored in various types of bags such as jute, polypropylene, polyethylene, *etc.* These are available in various capacities – 1, 5, 10, 25, 50 and 100 kg packages in the retail or wholesale market.

The following packaging materials are used in the packaging of pulses:

- Jute (gunny) bags: widely used by farmers and traders; maximum capacity 100 kg
- High-density polyethylene/polypropylene (HDPE) bags: available in various capacities; pulses packed in such bags are safe compared to jute or cloth bags
- Polythene/jute bags: jute bags blended with synthetics or impregnated jute bags
- Polyethylene individual retail pouches: generally available in 1, 2 and 5 kg pack sizes
- Cloth bags: also used for packing pulses.

11.6.2. Triple Bagging System

The triple bag consists of two layers of polyethylene bag which are expected to be as hermetic as possible, both included in a protective polypropylene woven bag. With triple bagging, cowpea seeds are sealed in a series of two heavy-grade polyethylene plastic bags.[3] Triple bagging of cowpeas for 7 months with two HDPE bags of 80 μm wall thickness placed inside an additional woven nylon bag, tightly sealed, has been shown to be effective in controlling *C. maculatus*. Moreover, the seeds remain undamaged and viable at the end of the storage period.[3]

Triple bagging is a quite simple method which can be used for the storage of cowpeas. The technique can be easily adopted by farmers for on-farm storage of pulse grains since heavy-grade polyethylene bags allowing low oxygen permeability are easily available and affordable.

11.7. COMMERCIAL STORAGE OF GRAINS IN SILOS

With several problems associated with traditional modes of grain storage, many changes evolved to offer improved grain storage structures to the farmers. For example, metal drums can be considered as the preliminary form of the present small metal silos used for storage of pulses, with a scale-up to average capacity 5 tonnes. For small-scale storage of grains similar structures, *viz.* the PAU bin, Pusa bin and Hapurtekka, are used in most parts of India. Of these, the Pusa bin is most popular and important as it is also employed in several other countries. In India, this storage structure, made of mud or bricks with a polythene film embedded within the walls, is the most common structure for storage of pulse grains for periods of up to 6–12 months.

The modern approach to storage of bulk commodities is based on metal or concrete silos. Silos storage technology well serves the world's highly developed economies, which are generally located in temperate climates.[44] Metal and concrete silo technologies originated in Europe and the USA where the climate permitted built-in ventilation systems to cool the grain, thus making it possible to keep a check over the insect activity. For tropical countries, however, there are multiple factors that affect efficient storage. These factors have been summarised by Navarro *et al.*[16] using a biosystems (silo ecosystem) approach in Figure 11.6. A stored-grain atmosphere is a man-made ecological system in which deterioration of the stored product results from interactions between physical, biological and other internal factors. The important factors influencing quality of stored pulses include temperature, moisture, carbon dioxide, oxygen, grain characteristics, microorganisms, insects, mites, rodents, birds, geographical location and granary structure.

The quality of stored grains is largely dependent on the water activity, the grain moisture content and grain temperature of storage. Ordinarily, a storage bin may be called a silo if its height is greater than twice its width. Grain silos can be made from cement and brick or metal. Silos made of cement or brick can have either a reinforced concrete or brick base and are suitable for pulse storage in both dry and wet tropics. Prefabricated concrete is also used for the manufacture of silos; this prevalent in developing countries and was quite widely employed previously. However, metal silos are cheaper than the concrete ones. A metal silo is a tall cylindrical structure, constructed from a galvanized iron sheet and hermetically sealed.[11] Silos are generally circular in cross-section, although different forms, such as square or rectangular cross-sections, are also adopted. In silos the grains in bulk are unloaded on to conveyor belts and, through mechanical operations, are carried to the storage structure. Pulses, being more hygroscopic than cereal grains, are usually stored in large-volume silos (as much as 25 000 kg capacity) in which they can be broken loose mechanically before being taken out.

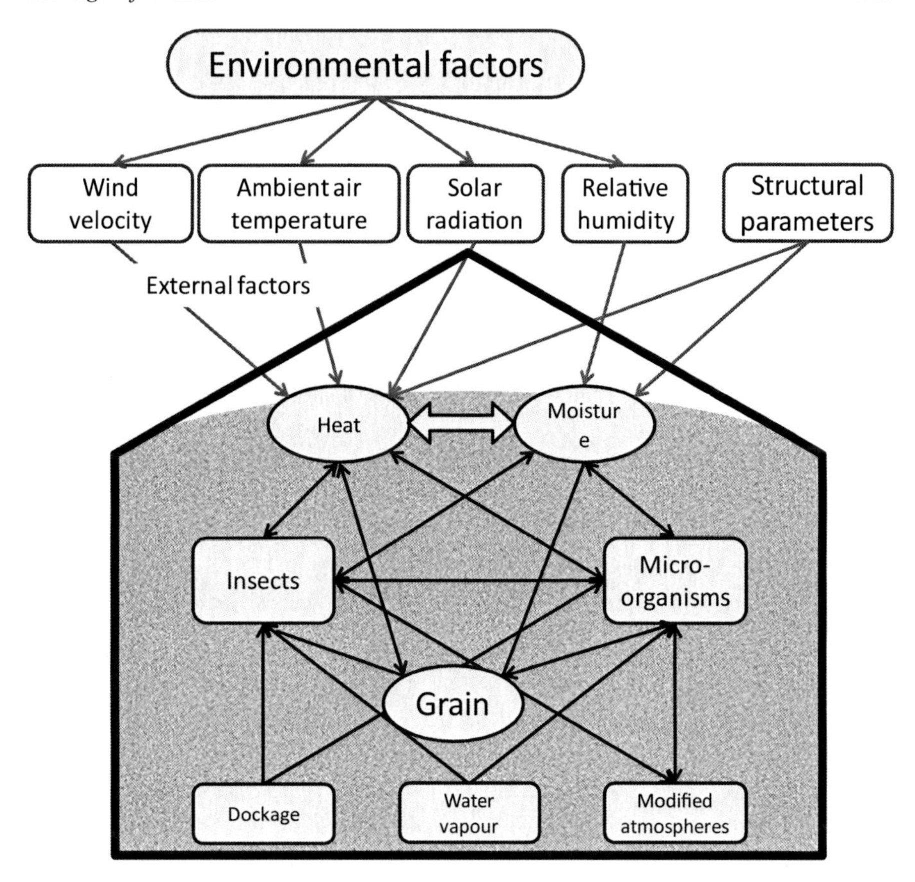

Figure 11.6 External factors and the interrelated grain bulk ecosystem components in a sealed bulk. (Adapted from Navarro *et al.*[16])

Metal silo technology has proved to be effective in protecting harvested grains from attack, not only from storage insects but also from rodent pests. A metal silo is airtight; it therefore eliminates oxygen inside, killing any insect pest that may be there. It also completely locks out any pest or pathogen that may invade the grains. Steel silos have high strength and can resist high loads.

Continued progress in solving post-harvest storage problems *via* metal silo promotion will require cooperation and effective communication among government organisations, non-governmental organisations, manufacturers and farmers.[11]

11.8. MOISTURE AND AIR MOVEMENT DURING STORAGE

11.8.1. Moisture Migration in Storage Structures

For satisfactory storage, preventing grain moisture migration and controlling moisture exchange is an important management process for farm-stored grain because grain adsorbs or desorbs moisture under changing environmental conditions.[17].The heat and mass transfer patterns within the grain bulk are primarily responsible for the development of undesirable properties throughout the grain mass. These changes in grain properties stem from diurnal and nocturnal variations as well as climatic seasonal changes.

Due to the hygroscopic properties of pulse grains they have good insulation characteristics; heat loss from grain is relatively slow in comparison to other materials. Therefore, it can be clearly visualised that when grains are placed in a bin during winter, the grain mass at the centre of the bulk tends to holds its original in-loading temperature (see Figure 11.7). On the other hand, grain mass near the silo wall approaches the average outside temperature. When the outside temperature falls, a temperature gradient is established between the grain at the centre and the grain at the walls. This temperature differential inside the grain mass produces natural convection currents, causing the cool air near the bin wall to fall since it is denser, thus forcing the warmer air up through the centre of the grain mass.

Further, as the cold air passes through the centre of the grain mass, it warms and picks up more moisture. As this air nears the top centre surface of grain, it cools to a point where it can no longer hold the moisture it has picked up. This moisture condenses on the surface of the grain mass, rapidly increasing the moisture content of the surface grain and creating a local environment that promotes mould or insect growth. This surface moisture change can occur even though the average grain moisture content is at or below recommended levels. The reverse situation occurs during the summer months. In this case, the moisture condenses near the bottom centre of the grain mass.

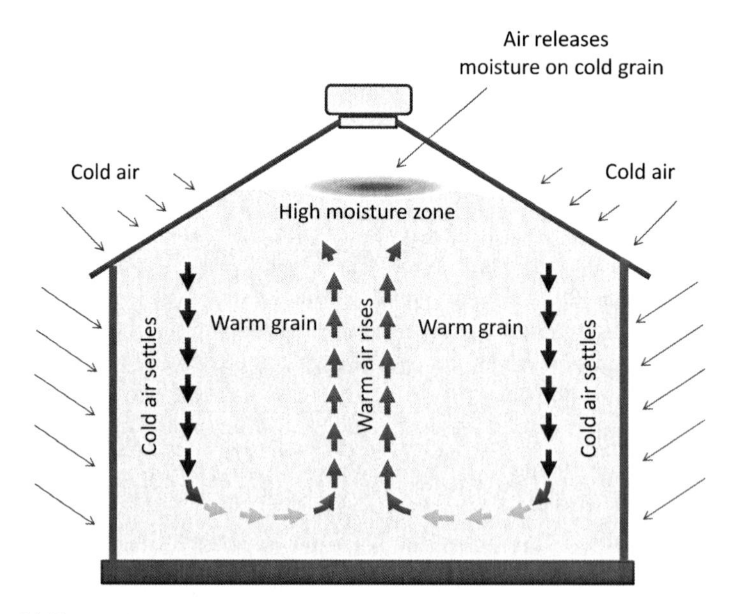

Figure 11.7 Moisture migration within warm grains and cool atmosphere

According to Navarro and Noyes[18] there are six mechanisms for moisture migration:

- Diffusion of moisture through interparticle contact conduction
- Leakage of water through openings in the structure
- Exchange of water vapour with the atmospheric air at the grain surface
- Diffusion of moisture to vapour gradients in the bulk
- Translocation of moisture due to convection currents
- Condensate that forms on the inside surfaces and downspout of the silo that falls on to the grain.

11.8.2. Aeration in Storage Structures

Grains are actually living mass and therefore biologically active. This biological activity yields moisture and heat, which in turn further enhance the biological activity. Thus, stored grains can self-heat (forming "hot spots") and rapidly deteriorate in quality if moisture and temperature are not managed within the storage structures. Aeration in the context of storage management of grains refers to forced movement of ambient air, which is conducted through a grain bulk to preserve or improve the physical conditions of the product. More precisely, the purpose of an aeration system is to preserve dry stored grain by cooling the grain and preventing moisture migration. The air is generally distributed using a ducting system so it enters the storage structure and passes through the grain before being able to exit the storage structure. A perforated floor in the silo produces a uniform air flow throughout the bulk and reduces the chance of unventilated spoilage pockets developing. Flat bottom silos with aeration are relatively common.

In general, aeration systems for grain storage structures use small, inexpensive fans and notably the airflow rates are too low to dry the grains. Aeration cooling is also useful after drying to allow thorough cooling and to "breathe off" any additional moisture mobilised in the drying process. In recent times, the use of this technique has receivedmore attention as a principal grain management tool since consumers, governments, processors and importers of grain are rejecting the use of residual chemicals on food grains.[19] Following are some of the important recommendations prior to loading of grains into a storage structure:

- The moisture content has a major impact on the extent to which pests infest stored crops and cause heating and spoilage of the grains. Therefore, it is essential to ensure proper drying of the food grains prior to storage. Painting or whitewashing of the storage system is encouraged to ensure closure of minute cracks and crevices.
- Insects and mites usually hide in the cracks and crevices of the storage bin even when the bins are empty, and manage to survive on the residues until a newly harvested crop arrives. Hence, it is of utmost priority to ensure

that the storage structures are repaired, properly cleaned and disinfested prior to loading of the freshly harvested grains.

- Insects rarely reproduce at temperatures below about 17 °C and mites below 3 °C. Therefore, as a proactive step it should be ensured that the storage structure should be kept in the coolest available area.
- Even a small amount of residue of grain pests in the storage can spoil the entire lot of grains during long-term storage. Therefore, mixing of non-uniform or different batches of grains during storage should be discouraged.

11.9. STORAGE PESTS

Pulses need continuous monitoring and protection at all stages of storage, for they are prone to attack by insects and rodents; in fact, the same applies to all food grains. Pulses stored in farm storage facilities have greater likelihood of pest infestation than storage at a processor's location. The major problem arises from the fact that most farmers use inadequate storage methods immediately after harvest and before processing. This problem is aggravated because infestation continues to increase during transportation and long-term seasonal storage before processing, causing an estimated overall loss of over 30%. Additionally, the situation is greatly magnified in regions where the relative humidity is high, even though temperature has the greatest influence on insect multiplication. At temperatures of about 32 °C, the rate of multiplication is such that a monthly compound increase of 50 times the original number is theoretically possible.[20] This means that 50 insects at harvest time could multiply to 312 million after just 4 months.

Storage losses incurred from storage pests are generally below 5% in traditional agriculture in tropical countries.[21,22] With the inception of high-yielding varieties to boost pulse production, storage loses owing to pests is yet another issue in focus. It is generally believed that high-yielding varieties are more susceptible than traditional low-yielding varieties.[23]

Broadly, storage pests may be classified into two different classes, namely primary and secondary pests. Primary pests are those which infest and cause damage to whole sound grain, while secondary pests are those that require grain that is going out of condition or is damaged to allow these insects to infest and thrive. It should be noted that primary pests, when left unattended, cause considerable damage and open channels for the dwelling of secondary pests. By common sense, the discovery of secondary pests will have less significance, because by that time the grains have already become unusable. These aspects of primary and secondary pests are presented in Figure 11.8.

The common primary and secondary insect pests of stored pulses, as identified by the Canadian Grain Commission, are listed in Table 11.2. Among microorganisms growth of bacteria is restricted due to low moisture content, whereas insects and moulds are major spoilage organisms in stored-grain ecosystems. The best approach therefore calls for regular inspection and

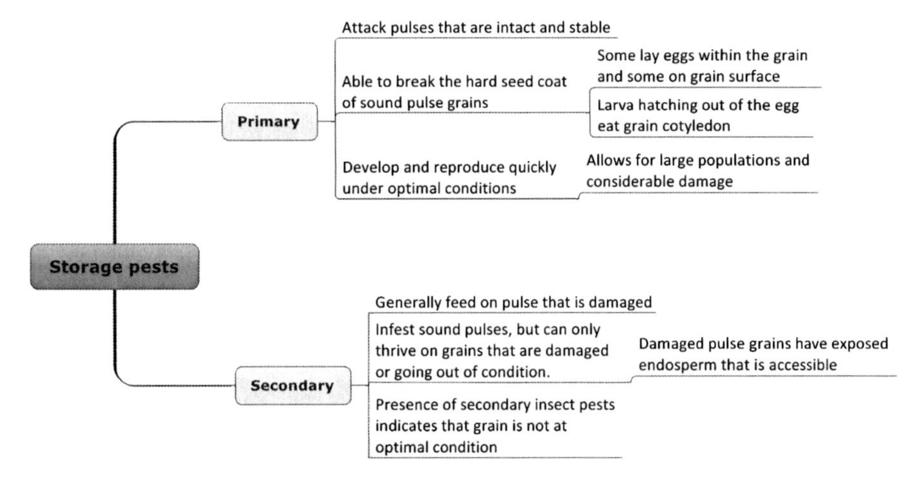

Figure 11.8 Classification of pulse storage pests

monitoring of railcar shipments, container and vessel loadings, and grain elevators for the presence of live insects. The only options after a primary infestation has been identified are either treatment or disposal of infested grain. Some studies show that pulses contain wide range of allelochemicals with toxic and deterrent effects against insect pests.[24,25] Fields *et al.*[25] reported repellence behaviours of pea fractions (fibres and proteins) against storage pests. Similarly, Bodnaryk *et al.*[26] observed toxicity, oviposition deterrence and reproduction inhibition of pea. Several other alternatives to synthetic chemicals were proposed, focusing on the use of natural pesticides such as plant volatile oils have been used in the protection of stored grains.[27–30]

11.9.1. Detection of Storage Pests

The first step to eliminate or control insect infestation in storage is to identify the infestation, if any. Early detection of infestation is important to implement necessary fumigation and pest control measures, thereby ensuring storage longevity, seed quality and food safety. Other important reasons for the timely check and detection of infestation include regulatory compliance and ascertaining the efficacy of control measures such as fumigation. Currently manual inspection, sieving, cracking–flotation and Berlese funnels are used to detect insects in grain handling facilities.[31] However, these methods are inefficient and time consuming[32] with poor accuracy for the developing life stages of pests. Infestation of grains by hidden insects can also be detected by staining of kernels to identify entrance holes for eggs, flotation, radiographic techniques, acoustic techniques, uric acid measurement, nuclear magnetic resonance (NMR) imaging and immunoassays.[32]

The International Organization for Standardization (ISO 6639-4, 1987) has identified five methods to determine the hidden insect infestations in pulses which are also applicable to cereals:[33]

Table 11.2 Primary and secondary pulse storage pests

Primary insect pests	
Lesser grain borer (*Rhyzopertha dominica*)	Flat grain beetle (*Cryptolestes pusillus*)
Bean weevil (*Acanthoscelides obtectus*)	Flour mill beetle (*Cryptolestes turcicus*)
Pea weevil (*Bruchus pisorum*)	Merchant grain beetle (*Oryzaephilus mercator*)
Cowpea weevil (*Callosobruchus chinensis*)	Sawtoothed grain beetle (*Oryzaephilus surinamensis*)
Broad-nosed granary weevil (*Caulophilus oryzae*)	Longheaded flour beetle (*Latheticus oryzae*)
Granary weevil (*Sitophilus granaries*)	Red flour beetle (*Tribolium castaneum*)
Rice weevil (*Sitophilus oryzae*)	Confused flour beetle (*Tribolium confusum*)
Maize weevil (*Sitophilus zeamais*)	Large flour beetle (*Tribolium destructor*)
Khapra beetle (*Trogoderma granarium*)	Cadelle (*Tenebriodes mauritanicus*)
Rusty grain beetle (*Cryptolestes ferrugineus*)	Angoumois grain moth (*Sitotroga cerealella*)
Secondary insect pests	
Cigarette beetle (*Lasioderma serricorne*)	Yellow mealworm (*Tenebrio molitor*)
Drugstore beetle (*Stegobium paniceum*)	Dark mealworm (*Tenebrio obscurus*)
Spider beetles (*Ptininae* sp.)	American black flour beetle (*Tribolium audax*)
Silken fungus beetles (*Cryptophagus* sp.)	European black flour beetle (*Tribolium madens*)
Black carpet beetle (*Attagenus unicolor*)	White shouldered house moth (*Endrosis sarcitrella*)
Larder beetle (*Dermestes lardarius*)	Brown house moth (*Hofmannophila pseudospretella*)
Glabrous cabinet beetle (*Trogoderma glabrum*)	Almond moth (*Cadra cautella*)
Mottled dermestid beetle (*Trogoderma inclusum*)	Mediterranean flour moth (*Ephestia kuehniella*)
Ornate carpet beetle (*Trogoderma ornatum*)	Indian meal moth (*Plodia interpunctella*)
Warehouse beetle (*Trogoderma variabile*)	Meal moth (*Pyralis farinalis*)
European larger cabinet beetle (*Trogoderma versicolor*)	Nocturnal butterfly (*Haplotinea ditella*)
Minute brown scavenger beetle (*Cartodere constricta*)	European grain moth (*Nemapogon granella*)
Hairy fungus beetle (*Typhaea stercorea*)	Large pale clothes moth (*Tinea pallescentella*)
Sap beetle (*Carpophilus* sp.)	Case-making clothes moth (*Tinea pellionella*)
Foreign grain beetle (*Ahasverus advena*)	Psocid (*Liposcelis bostrychophilus*)
Lesser mealworm (*Alphitobius diaperinus*)	Larger pale booklouse or deathwatch (*Trogium pulsatorium*)
Black fungus beetle (*Alphitobius laevigatus*)	Grain mite (*Acarus siro*)
Small-eyed flour beetle (*Palorus ratzeburgi*)	

(i) Determination of carbon dioxide production method
(ii) Ninhydrin reaction with amino acids
(iii) Whole grain flotation method
(iv) The acoustic method
(v) The X-ray method

Besides these, recent approaches under investigation include use of thermal imaging to detect post-embryonic stages in grains. The underpinning principle of this relies on the temperature difference between grains and respiring insect-infested pulses.

11.9.2. Pest Control Measures

Dusting of crops in field is usually a prophylactic measure to control infestation. Favourable conditions for insect growth during storage, such as high temperature and high humidity, must be avoided as primary measures to prevent infestation. Insects and moulds require aerobic conditions for their growth and development. Creating an anaerobic atmosphere in the stored-grain ecosystems therefore has a lethal effect on insects and moulds and extends the storage ability of grains.[34] Damage during harvest may increase the vulnerability of some crops to insect attack. When grain legumes are husked and split into dhal they become vulnerable to infestation by other stored-product insects and moths. Pre-harvest prophylaxis is the insecticidal treatment of the mature grain before harvest in order to prevent carry-over of field infestation to the warehouse. Until recently several insecticides and fumigation techniques were used for storage pest management. Currently, commonly used pesticides (fumigants) for grain storage include aluminium phosphide, methyl bromide and phosphine.[41] Among these, the use of methyl bromide is nearly phased out as agreed in the Montreal Protocol. The use of chemicals is now restricted due to the growing concern over the use of harmful pesticides to kill storage pests. The persistent use of pesticides has been reported to disrupt biological control systems by natural agents, leading to outbreaks of insect pests, widespread development of resistance, undesirable effects on non-target organisms, and environmental and human health concerns.[35–39] Storage of grains including pulses at higher level of carbon dioxide (>60%) and low level of oxygen (<10%) with moderate humidity (<50%) and high temperature (>27%) is lethal to most of the storage pests responsible for qualitative and quantitative losses. Ozone as a fumigant is reported to kill stored-grain insects such as *Tribolium castaneum*, *Rhyzoperthadominica*, *Oryzaephilussurinamensis*, *Sitophilusoryzae* and *Ephestiaelutella*.[45] Laboratory and field studies report the efficacy of ozone in controlling both phosphine-susceptible and phosphine-resistant strains of *Silophiluszeamais*, *S. oryzae*, *R. dominica*, and *T. castaneum*.[40] However, the effectiveness of ozone depends on several factors including the amount of ozone applied as well as various environmental factors such as grain mass temperature, moisture and the surface characteristics.[41]

11.10. EFFECT OF STORAGE ON QUALITY AND NUTRITIONAL PROPERTIES

Grain deterioration during storage is caused mainly through (a) biodeterioration, (b) insects and pests and (c) moulds and fungi. Biodeterioration is caused by the activity of enzymes present in the seed. The extent of deterioration depends upon the level of enzyme activity, which in turn is determined by moisture and temperature. Growth of insect pests and moulds raises both temperature and moisture, and thereby accelerates the activity of the enzymes, which would remain at a low level if conditions of storage were favourable. All these circumstances lead to grain deterioration during storage resulting in weight loss. Apart from quantitative losses (weight loss) during improper storage, biodeterioration results in an enhanced loss of nutrients and possibly formation of antinutritional factors. The nutritional value of beans is affected by a variety of factors during storage and processing. The storage of pulses under poor conditions lead to development of hard-to-cook defect. High moisture in the grain favours hardening as storage time increases. The hard-to-cook phenomenon that develops in pulses under improper storage condition is not fully understood. Available evidence suggests that an increase in the bound protein takes place in the seed coat and aleurone layer, although the cotyledons also lose their capacity to absorb water because of changes in pectin and calcium ions. The grey colour that very often develops is suggestive of carbohydrate–protein reactions.[42] In beans, storage problems result in the loss of grade, including crusting and discoloration of the top layer caused by moulds. Moulds cause spoilage, discoloration and heating. Safe storage of a commodity depends largely on its moisture content (more strictly, the relative humidity of the intergranular atmosphere), its temperature, the period of storage, and other factors. Storage at lower moisture contents is suggested to minimise darkening and colour loss from the seed coat. Lower seed moisture influences the susceptibility of chickpeas to break during handling.

Storage conditions have a large bearing on the behaviour of the starch during cooking of adzuki beans. Fresh beans (11% moisture content) attain 50% starch gelatinisation after 60 minutes of cooking, compared with only 18% for beans stored for 1 year (8% moisture content) and cooked for the same period of time. Yet another example is the gelatinisation temperature of isolated cowpea (*Vigna unguiculata*) starch, which remains constant during storage for 18 months at 30 °C and 64% RH. As beans harden, water is excluded from entering the cell due to a decrease in the permeability of the cell wall; thus, water needed for starch gelatinisation during cooking may be restricted. Water absorption capacity of starch from common bean (*Phaseolus vulgaris*) usually increases during storage under unfavourable conditions.[43] This is due to physical breakdown of starch granules during storage as well as degradation of starch by α-amylase. The starch content of all pulses decreases as the period of storage increases: the reduction in starch content after 12 months was maximum in pigeon peas (7.24%), followed by green gram (6.19%), black gram (3.49%) and chickpeas 12.92%). The amylose content also

Table 11.3 Some of the effects of storage condition on cooking quality

Storage condition	*Effect on cooking quality*
High-temperature storage (30–35 °C)	Increase in hardness of cooked grains relative to freshly harvested grains increase in cooking time
Stored grain moisture	Hardening of grain slower water uptake during cooking
High relative humidity (RH>80%) storage	Increase in grain moisture cause mould growth
Low relative humidity (12–35% RH) storage	RH decrease the moisture content Increase in cooking time Decrease in starch gelatinisation during cooking

decreases with increase in storage period but this decrease is usually much slower.

Protein quality and quantity of pulses decreases during storage. Various factors are responsible for this reduction, namely storage temperature, moisture, RH. Insect infestation is reported to decrease protein quality. Therefore, insect control is also desirable to maintain nutritive value and ensure increased efficiency of processing. Decrease in water-soluble protein with increase in storage and decrease solubility is reported for cowpeas stored at 30 °C/64% RH. The protein solubility decreased from 76.5% to 11.2% after 18 months storage. Therefore, in the aged cowpea seed, most proteins become water insoluble and the remaining soluble proteins readily coagulate upon cooking. Storage of legumes at high temperature (30 °C) and high RH (85%) also exhibits qualitative and quantitative changes in the protein electrophoretic patterns. Storage duration, temperature, relative humidity and moisture content are important parameters for pulse storage. These parameters influence the cooking quality (defined by cooking time and textural softening) of pulses during storage. Unfavourable storage conditions results in a lower water imbibition, which in turn results in reduced cooking quality and increase in cooking time. Some of the effects of storage condition on cooking quality are listed in Table 11.3.

REFERENCES

1. J. M. Thamaga-Chitja, S. L. Hendriks, G. F. Ortmann and M. Green, *J. Fam. Ecol. Consumer Sci.*/*Tydsk.. Gesinsekolog.. Verbruikerswetensk.*, 2004, **32**, 8–15.
2. A. P. Ouédraogo, S. Sou, A. Sanon, J. P. Monge, J. Huignard, M. D. Tran and P. F. Credland, *Bull. Entomol. Res.*, 1996, **86**, 695–702.
3. A. Sanon, L. C. Dabiré-Binso and N. M. Ba, *J. Stored Prod. Res.*, 2011, **47**, 210–215.
4. Q. Mohy-ud-Din, *Econ. Rev.*, 1998.
5. J. Singh and P. Verma, Central Institute of Post Harvest Engineering and Technology, Ludhiana, India, 1995.

6. K. E. Ileleji, D. E. Maier and C. P. Woloshuk, *J. Stored Prod. Res.*, 2007, **43**, 480–488.
7. M. T. Talbot, *Management of Stored Grains with Aeration*, Circular 1104, Institute of Food and Agricultural Sciences, University of Florida, Agricultural and Biological Engineering Department, Florida Cooperative Extension Service, 1999.
8. J. T. Mills and S. M. Woods, *J. Stored Prod. Res.*, 1994, **30**, 215–226.
9. J. Cassells and E. Armstrong, CSIRO Stored Grain Research Laboratory, 1998.
10. CGC, *Protection of Farm-Stored Grains, Oilseeds and Pulses from Insects, Mites and Moulds*. Canadian Grain Commission, Agriculture and Agri-Food Canada Publication, 2001. (http://www.grainscanada.gc.ca/storage-entrepose/aafc-aac/pfsg-pgef-8-eng.htm)
11. T. Tefera, F. Kanampiu, H. De Groote, J. Hellin, S. Mugo, S. Kimenju, Y. Beyene, P. M. Boddupalli, B. Shiferaw and M. Banziger, *Crop Protect.*, 2011, **30**, 240–245.
12. FTIC, *Hermetically Sealed "GRAINPRO COCOONS™" for Bagged Grain Storage*, http://ftic.biz/index.php?option=com_k2 and view=itemlist and task=category and id=15 (Accessed 12 August 2011).
13. FTIC, *Bunker Storage of Cereal Grains*, http://ftic.biz/index.php?option=com_k2 and view=itemlist and task=category and id=20 (Accessed 12 August 2011)
14. S. Navarro, E. Donahaye, Y. Kashanchi, V. Pisarev and O. Bulbul, in *Controlled Atmosphere and Fumigation in Grain Storages*, ed. B. E. Ripp, Elsevier, Amsterdam, 1984, pp. 601–614.
15. M. Khan, *Anzeiger für Schädlingskunde*, 1983, **56**, 65–67.
16. S. Navarro, T. De Bruin, A. R. Montemayer, S. Finkelman, M. Rindner and R. Dias, *IOBC/WPRS Bull.* 2007, **30**, 197–204.
17. F. Jian, D. S. Jayas and N. D. G. White, *J. Stored Prod. Res.*, 2009, **45**, 82–90.
18. S. Navarro and R. T. Noyes, *The Mechanics and Physics of Modern Grain Aeration Management*, CRC Press, Boca Raton, FL, 2001.
19. D. d. C. Lopes, J. H. Martins, A. F. L. Filho, E. d. C. Melo, P. M. d. B. Monteiro and D. M. d. Queiroz, *Comput. Electron. Agric.*, 2008, **63**, 140–146.
20. P. Pushpamma and V. Vimala, eds, *Storage and the Quality of Grain: Village-Level Studies*, United Nations University Press, Tokyo, 1984.
21. J. A. F. Compton, S. Floyd, P. A. Magrath, S. Addo, S. R. Gbedevi, B. Agbo, G. Bokor, S. Amekupe, Z. Motey, H. Penni and S. Kumi, *Crop Protect.*, 1998, **17**, 483–489.
22. E. Getu and A. Gebre-Amlak, *Pest Manage. J. Ethiop.*, 1998, **2**, 26–35.
23. T. Abate, A. van Huis and J. K. O. Ampofo, *Annu. Rev.Entomol.*, 2000, **45**, 631–659.
24. J. B. Harborne, D. Boulter and B. L. Turner, eds, *Chemotaxonomy of the Leguminosae*, Academic Press, London, 1971.

25. P. G. Fields and N. D. G. White, *Annu. Rev. Entomol.*, 2002, **47**, 331–359.
26. US 5955082, 1999.
27. A. K. M. El-Nahal, G. H. Schmidt and E. M. Risha, *J. Stored Prod. Res.*, 1989, **25**, 211–216.
28. B. P. Saxena and A. C. Mathur, *Experientia*, 1976, **32**, 315–316.
29. E. M. Risha, A. K. M. El-Nahal and G. H. Schmidt, *J. Stored Prod. Res.*, 1990, **26**, 133–137.
30. A. A. Gbolade and T. A. Adebayo, *Int. J. Trop. Insect Sci.*, 1993, **14**, 631–636.
31. A. A. Gowen, B. K. Tiwari, P. J. Cullen, K. McDonnell and C. P. O'Donnell, *Trends Food Sci. Technol.*, 2010, **21**, 190–200.
32. S. Neethirajan, C. Karunakaran, D. S. Jayas and N. D. G. White, *Food Control*, 2007, **18**, 157–162.
33. A. Manickavasagan, D. S. Jayas, N. D. G. White and J. Paliwal, *T Asabe*, 2008, **51**, 649–651.
34. D. S. Jayas and S. Jeyamkondan, *Biosyst. Eng.*, 2002, **82**, 235–251.
35. M. A. G. Pimentel, L. R. D. A. Faroni, R. N. C. Guedes, A. H. Sousa and M. R. Totola, *J. Stored Prod. Res.*, 2009, **45**, 71–74.
36. M. A. G. Pimentel, L. R. D. A. Faroni, M. R. Totola and R. N. C. Guedes, *Pest Manage. Sci.*, 2007, **63**, 876–881.
37. P. J. Collins, G. J. Daglish, H. Pavic and R. A. Kopittke, *J. Stored Prod. Res.*, 2005, **41**, 373–385.
38. M. S. Islam, M. M. Hasan, W. Xiong, S. C. Zhang and C. L. Lei, *J. Pest. Sci.*, 2009, **82**, 171–177.
39. S. A. Kells, L. J. Mason, D. E. Maier and C. P. Woloshuk, *J. Stored Prod. Res.*, 2001, **37**, 371–382.
40. Z. Qin, X. Wu, G. Deng, X. Yan, X. He, D. Xi and X. Liao, *Adv. Stored Prod. Prot*, 2003, 846–851.
41. B. K. Tiwari, C. S. Brennan, T. Curran, E. Gallagher, P. J. Cullen and C. P. O'Donnell, *J. Cereal Sci.*, 2010, **51**, 248–255.
42. R. Bressani, L. Elias, A. Wolzak, A. Hagerman and L. G. Butler, *J. Food Sci.*, 1983, **48**, 1000–1001.
43. C. Reyesmoreno and O. Paredeslopez, *Crit. Rev. Food Sci. Nutr.*, 1993, **33**, 227–286.
44. P. Villers, S. Navarro and T. De Bruin, *Julius-Kühn-Archiv*, 2010, S. 446.
45. A. H. Sousa, L. R. D. Faroni, R. N. C. Guedes, M. R. Tótola and W. I. Urruchi, *J. Stored Prod. Res.*, 2008, **44**, 379–385.

CHAPTER 12

Processing of Pulses

12.1. INTRODUCTION

Pulses are consumed in various forms after suitable processing. Pulse processing is one of the earliest and most important of all food technologies and forms a large and important part of the food production chain. Unlike cereal processing not much has changed for pulse processing, the reason being the processing challenges which are associated with the processing and subsequent utilisation of pulses. These challenges include (1) low profitability of pulse production; (2) post-harvest losses, primarily during storage; (3) inadequate supply of high quality and reasonably priced raw materials; (4) lack of sustainable and efficient processing and packaging technologies and (5) lack of internationally recognised quality standards and accepted nomenclature. Interest in the utilisation of whole pulses and their milled constituents in food formulations is growing in many developed countries. The processing of pulses into ingredients such as flours and fractions (*e.g.* protein, starch and fibre) and utilising them in food products is virtually non-existent in Western-style food products, apart from a few specialty or niche markets, and exists only in a limited way in a few other countries. However, more recently pulse flour and fractions have been used successfully as ingredients in the formulation of several meat products to improve functionality.

The earliest and most common home-scale technique for hulling pulse grain is to pound it in a mortar with a pestle, either after spreading the grains in the sun for a few hours, or after mixing them with a little water. The hull is then winnowed off to produce clean cotyledons. The traditional stone *chakki* design was used as a model for attrition-type mills in commercial-scale dehulling and splitting of pulses. Large-scale industries have evolved to some extent from these traditional processing methods. The processing of pulses by various means has been used for eliminating toxic substances and non-nutrients, removing the seed coat, softening the cotyledons and making them palatable.

Pulse Chemistry and Technology
Brijesh K. Tiwari and Narpinder Singh

Published by the Royal Society of Chemistry, www.rsc.org

12.2. TYPES OF PROCESSING

Processing operations could be divided into primary, secondary and tertiary processing as shown in Figure 12.1. Primary processing is generally defined as conversion of raw material (grains) to foods that can be eaten, or to ingredients that can be used to make edible food products. Primary processing of pulses involves cleaning, grading, dehulling, milling, splitting, *etc.* Secondary processing is defined as the conversion of ingredients (end or finished product of primary processing) to edible food products or finished food. This generally involves a number of unit operations which modify or alter the properties. Secondary processing of pulses involves utilisation of the primary products (dehulled grains, splits, *etc.*). Secondary processing includes grinding (*e.g.* dehulled beans are ground to obtain flour), roasting (clean and graded chickpeas are roasted), fermentation, frying and extrusion fortification. Tertiary processing includes those processes which utilise end products of primary and secondary processing to obtain products that are more interesting and add variety to the diet. These can include ready meals or ready-to-eat snacks prepared by mixing a variety of pulses, *e.g.* roasted chickpea, deep fat fried snacks, fried lentils, *etc.* (commonly known as Bombay mix). However, there is no clear distinction between primary and secondary processing of legumes and most of the unit operations overlap each other. In the following section some of the important primary processing methods are discussed in detail.

12.3. PRIMARY PROCESSING OF PULSES

12.3.1. Pulse Dehulling

The term "dehulling" and "milling" of pulses are often used interchangeably. Milling is a general term and usually refers to a process where grains are reduced to meal or flour, as in the case of wheat milling. However, in the case of rice milling and pulse milling the finished product is either polished rice (rice milling) or dehulled splits (pulse milling). In this chapter the term dehulling

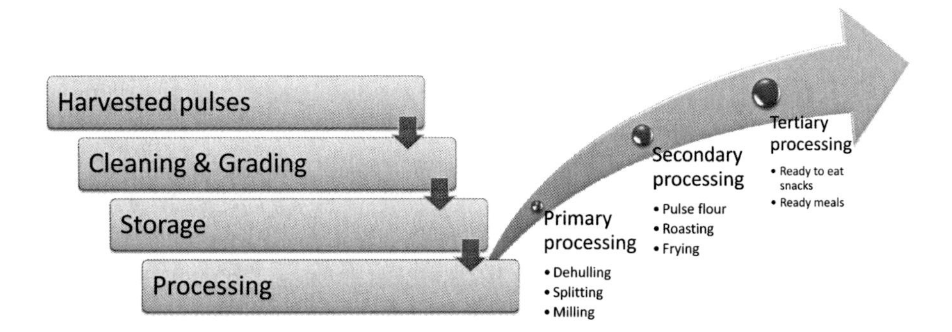

Figure 12.1 Flowchart showing the processing of pulses

refers to the removal of hull to obtain dehulled whole or splits and milling refers to a process for obtaining pulse flours. Dehulling is the most common and widely practiced primary processing method followed in various countries of the Asian subcontinent and Africa, either on a home-scale or as a cottage industry, in order to produce dehulled cotyledons (whole or splits) also known (in India) as dhal, with good appearance, texture and cooking qualities. Dehulling of pulses is the third largest food processing industry in India after rice milling and wheat milling; more than 75% of pulses produced are converted to splits. Dehulling of pulses involves the removal of the outer hull and splitting the grains into two equal halves (splits). The dehulled splits thus obtained are known as dhal. The dehulling process probably originated in the kitchen by coarse grinding on a kitchen stone or grinding with a pestle and mortar after slight heat treatment (roasting) or sun-drying and winnowing off the hull. This practice is still followed in many tribal and rural areas. Other methods commonly practiced in rural areas are trampling or grinding in a wet grinder to remove the hull from pulses previously soaked in water. The hull envelops the endosperm tightly, usually through a thin layer of gums and mucilages, which are loosened up on slight heat treatment, or by soaking in water. The dehulling of pulses on a small scale or cottage level involves laborious procedures. The most common and oldest home-scale technique for dehulling pulses is to pound them in a mortar with a pestle, either after spreading the grains in the sun for a few hours, or after mixing them with a little water. The hull is then winnowed off to get the clean cotyledons. Precisely, dehulling of pulses involves three principal steps: (1) loosening of the hull by pre-milling treatments, (2) removal of the hull and (3) splitting the dehulled whole grains.

Methods followed in the home or cottage industry or in commercial pulse dehulling industries are usually similar in principle, but differ in the use of techniques for better yield, operational efficiency and large-scale application. The household and small-scale practice of preparing dehulled split pulses involves loosening the husk and subsequent removal and splitting into two cotyledons. Home-scale milling consists generally of two steps. The first step is achieved by sun-drying the raw mature grains as they are, or after they have been treated with edible oil and/or water to loosen the hull. In some areas, grain is steeped in water for 2–8 hours prior to sun-drying. Grain varieties whose hulls are tightly attached to the cotyledons are soaked and then treated with red-earth paste before being sun-dried. The steeping technique to loosen the hull is also practiced in several South East Asian and African countries. Cleaning, grading, drying, pre-milling treatment, dehulling, packaging and storage of processed pulses are major operations performed for dehulling of pulses. Figure 12.2 shows a general dehulling process for pulses; however, there is no common dehulling process applicable for all pulses. There are variations in the process of dehulling depending on the type of pulses. Hulling of pulses on a commercial scale is generally based on dry-processing techniques. Table 12.1 shows some of the dehulling techniques adopted for pulses.

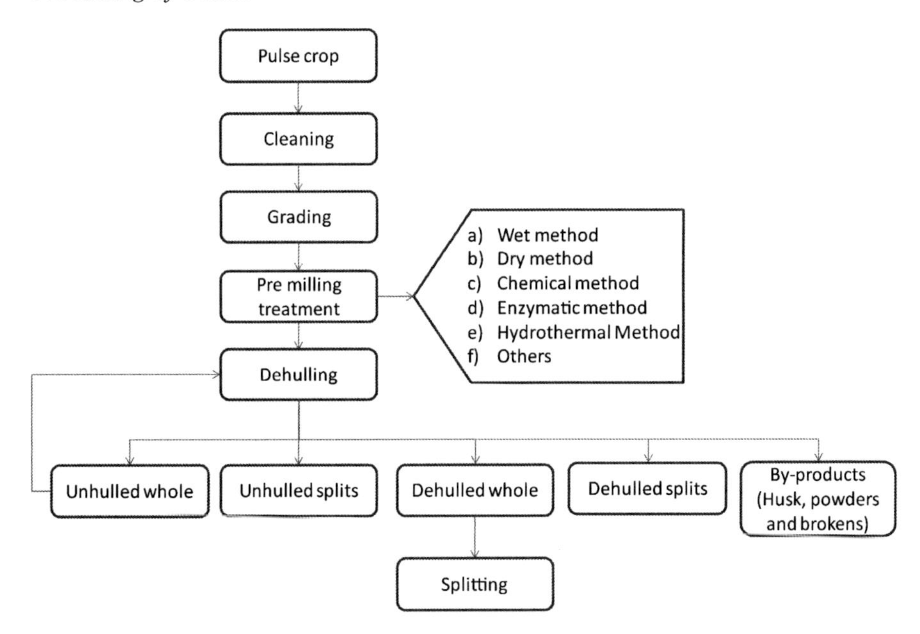

Figure 12.2 General dehulling procedure for pulses

Many of the operations, particularly dehulling and splitting, are mechanised. The drying is done in large yards, and is completely dependent on sunshine. Pulses such as pigeon peas, black gram, and mung bean, which are more difficult to hull, require more of the oil or water treatments followed by prolonged sun-drying (pre-milling treatments), while grains such as chickpeas, lentils, peas or *Lathyrus sativus*, with more easily removable hulls, require short periods of drying and fewer oil or water treatments. Sometimes these grains are given an initial "pitting" in the roller mill to crack the hull or induce scouring on the grain surface (*e.g.* black gram, pigeon peas) which facilitates absorption of water or oil. Hulling and splitting are done either in a single operation or more advantageously, as independent operations. Moisture addition adversely affects hulling, but it helps to split the grain. Addition of water prior to dehulling helps to induce simultaneous splitting, but this often leaves patches of hull on the split cotyledons (dhal) that have to be removed by scouring in polishing machines. As the hull forms 10–16% of the grain, a maximum theoretical yield of 84–90% of kernels should be possible during hulling. In practice, the yields vary from 68 to 76% because of breakage, powdering and other milling losses. During splitting, the germ, which forms about 2–5%, is also lost. In wet methods, water-soluble nutrients are also lost in the soak water. Table 12.2 shows various pulse dehulling fractions by some traditional methods.[1]

12.3.2. Cleaning and Grading of Pulses

Cleaning and grading is an essential step prior to dehulling or any other processing. During harvesting and post-harvest handling, including drying and

Table 12.1 Dehulling process for some pulses

Pulses	*Dehulling characteristics*	*Pre-treatment*	*Dehulling process*
Chickpeas, peas and lentils	Easy to dehull	Dry and wet methods	Pitting of clean and size-graded grains in smooth rollers at low peripheral speed. Mixing of grains with about 5–10% water in a screw conveyer-type mixer and tempering for a few hours for equilibration Drying of grains Dried grains are passed through either a horizontal or vertical emery roller for simultaneous dehulling and splitting. Separation of milling fractions using appropriate screening and grading system coupled with aspirators
Lentils[3]	Easy to dehull	Wet method	Clean and size-graded lentils are immersed in water for 1–5 min followed by tempering (1–6 h). Tempered grains are dried in mechanical dryers followed by tempering up to 24 h before dehulling in suitable dehulling machine The mixture of whole and split lentils, hulls, broken seed and fine particles is separated through an inclined gravity separator and aspiration/suction system to remove fine particles
Black gram	Difficult to dehull	Dry method	Pitting of clean and size-graded grains in using emery rollers in 2–3 passes for effective surface scarification Pitted grains are mixed with 0.5–1.0% edible oil and tempered overnight for uniform absorption of oil Drying of grains using mechanical dryers or sun-drying After drying 2–5% water is sprinkled and kept for tempering Grains are dehulled using emery rollers or cone. Separation of milling fractions using appropriate screening and grading system coupled with aspirators. Dehulled splits are separated using screening system, dehulled whole are either passed through splitting machine or sold as dehulled whole depending on the demand. Dehulled splits obtained from whole dehulled pulses are considered as grade 1 splits whereas dehulled splits obtained along with dehulled whole are considered as grade 2.

Table 12.1 Continued

Pulses	*Dehulling characteristics*	*Pre-treatment*	*Dehulling process*
			Dehulled splits or whole are polished using oil and water or soapstone in a screw conveyor to enhance appearance
Green gram	Easy to dehull	Dry method	Dehulling method is similar to black gram, however, green gram hull is thin, soft and slippery compared to black gram. Unlike black gram, splitting occurs during pitting and may result in excessive scouring of the cotyledons, resulting in large losses in the form of brokens and powder
Pigeon peas	Difficult to dehull	Dry method	Cleaned and size-graded grains are pitted as in the case of black gram. Pitted grains are mixed with edible oil (0.2–1.0%) followed by tempering for 12–24 h.
			Oil-coated pulses are mechanically dried or sun-dried for 1–3 days, followed by spraying with water (2–6%) and tempering overnight
			Tempered grains are dehulled using appropriate dehulling machines and milling fractions are separated as in the case of black gram. During dehulling splits with sharp edges are considered as grade 1 whereas splits obtained along with dehulled whole do not have sharp edges and are considered as grade 2. Dehulled splits obtained by splitting dehulled whole have sharp edges. Unlike black gram, dehulled whole grains are not sold in market

storage of pulses at the farm, pulses contain a range of foreign matter including pod shells, soil dirt, stones, mudballs, plant materials, weed seeds and grains of other crops. During mechanical or manual harvesting the unthreshed broken, damaged, shrunken and immature pulse kernels must be adequately removed prior to subsequent processing. These immature, broken and inferior quality grains affect the dehulling efficiency and subsequent processing. Non-synchronous maturity of pulses leads to the presence of immature grains. Mechanical harvesting also yields broken grains which must be removed prior to processing. These problems are often encountered when pulses are stored without cleaning. However, large pulse growers and processors store their product after adequate cleaning and grading operations to secure a better price for their produce. Storage of uncleaned pulse grains with broken and shrivelled grains harbours storage pests and has a greater chance of microbial

Table 12.2 Pulse dehulling fractions by various traditional methods

Pulses	*Theoretical yield (%)*	*Dehulling method*	*Dehulled splits (%)*	*Brokens (%)*	*Powdering loss (%)*	*Hull (%)*
Chickpeas						
large size	88.5	Dry methods	74	7	10	8
small size	87.5	Dry methods	72	6	13	9
Pigeon peas						
large size	88.5	Dry methods	75	5	11	8
small size	85	Wet method	68	8	14	9
		Shellers	75	7	3	14
Blackgram	87.5	Dry methods	71	7	13	8
		Stone powder method	74	10	7	8
Mung gram	89.5	dry methods	65	13	17	4
		Stone powder method	74	10	7	8
Lentils	88.5	Dry methods	76	5	12	6
Lathyrus	88	Dry method	76	6	8	9

Source: Adapted from Kurien *et al.*[22]
These estimates are based on the survey of pulse processors in India.

contamination and insect infestation that may cause concern to the processors and may render the pulses unsuitable for processing. However, stored clean pulses may also acquire foreign matter during loading/unloading and transportation or due to pest infestation during storage. Hence, cleaning and grading become essential prior to dehulling process.

The *Codex Alimentarius* clearly outlines that pulses shall be free from abnormal flavour and odours, and free from filth including living or dead insects, impurities of animal origin and extraneous matter in amounts that may represent a hazard to human health. Extraneous matter represents mineral or any organic matter including dust, twigs, seed coats, seeds of other pulses, dead insects, fragments or remains of insects, other impurities of animal origin. Although regulation varies from country to country, however, the *Codex Alimentarius* outlines that pulses shall have not more than 1% extraneous matter of which not more than 0.25% shall be mineral matter and not more than 0.10% shall be dead insects, fragments or remains of insects, and/or other impurities of animal origin. In Canada and USA a term "dockage" is used in the context of grading of pulses. Pulse dockage is described as weed seeds, weed stems, chaff, straw, or grain other than pulses, which can be readily removed from the pulse grain by the use of appropriate sieves and cleaning devices. Dockage also includes immature, shrivelled and broken grains, which should be removed by following the approved cleaning processes. The basic principle of cleaning is to remove all foreign materials that come along with the pulses before subsequent processing based on physical properties (Table 12.3). Cleaning equipment used to clean pulses varies from simple screening devices

Table 12.3 Principle of cleaning and equipment

Foreign matter properties	*Types of foreign material*	*Equipment in use*
Size		
Larger	Large size grains of other crop or any material larger than the size of pulses to be cleaned	Grain separators, cleaning systems with screens and sieves Cleaning equipment based on mechanisms of gyration and use of forced air to separate large and lightweight grain
Smaller	Small size grains of other crops, *e.g.* weed seeds. Sand, dust and any material smaller than the size of pulses to be cleaned	
Specific gravity		
Lighter	Barley, oats, wild oats, and ergot	Table separators, gravity selector, and combination
Denser	Stones, metals, and mud balls	Destoner
Shape		
Shorter, longer and round	Broken wheat, cockle, and wild buckwheat	Disc separator, indented cylinders and spiral seed separator
	Oats, barley, wild oats, and ergot	
	Cockle	
Magnetic properties	Ferrous and ferrous alloy objects	Electrical and permanent magnets
Differences in terminal velocity	Chaff, husk, straws, dust, and shrivelled kernels	Aspiration channels, air-recycling chambers

to aspiration type cleaners. Gravity type cleaners used in grain cleaning employ vibration and forced air to clean and grade grains according to density. Sound grains settle to the bottom of the layer while less dense or lighter, shrivelled grains separate to the top. Spiral types of cleaners are also employed for cleaning of pulses and they separate grains efficiently based on shape. Some pulse processors employ complex cleaners which can clean grain based simultaneously on physical properties such as shape, size, density and weight. These types of cleaners are equipped with a series of screens of varying mesh size that shake back and forth to separate the foreign matter and aspiration system to remove lighter impurities (Figure 12.3). Foreign matter larger or smaller than the pulse grain to be cleaned can be removed efficiently without any difficulties. Destoners are also employed to remove heavy foreign contaminants like stone, pieces of metal or glass, mud balls, *etc.* The working principle of the destoner can be either based on vacuum (negative pressure) or on weight difference.

Magnetic separators are also used to remove any ferrous metal or its alloy from the pulse grains. A grain separator or classifier is generally used with an

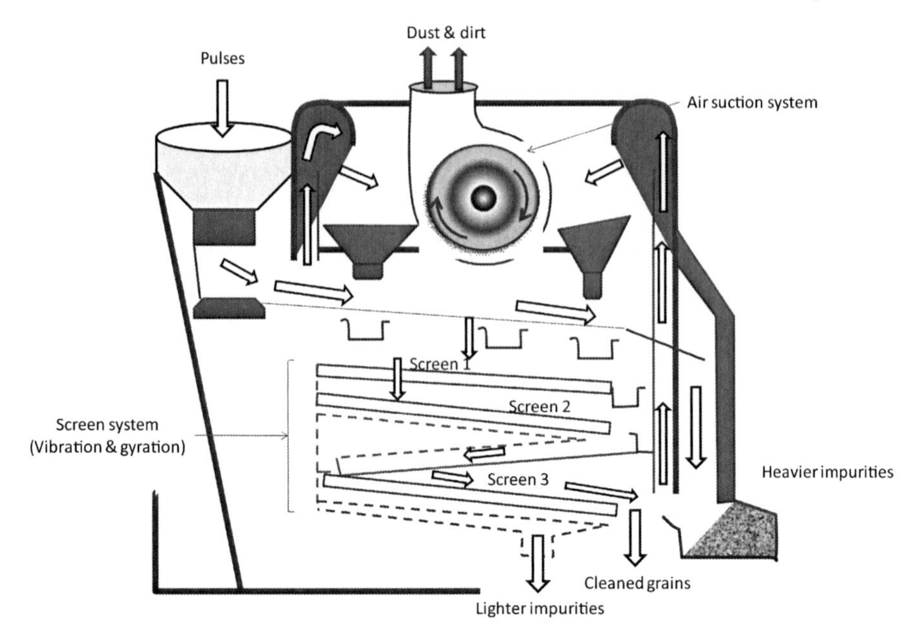

Figure 12.3 Cleaning equipment

air-recycling aspiration system for simultaneous grading and cleaning. Aspiration helps to remove larger, finer and lighter foreign material from pulses. A destoner is used to remove stones, mud balls, and other heavy material. Impurities commonly found in pulses include pod walls, broken branches, soil, cereals, oilseeds, weed seeds, diseased and deformed seeds, and stones, dust, dirt, foreign material, off-sized, immature, and infested grains, and dead or living insects. Various types of cleaning and grading equipment used for pulses are listed in Table 12.4. The sequence for cleaning and sorting pea, beans and lentil grains is shown in Figure 12.4.

12.3.3. Pre-milling Treatment

Pre-milling treatment is also known as grain conditioning. Conditioning is necessary prior to dehulling of pulses. Pre-milling treatment is the most important step in processing of pulses, it is essentially required to loosen the hull from the cotyledon, which is tightly bound by gums and mucilages, and facilitate the easy removal of husk from the cotyledon. Pre-milling treatments are used with the following objectives: (1) to loosen the husk, (2), to increase ease of milling, (3) to reduce breakage and (4) to improve the quality of split. Various types of pre-treatments are currently in practice depending upon the type of the grain and varying from region to region. Pre-treatments of grains can be classified based on the technique involved: wet and dry methods, and chemical, thermal and enzymatic treatments are used.

12.3.3.1 Dry Milling

The dry method includes pitting of grains (scarification) and application of oil and water. In this method, cleaned and graded grains are subjected to pitting in an emery roller; usually one pass is given to scour over the grains. Then pitted grains are mixed thoroughly with 0.5–1.5% of oil followed by sun-drying in a thin layer for 2–3 days. Nowadays, mechanical driers are used for drying of grains. After drying, about 2–3% of water is sprayed and mixed thoroughly. For tempering, grains are heaped overnight to equilibrate the moisture in grain. Spraying of water over grains is advantageous in that it reduces powdering losses during milling. Table 12.5 shows the effect of various edible oils on the milling fractions of black gram. The table shows that use of oil improves the yield of dehulled grains.

Home-scale methods are generally laborious and not always hygienic. In village industries, the techniques employed for loosening the hull are the following: (1) prolonged sun-drying until the hull is loosened; (2) application of small quantities of oil, followed by several hours or even days of sun-drying and tempering; (3) soaking in water for several hours, followed by coating with red-earth slurry and sun-drying; (4) soaking in water for several hours to loosen the hull before processing into food products; or (5) a combination of these techniques. There are no standard procedures developed for any specific variety of pulses. Removal of the loosened hulls from the grain in the dry milling technique is commonly done in small machines. Hand-or power-operated under-runner disc-shellers or grinders with emery or stone contact surfaces are used. A plate mill with a blunt contact surface is sometimes used both to hull and split soaked and dried grains. After aspirating or winnowing off the hull, the split cotyledons are separated by sieving. The remaining unsplit whole grains are similarly processed until almost all the grain is hulled. In certain parts of India, oil-treated and sun-dried grains are hulled in an Engelberg-type rice-huller after being mixed with 2–3% stone powder. Sound kernels are removed by sieving, while the hull, powder, and small brokens remain in the stone powder. About 50% removal is achieved in the first operation. After separation of the hulled, split cotyledons (dhal), the process is repeated several times until almost all the grain is converted into dhal. In the process, excessive breakage with powdering of grains occurs because of repeated splitting and hulling operations.

12.3.3.2 Wet Milling

In wet milling, the main objective is to soften the hull prior to its removal. Traditionally, water soaking of pulses is carried out to hydrate and swell the seed coat. In some instances the grain is merely sprinkled with water and, being hygroscopic in nature, absorbs moisture. Commercially, the wet method employs mixing of grains with red earth after soaking in water for about 5–12 hours. The red earth is mixed with pulses at about 10–15% and kept in the form of a heap for about 16–24 hours. Following this, grains are spread in a thin layer in drying yards for 2–4 days and the dried earth is removed by sieving. The grains are then milled

Table 12.4 Cleaning equipment used for pulses

Scalpers	These are used to remove large unwanted trash and fines. The separation is done on reciprocating or rotating screens Helps to remove large, unwanted trash and fines The scalper is based on the overall size of the seed and airflow This pre-cleaning makes the workings of the subsequent machines easier
Air-screen machines	These are used as scalpers for size separation as well as weight separation. The material is aspirated to remove light materials, then passed over screens to remove large materials as overs. The unders are passed on a second screen to separate fines from desired size seeds Based upon seed size and airflow, but the motion of the mix through the machine is more precisely controlled Several sieves may be required for a complete clean out of the seed. Pea seeds might be completely cleaned on these machines
Disc and cylinder separator	These consist of series of cast iron discs mounted on a shaft revolving at a very precise speed relative to disc diameter within a cylindrical housing. The discs have precise undercut pockets with size and shape variations which allow the smaller seeds to be lifted and to reject larger seeds Disc separators and indent cylinders are more precise machines than scalpers and air screens. They separate and grade seeds based on their length. Disc separators and indent cylinders can be adjusted in many ways and with a proper adjustment a high degree of cleaning can be expected
Indent cylinders	These are the machines of choice for length separation. They are almost horizontal rotating cylinders lined with hemispherical depressions. Short seeds are picked up in the indent and lifted up and thrown into an auger to be carried out of the rotating cylinder
Width and thickness separators	These are typically rotating, cylindrical, perforated shells. Larger seeds will not pass through and are discharged from the end of the cylinder
Gravity separators	These separate seeds based on a combination of shape, size, specific gravity, and surface characteristics. The seed mixture is fed onto an oscillating deck with a carefully controlled air movement to fluidize the material. The mixture is stratified – lighter up and heavier down; the seed layers along the deck in different directions toward discharge ends Contaminants in a lot that have a size similar to the seed, but are of different specific gravity, can be separated on a gravity table. The seed mixture is fed onto a perforated table, the air blowing through perforations keeps the mixture fluidized, while an oscillating/vibrating motion stratifies the material and separates the heavy seeds from light ones The material is divided into several fractions, each individually collected at the other end or at the side of table. Gravity tables are quite versatile machines, but require experienced operators to run at their peak efficiency The most important variables are degree of oscillation; rate of vibration; airflow; slope and direction of the deck; decking configuration and decking material; number of output spouts, loading rate on the deck, and the location of the unloading dividers
Spiral separators	These work on the basis of the shape of the seed to roll at different rates. The angle of the flights of the spiral has to be adjusted based on the type of crop

Table 12.4 Continued.

Colour sorters	Colour sorters have proved to be very effective but need considerable adjustment andcapital investment, and are slow, although it is expected that improvements in speed will be made. Colour sorters are sometimes used as final equipment to separate seeds based on their reflective (colour) characteristics. In these devices the seed is fed into the machine in single file (or channel). One or a series of light diodes emits a filtered light onto the seed as the seed passes through the detector ring. The reflected light from the seed is compared electronically to the reflected light from a reference plate. The seed is ejected from the stream by mechanical means, if its reflected light is different from the reference. The mechanical ejector is usually a pressurized air nozzle and may pulsate at a maximum rate of 70 ejections per second. The amount of air per ejector may be as much as 0.75 cubic feet per minute. Factors that affect the operation of a colour sorter are: loading rate, *i.e.* the rate seed flows against the sensor light; electronic settings; ambient light (or background light); dust accumulation on lights and reflector plates; vibrations, and type and degree of discoloration on the seed. The capacity of colour sorters can be increased by adding channels and the size of a channel controls the grain size. For larger seeds, such as Laird lentil, rollers are used to guide the seed through the detector ring. Slider channels are used for small seeds (Figure 12.5)

Source: Adapted from Sokhansanj and Patil[23] and Milne.[24]

on a power-operated stone or emery-coated vertical *chakki*.[2] The cooking quality of dhal prepared by wet methods is usually poor. This is especially true for pigeon peas, for which cooking time increases with duration of soaking. However, such dhal has an attractive appearance and a more desirable flavour. The wet method may be useful where the pulse is to be ground to a paste for further processing. If it is to be dried as dhal, the wet method is laborious and entails the loss of soluble solids in the soak water. The use of red earth was primarily employed for dehulling of pigeon peas in India, but rarely used commercially.

Another wet milling method involves dehulling by pounding with a pestle and mortar after soaking grains for a brief period. Wetting the grains followed by heat treatment also loosens the hull, makes its removal easier, and it can be removed by scouring action during dehulling. Soaking time of pulses is also critical for obtaining best dehulling yield. Figure 12.6 shows the effect of immersion time followed by tempering and drying. The best dehulling yield in case of lentils using the wet method has been found with the combination of shortest drying time, least immersion time and longer tempering time.[3] Longer periods of soaking increase grain moisture content and reduce the dehulling efficiency with an increase in split lentils with hulls.[3] Soaking and drying is one of the effective methods for dehulling of pulses because the shrinkage of cotyledons on drying is more than that of the hull, which results in a bubbled hull that can be removed easily by a shearing or scouring action. Table 12.6 shows the various features of wet and dry methods of pre-treatment.

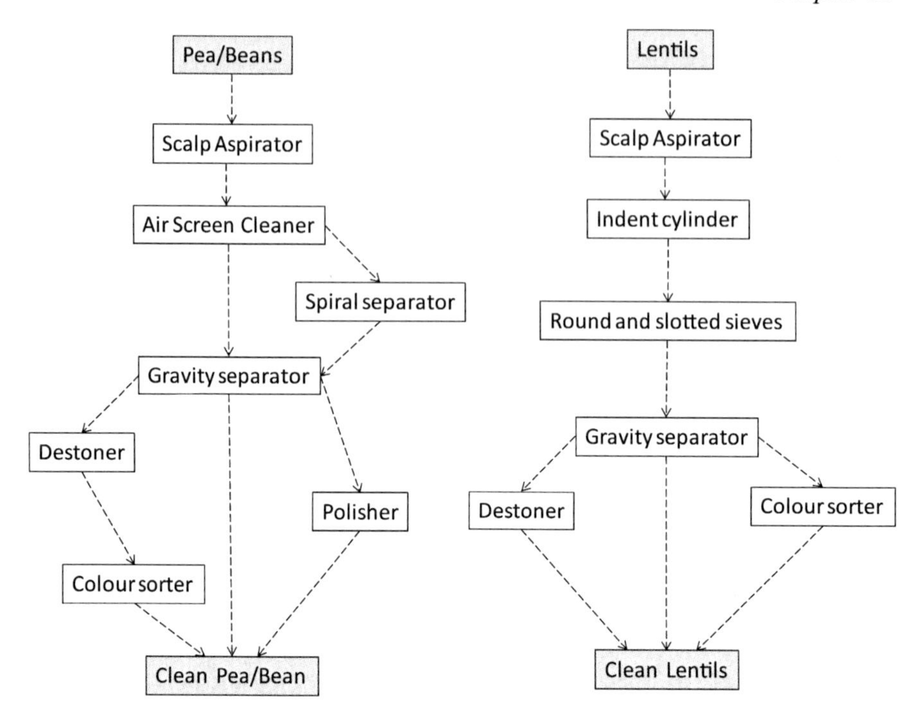

Figure 12.4 Sequence of cleaning and sorting pea, beans, and lentil grains (Adapted from pulse production manual)[26]

12.3.3.3 Thermal Treatment

In this method of pre-treatment, cleaned pulse grains are conditioned in two passes in a dryer using hot air at about 120 °C (grain temperature 80 °C) for a certain period. After each pass the hot pulses are tempered in the tempering bins for about 6 hours. Such hot air preconditioning of grains helps in loosening the hull significantly. Application of heat treatment to moistened pulses makes the hull easier to remove as hull becomes brittle and cracks. Hulls from heat-treated grains can be removed easily by a dehuller. During heating the cotyledons shrink more than the hull and this also assists in easy removal of hull. Drying of pulses using heat hardens the grains and increases their resistance to scouring. Pigeon peas are dried from a typical moisture content of 10–11% to 6–7% to provide increased resistance to peripheral scouring during dehulling.

12.3.3.4 Chemical Treatment

Some chemical treatments have demonstrated the potential to replace oil treatment given before milling. These treatments have shown the same hulled grain recovery as traditional methods and in some cases more. Sodium bicarbonate, sodium chloride, urea, vinegar (acetic acid) and molasses have been used to treat pigeon pea grains. This treatment involve soaking of grains

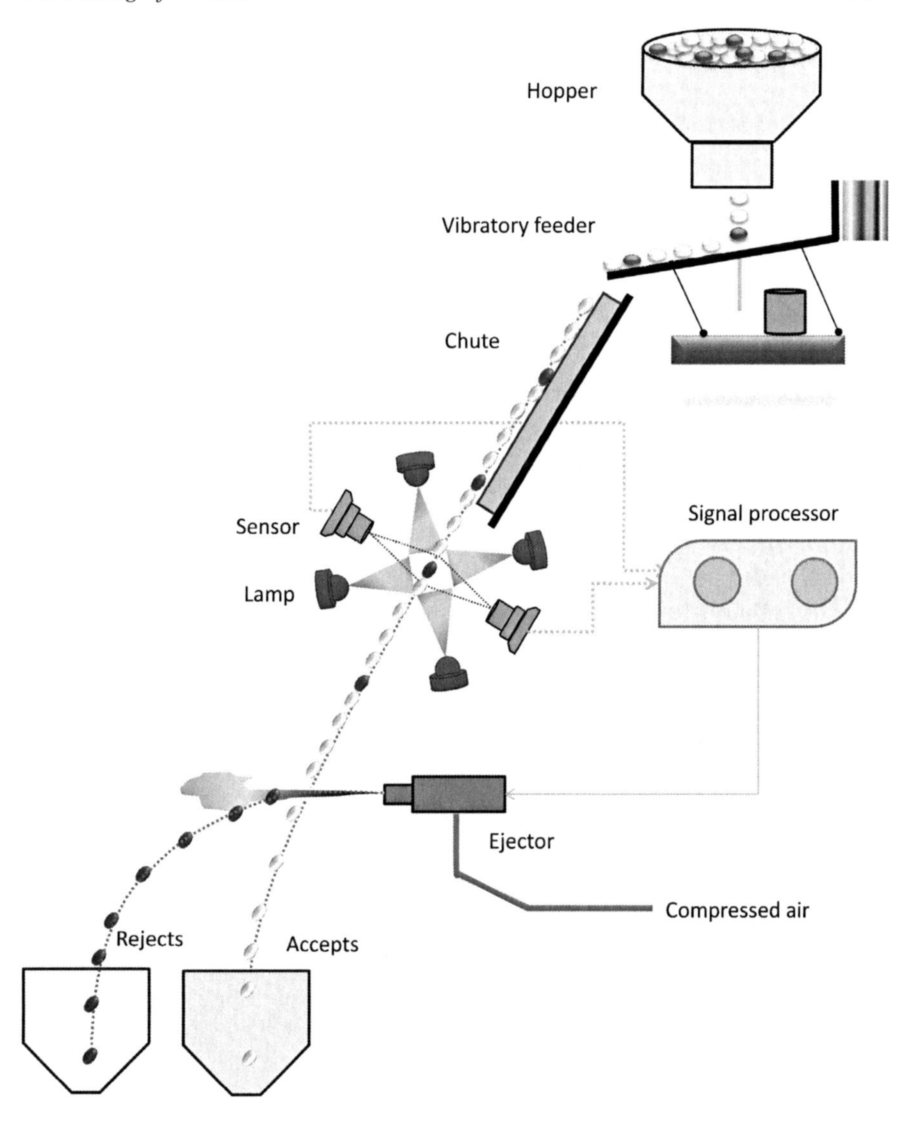

Figure 12.5 Working principle of colour sorter

in sodium bicarbonate solution (4–6%) for 0.5–1.0 hours followed by drying at 65 °C to 10–15% moisture content. This treatment gave hulling efficiency up to 95% and dhal yield of about 80%. Chemical treatments on pitted pigeon peas enhances dehulling rates by acting on the uronic acid present in gums and mucilages, which are known to be present between seed coat and cotyledon.[4] Studies have shown that acetic acid (vinegar) is effective for loosening the hull of pigeon peas. Application of vinegar followed by thermal treatment at 60 °C for 30 min results in dehulling of 77.5% of pigeon pea grains.[5] Use of chemicals

Table 12.5 Effect of various edible oils on milling fractions of black gram

	Dehulled (%)		*Unhulled (%)*			
Edible oil	*Whole*	*Splits*	*Whole*	*Splits*	*Broken%*	*H&P*
Control	10.5	13.1	23.5	39.2	3.1	5.2
Sesame oil	44.3	38.8	83.0	4.9	3.4	7.9
Groundnut oil	49.5	37.0	86.4	0.4	3.3	7.7
Palm oil	44.4	38.3	82.8	5.4	3.5	8.0
Coconut oil	44.8	37.1	81.9	5.3	3.3	7.7
Sunflower oil	41.8	35.6	77.4	7.8	3.3	8.1
Mustard oil	44.2	38.7	82.9	5.4	3.4	8.0
Castor oil	43.3	35.3	78.6	6.5	3.3	7.7

Source: Adapted from Tiwari *et al.*[25]
Control indicates no pre-treatment
H&P indicates hull and powdering losses

as pre-treatment process has been investigated but has not been commercially adopted and is yet to be developed as complete package. Chemical treatment with sodium bicarbonate, sodium carbonate, sodium hydroxide or acetic acid has already been tried to replace vegetable oil. Use of chemicals for pre-treatment of pulses is not advisable because a higher concentration of chemicals is required for improving the yield of dehulled grains, which may have an adverse corrosive effect on the milling machinery and also may affect the grain quality and feed quality of the hull (Tiwari *et al.*[25]).

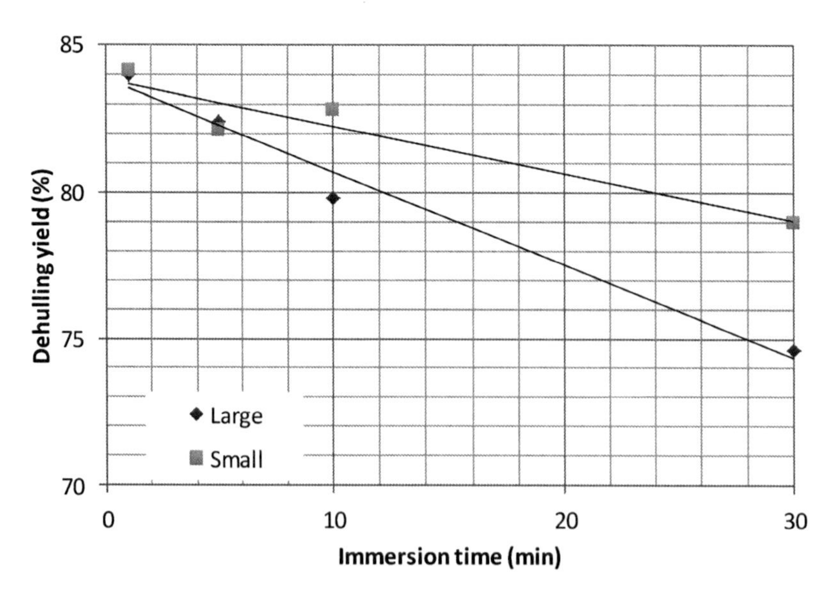

Figure 12.6 The effect of immersion time on dehulling efficiency of large (4.5 mm) and small (4 mm) lentil grains (Erskine *et al.*[3]).

Table 12.6 Basic features of wet and dry methods of pre-treatment

Advantages	*Disadvantages*
Dry method	
• Cooking quality of processed dhals is better	• More costly due to higher percentage of brokens
• Dehusked splits possess good appearance	• Requires longer processing time due to repeated drying and milling steps
• Cotyledons soften easily and quickly as compared to wet method	• This process involves application of oil which is dearer commodity
• Large quantity of dhal can be processed	• Requires less processing time (2–3 days) than wet method
• More advantageous process for dehusking	
• This process can be used for pigeon peas, green gram and black gram	
Wet method	
• 15 to 20% better dhal yield	• Cooking quality is poorer and takes a long time to cook
• Cooked dhals have better taste and flavours	• Dehusked splits give cupped shaped appearance due to small depression in the centre of dehusked splits
• Easy husk removal due to pre-soaking	• This process is laborious and mostly depends on climatic conditions
• Traditional *chakki* (sheller) is more suitable for wet processed grains	• Causes lot of processing losses during outdoor drying of grains
• More advantageous process for splitting of grains	• This process can only be applied to pigeon peas and not to other grains
	• Requires more processing time, usually 5–6 days

12.3.3.5 Enzymatic Pre-treatments

Enzymatic processing involves use of enzymes, which may break down hemicelluloses, pentoses, hexoses, *etc.* present in germ and seed coat, thus enhancing the ease of husk removal. The enzymatic approach for dehulling is yet to be developed as a complete package. Enzymes such as xylanase and protease can be used to dehull pulses such as green gram, black gram, red gram and horse gram. Partial hydrolysis of non-starch polysaccharides, gums and mucilages can be achieved by employing enzymes.[6,7] Xylanases aid in the degradation of non-starch polysaccharides and hence facilitate the easy dehulling of green gram, black gram and horse gram, whereas proteases act on cell wall proteins and the partial hydrolysis of these proteins facilitates the easy removal of seed coat.[8] Table 12.7 shows comparative yield of milling fractions subjected to enzyme treatment and oil pre-treatment. Green gram, black gram, red gram and horse gram were treated with xylanase in sodium acetate buffer at pH 5.0 followed by incubation period of 3 h at 50 °C. For protease treatment green gram, black gram, red gram and horse gram were treated in sodium phosphate buffer, pH 7.8, followed by an incubation period of 3 h at

Table 12.7 Comparative analysis of the yield of milling fractions of pulses subjected to oil and enzyme treatments

Dehulling treatment	*Dehulled kernels (%)*	*Undehulled kernels (%)*	*Fines (%)*	*Hulls (%)*
Green gram				
Oil treatment (0.3% peanut oil)	76.8	5.7	4.7	12.8
Protease treatment	83.3	1.9	1.5	13.4
Xylanase treatment	78.4	4.5	5.0	12.5
Black gram				
Oil treatment (0.3% peanut oil)	74.1	8.3	6.0	11.6
Protease treatment	82.8	3.3	2.8	11.2
Xylanase treatment	75.7	5.5	6.5	11.9
Red gram				
Oil treatment (0.3% peanut oil)	79.5	3.0	2.6	14.9
Protease treatment	78.4	4.7	4.7	12.1
Xylanase treatment	58.9	23.4	7.1	10.6
Horse gram				
Oil treatment (0.3% peanut oil)	69.4	13.0	3.8	13.7
Protease treatment	51.3	34.8	5.1	8.7
Xylanase treatment	84.4	1.6	1.5	12.5

Source: Adapted from Sreerama *et al.*[8]

37 °C. After equilibration, enzyme-treated grains are dried at 70 °C for 7–8 h until the grain moisture reaches 8–10%.[8]

12.3.4. Dehulling

Pre-milling treatment allows loosening of the hull and subsequent removal of hull and splitting into two cotyledons, which can be achieved either by shearing type disc shellers or the abrasion type roller machines. Various types of disc shellers are employed for dehulling of pulses. It should be noted that disc shellers cause excessive breakage in case grains which are not graded prior to dehulling; this also depends on uniformity of grains used for dehulling. Splitting can be achieved by using attrition or impact. Dehulling and splitting can also be achieved simultaneously. Splitting of pulses results in excessive losses in terms of powder and broken pulses due to scouring at the edges of pulses. Roller machines are normally used for dehulling of pulses. Roller machines work on the principle of abrasion. Commonly used roller machines have tapered or cylindrical emery coating revolving inside a metal cage fitted with a perforated sheet (Figure 12.7). Generally, cylindrical rollers have an inclination with a slope whereas tapered rollers have a horizontal setup. The horizontal or inclined arrangement helps in movement of grains for effective dehulling. Various factors including roller speed, emery grit size (*e.g.* surface of the roller), inlet and outlet rate of grains, annular gap between perforated sheet and the roller must be considered for effective dehulling of pulses. These machine parameters should be adjusted based on the physical properties of

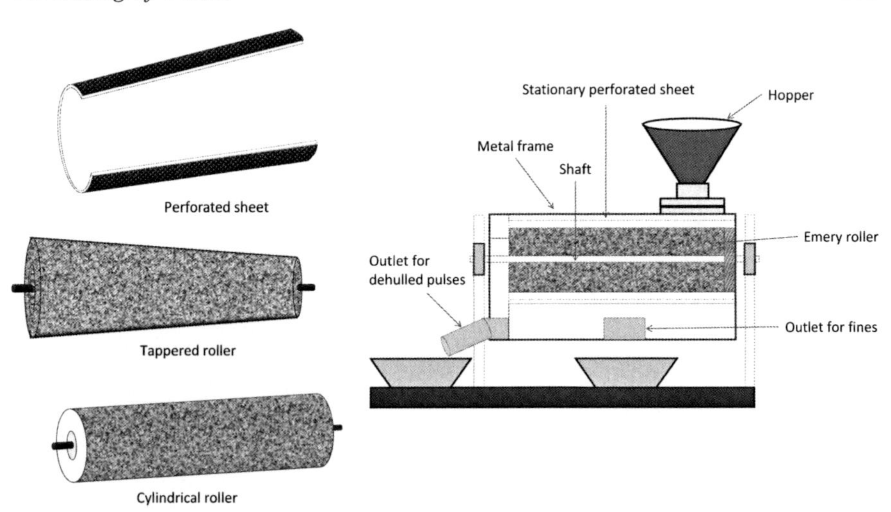

Figure 12.7 Abrasive roller machine along types of roller and perforated sheets housed inside the metal cage

grain. These emery rollers are also used for pitting or scarification of pulses (see dry milling). Optimum machine parameters should be adjusted based on the pulses. The difference between various dehulling and splitting approaches for pulses can be explained in terms of orientation of roller, type of roller and feed inlet. With a horizontal roller, pulses move in between perforated sheets and the emery roller. In case of the conical emery roller designed and developed by CFTRI, India, an emery-coated metal cone is fixed to a metal shaft inside a fixed conical perforated sheet, and both emery cone and sheet are concentrically housed in a metal cage (Figure 12.8). Grains are dehulled based on abrasive force. These machines are also known as pearling machines, working on the principle of abrasion. Surface and length of emery roller, peripheral speed of roller, clearance between perforated sheet and emery roller, inlet feed and discharge rate govern the dehulling efficiency. Pulses with a thin and loose seed coat require mild abrasive force to dehull which can be achieved by using rollers with slow speed and a fine emery-coated roller or cone. The tangential abrasive dehulling device (TADD) was developed at the Prairie Regional Laboratory (PRL) of the National Research Council of Canada in Saskatoon. TADD is generally used for testing and dehulling properties of pulses. A schematic of TADD is shown in Figure 12.9. It consists of a sample cup and abrasive disc mounted onto a motor. Samples are placed in a cup and closed using a lid provided.

12.3.5. Splitting of Pulses

Dehulling often results in a mixture of whole dehulled pulses and dehulled splits. Whole dehulled pulses are also known as "footballs" or "gota". Splitting of dehulled whole pulses is aimed at the production of perfect splits

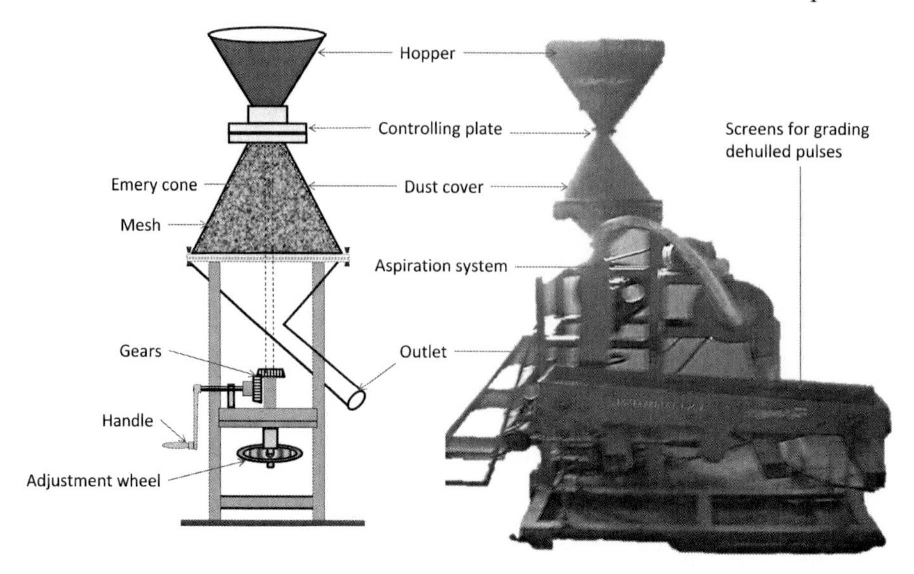

Figure 12.8 Manual and motorised pulse dehulling machine (CFTRI design) equipped with emery cone

without losses. Simultaneous dehulling and splitting can be achieved in the same milling process for most pulses. However, in some cases pulse processors prefer to split pulses depending on the demand. For example, in India black

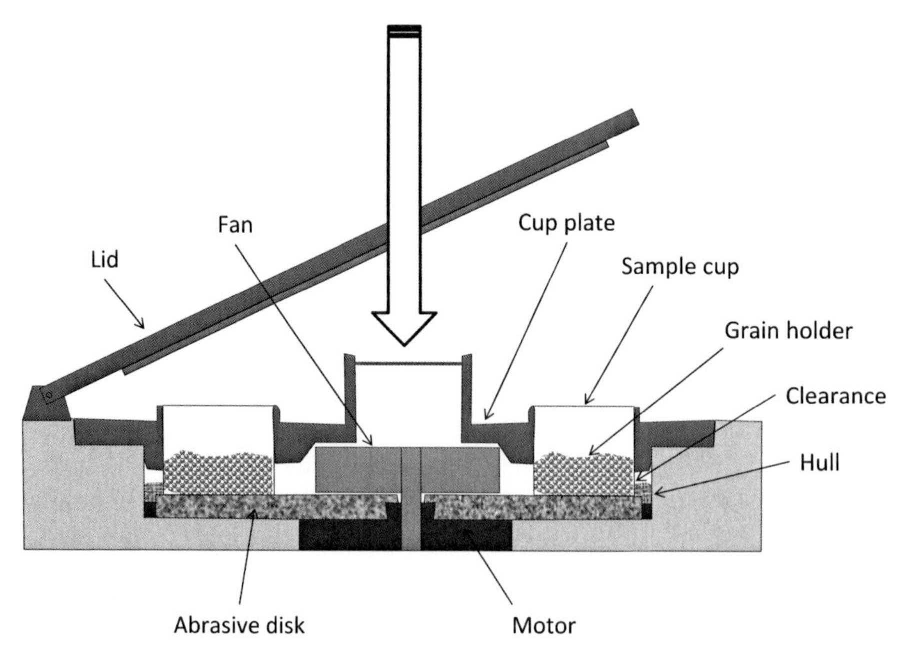

Figure 12.9 Tangential abrasive dehulling device (TADD)

gram is sold in various forms including whole dehulled (gota), dehulled splits (dhal), unhulled whole and splits (splitting without dehulling). Splitting can be achieved by using a roller machine, under-runner disc sheller, vertical or horizontal attrition mill and impact sheller. Roller milling machines similar to those for dehulling are used for splitting of pulses. A coarse emery roller is required for splitting of pulses. Horizontal or inclined rollers are used for splitting green gram, pigeon peas, lentils, chickpeas, *etc.* Machine parameters influencing dehulling of pulses also affect the splitting efficiency. Under-runner disc shellers are also used for splitting of pulses. In case of horizontal discs, the upper disc is stationary and the lower disc rotates to cause splitting of dehulled whole pulses. Under-runner disc shellers and horizontal or vertical attrition mills (Figure 12.10) are used for splitting of black gram, green gram, pigeon peas, lentils, chickpeas (mainly desi type), *etc.* Grading of pulses and clearance between stationary and rotating disc governs splitting efficiency. Breakage up to 40% can occur if grains are not graded according to size.

12.3.6. Separation and Grading

Various fractions are obtained during dehulling of pulses, namely dehulled splits, dehulled whole, unhulled splits, unhulled wholes, brokens, fines (powdering loss) and hull (Figure 12.11). Hulls and fines are separated by aspirators attached to the dehulling machine. Brokens can be separated by using sieve or disc separators. Brokens are generally mixed with powder and hull are used as cattle feed. Hull and fine particles can be separated by using gravity separator. In modern pulse mills a cyclone type separating system is used for improving separation efficiency and reducing dust pollution in milling premises. Unhulled splits and wholes are again passed through the dehulling system. However, efficient size grading and adequate pre-treatments ensure lower chances of unhulled fractions. Separation of the dehulled whole and

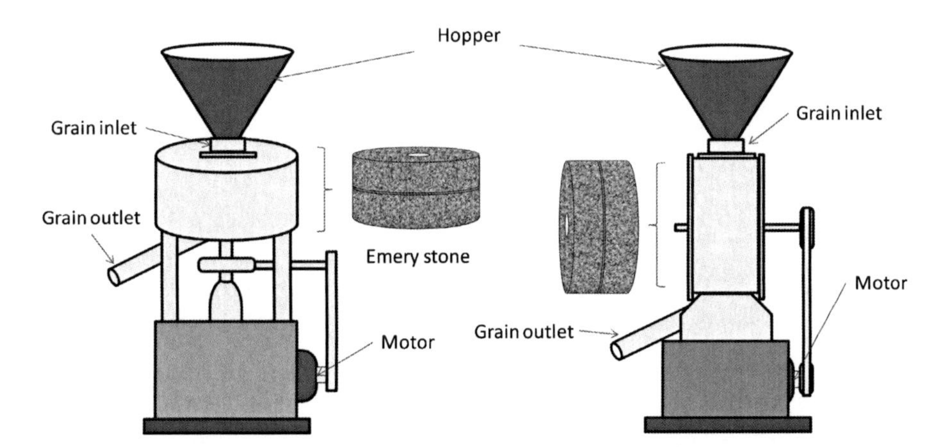

Figure 12.10 Motorised stone grinders with horizontal and vertical arrangements

Figure 12.11 Various dehulling fractions of pigeon pea

unhulled whole grains is quite tricky because they are the same size. Repeated exposure of unhulled and dehulled pulses causes excessive scouring on dehulled grains leading to powdering losses. Dehulled wholes have a tendency to bounce more than unhulled pulses, and this allows effective separation of grains. Grading and separation of various fractions of pulses is achieved by using machines similar to those used for grading of pulses (see section cleaning and grading). Appropriate combinations of screening sieves can be used for the separation of dehulling fractions depending on the grain dimensions.

12.3.7. Polishing

Sometimes dehulled splits or whole pulses are polished for value addition and trade, depending on consumer preferences. Generally, pulses are polished by mixing small amount of water, edible oil or both with pulses in a screw conveyor of various shapes as shown in Figure12.12. It is considered as the final stage in pulse dehulling prior to packaging.

12.4. FACTORS AFFECTING DEHULLING OF PULSES

The milling characteristics of pulses are influenced by a myriad of factors such as variety, climatic conditions and region. Larger or bold-grain varieties are easier to hull, give a higher yield and are preferred by millers, while the smaller varieties require repeated and severe pre-treatments and complex procedures. Freshly harvested grain and winter crops are more difficult to dehull, possibly because of their higher moisture content. Apart from pre-treatments and machine parameters, other factors such as seed characteristics also play important role in dehulling, in terms of loosening of hull, which in turn influences milling yield. Dehulling efficiency largely depends on physical properties and seed structure apart from the machine parameters employed and type of dehulling equipment used. The following factors have an impact on the dehulling yield:[9]

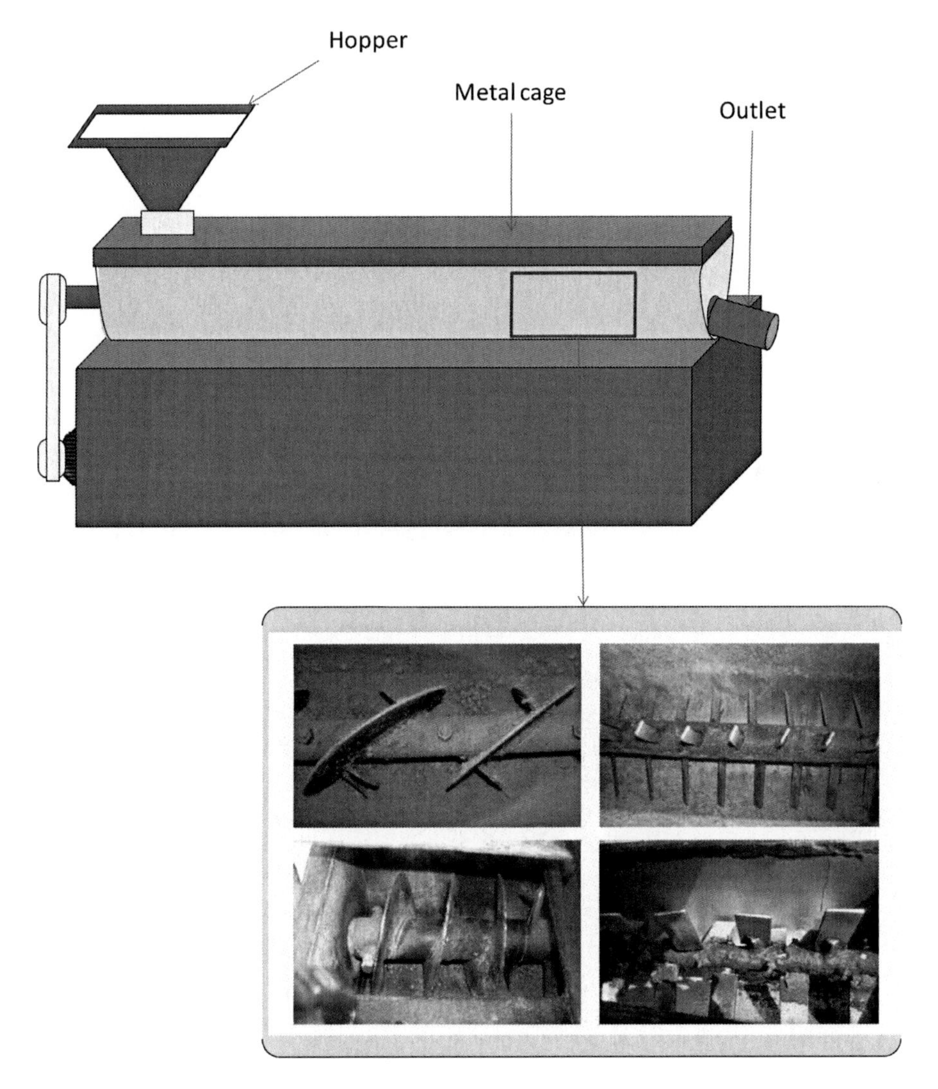

Figure 12.12 Conveyor machine along with various types of auger used for polishing of pulses

- The type of grain and its properties
- Bond strength between kernel and hull, strength of kernel, and strength of hull
- The proportion of hull in the kernel
- Grain size and uniformity
- Moisture content of the grain and the difference in moisture content between the hull and the rest of the kernel
- Pre-treatment given to the grain
- Proportion of dehulled kernel in the grain
- The ease of classification of dehulled kernels and hulls
- Dehulling machine parameters such as such as peripheral speed, abrasive surface, annular space.

12.4.1. Raw Material Characteristics

Large variability in dehulling characteristics is observed among pulses including green gram, cowpeas, chickpeas, lentils and pigeon pea cultivars. These variations are mainly due to grain properties (see Chapter 3). Variability in dehulling characteristics can be reduced by using appropriate grading systems and largely by using cultivars with fewer variations in grain characteristics. Plant breeders are constantly developing cultivars with synchronous maturity to avoid intravarietal variations.

12.4.2. Size and Shape of Grains

As discussed earlier in Chapters 2 and 8, there is wide variation in grain shapes and sizes. Inter- and intravarietal variations are usually due to unsynchronised maturity of pods. Variation in pulses is mainly due to their unique genetic characteristics such as annular, oblong, round, trapezoidal or elongated shapes and various sizes of grains such as small, medium and large. Such variations in grain size pose problems in designing a suitable dehulling machine. Seed size and shape is a varietal characteristic, but may be influenced by agronomic practices, growing season and harvesting time. Dehulling and splitting are both affected by size and shape of grain; bold grains split more efficiently than smaller grains whereas bolder grains are more prone to milling losses in terms of powder and brokens, if milling machinery is not properly adjusted.[10] Shape is a varietal characteristic and unlike grain size, it is not much influenced by agronomic practices. Spherical or globular grains are easy to dehull and easy to split as compared to oblong, irregular or angular shaped grains because sharp edges in grains may lead to damage to grain kernel and yield into brokens and powders. Kibbling is often the result of irregular shapes in case of splitting of black gram.

12.4.3. Hull Characteristics

Generally variation in hull content and thickness of hull is not seen within a variety, but there are wide differences between species and varieties, *e.g.* pigeon peas (14–15.5%), black gram (12–12.5%), green gram (10–12%). Dehulling yield is directly proportional to hull content and thickness.

12.4.4. Adhesion Between Husk and Cotyledons

Dehulling of pulses such as pigeon peas, black gram and green gram is more difficult because of the presence of a gum layer between the seed coat and the cotyledons, which causes strong bond attachment between the hull and cotyledons. The nature of gums and mucilages present in this layer is the main factor for strong bonding. Layer of gums and mucilages are characterized by the presence of water-soluble and water-insoluble non-starch polysaccharides, lignin, hemicelluloses, celluloses, sugars such as pentose and hexoses, uronic acid. During germination or soaking, the vitreous layer of gums and mucilages weakens due to solublisation of water-soluble polysaccharides and sugars. The gum layer is very strong in pigeon peas and black gram as compared to bengal gram and green gram.

12.5. DEHULLING CALCULATIONS

Dehulling efficiency

Over time, several approaches have been developed to calculate the milling efficiency. However, the descriptive terms of the calculations often vary. Dehulling efficiency (*DE*, Equation 12.1),[11,12] sometimes called degree of dehulling or dehulling index, is essentially the yield of dehulled product material (dehulled whole seed and dhal) as a percentage of original seed weight:

$$DE(\%) = \frac{DWS + D}{w_t} \times 100 \tag{12.1}$$

where DWS is the weight of dehulled whole pulses (g), D is the weight of dehulled splits (g) and W_t is the original weight (g).

The dehulling index, sometimes called the pearling index, varies from a maximum of +1 (100% seed is dehulled with no fines or undehulled grain) to a minimum of -1 (indicates 100% unsuccessful result, with all seed undehulled or broken into fines).[13] A value of -1 indicates that the dehulling is not complete, thus the grains are either not dehulled properly or yielded to broken and powdering loss.[13]

Dehulling index (η)

The dehulling index may vary from a maximum value of +1 to a minimum of -1. A value of +1 indicates that all the samples are completely dehulled.

$$\eta = \frac{(w_c + w_h) - (w_{uh} + w_f)}{w_t} \tag{12.2}$$

where W_c is the weight (g) of cotyledons and broken cotyledons, W_h is the weight (g) of removed hulls, W_{uh} is the weight (g) of kernels that remained undehulled, W_f is the weight (g) of fines in the final product, and W_t is the total weight (g) of original grain fed into the dehuller.

The dehulling index can also be calculated by using the following Equation 12.3:[9,12]

$$DI = \frac{100 \times HR}{AF} \qquad (12.3)$$

where AF denotes abraded fines (Equation 12.4) and HR is hull removal (Equation 12.5):

$$AF = \frac{100 \times (W_1 - W_2)}{W_1} \qquad (12.4)$$

$$HR = 100 - \frac{100 \times (W_3)}{W_4} \qquad (12.5)$$

Here W_1 is the weight (g) of original sample placed in each TADD cup, W_2 is the weight (g) of partially dehulled grains, W_3 is the weight (g) of hull and W_4 is the weight (g) of fully dehulled grains.

The splits yield (SY) (Equation 12.6), also known as dhal yield, is essentially the yield of dehulled and split product material (dhal) as a percentage of original grain weight:[11,12]

$$SY(\%) = \frac{100 \times D}{W_t} \qquad (12.6)$$

where D is the weight (g) of dehulled splits and W_t is the original weight (g) of grain. This calculation can be adapted to calculate percentage of other desired milled products such as dehulled whole seeds.

The broken percentage or kibble (fines or fragment of cotyledon) can be calculated by using Equation (12.7). It is essentially the yield of kibble or broken product material as a percentage of the original seed weight.[11,12]

$$B(\%) = \frac{100 \times K}{W_t} \qquad (12.7)$$

where K is the weight (g) of kibble or brokens and W_t is the original grain weight (g).

Milling parameters can also be calculated by using the following equations

Coefficient of Hulling (C_h):

$$C_h = 1 - W_{uh}/W_{th} \qquad (12.8)$$

Coefficient of wholeness of kernel (C_{wk})

$$C_{wk} = W_{fp}/\left(W_{fp} + W_{br} + W_{po}\right) \qquad (12.9)$$

Degree of dehulling (M_h): The degree of hull removal is the ratio of weight of hull removed during dehulling to the initial weight of sample used for the dehulling process.

$$M_h = W_h / W_{th} \quad (12.10)$$

Effectiveness of dehulling (E_d): Effectiveness of dehulling is the ratio between the weight of the material remaining unhulled and the initial weight of material taken for dehulling.

$$E_d = W_{uh} / W_{th} \quad (12.11)$$

Hulling efficiency (HE):

$$HE = C_h \times C_{wk} \times 100 \quad (12.12)$$

Efficiency of dehulling (η_d):

$$\eta_d = \left(W_{fp} / W_{ty}\right) \times 100 \quad (12.13)$$

Overall dehulling efficiency (η_o):

$$\eta_o = (M_h + Q_d) \times C_h \times 100 \quad (12.14)$$

where Q_d is the quality of dehulling, which is expressed as a ratio between the weight of dehulled kernel (both split and whole) and initial weight of material taken for dehulling. A higher value of overall dehulling efficiency is considered as the index of a better process.

In the above Equations 12.8–12.14, W_{uh} is the weight of unhulled grain after milling, W_{th} is the weight of grains used for milling, W_{fp} is the weight of finished product (splits and whole dehulled kernel), W_{br} is the weight of brokens, W_h is the weight of hull, W_{po} is the weight of powder and W_{ty} is the theoretical yield of the test sample.

The efficiency of dehulling is a ratio between amount of dehulled grains obtained, both splits and whole, to that of theoretical yield. Theoretical yield varies depending upon type of grain, hull, cotyledon and germ proportion in a grain. This parameter can be obtained by taking 10 g of sample, manually removing the hull using sharp knife and weighing each fraction separately. Value of efficiency of dehulling can vary from 0 to +1: 0 means there is no dehulling and value near to 1 shows complete dehulling without any broken and powdering loss. When manual dehulling of the grains is done, it is usually reported as the mean value of triplicate. If in a milling trial, manual dehulling of pigeon peas as a test grain sample shows 14.5% hull, 83.2% cotyledon and 2.3% germ, the theoretical yield of pigeon pea will be 85.5%.

12.6. EFFECT OF DEHULLING ON QUALITY OF PULSES

12.6.1. Nutritional Quality

The main objective of pulse milling is to remove the hull. This has great significance, considering that removal of the hull helps to reduce antinutritional factors, such as tannins and insoluble fibre (non-nutrients that can bind protein and other nutrients), thereby improving nutritional quality, protein digestibility, texture and palatability. Further, dehulling also removes astringent taste caused by tannins and allows the production of higher quality flours, without browning/speckling. Soaking of grains in water or acidic media may cause formation of insoluble higher oligomeric polymers. Water soaking has been shown to have no effect on total extractable phenolics in mung bean, while alkali and acidic soaking, show no effect on total extractable phenolics in brown kidney bean.[14] It should be noted that polyphenols have contradictory positive effects on human health and it has been reported that these have anticarcinogenic and antioxidant properties. Looking at the other side of the coin, even if thye are considered antinutritional, the notable fact is that cooking results in significant reduction in phytic acid and tannins in pulses.[15] Anton *et al.*[16] have observed higher antioxidant activity associated with the seed coats of heat-treated pinto beans, which was in turn correlated with the total phenolic content. The digestibility of pulse proteins also increases during the heat treatment process, though to a lesser extent. To cite an example, the true protein digestibility (TPD) and net protein utilisation (NPU) of dehulled components significantly increases, when compared to whole seeds of pigeon pea.[17] On the other hand, the dehulling process, without exception, removes a major portion of the germ along with the hull, resulting into the loss of proteins and vitamins. In some studies it has also been seen that dehulling causes a decrease in the calcium and iron content of the splits. It is worthwhile mentioning that even milder approaches, such as scouring, lead to such losses. To justify this, Singh *et al.*[18] have observed that protein, soluble sugars, and ash of scarified cotyledons decreased with increased scarification time. In wet methods, water-soluble nutrients are also lost in the soak water. Excessive abrasive forces can cause surface scouring of cotyledon and may result in protein losses (Figure 12.13). This is because proteins in pulses are mainly concentrated in the peripheral layers of cotyledons. Figure 12.9 shows a decrease in protein content of dehulled pigeon pea splits and an increase in protein content of powder using the TADD.[18]

Additionally, optimum hydrothermal treatment is necessary to obtain good quality dhal. Excessive water treatment has been reported to induce cupping in pulses, which lowers the quality drastically. To summarise, dehulling results in a significant changes in protein, starch, resistant starch, K, P, phytic acid, stachyose and verbascose content. However, significant decrease in soluble dietary fibre, insoluble dietary fibre, Ca, Cu, Fe, Mg, Mn and tannin content can also be observed.[19]

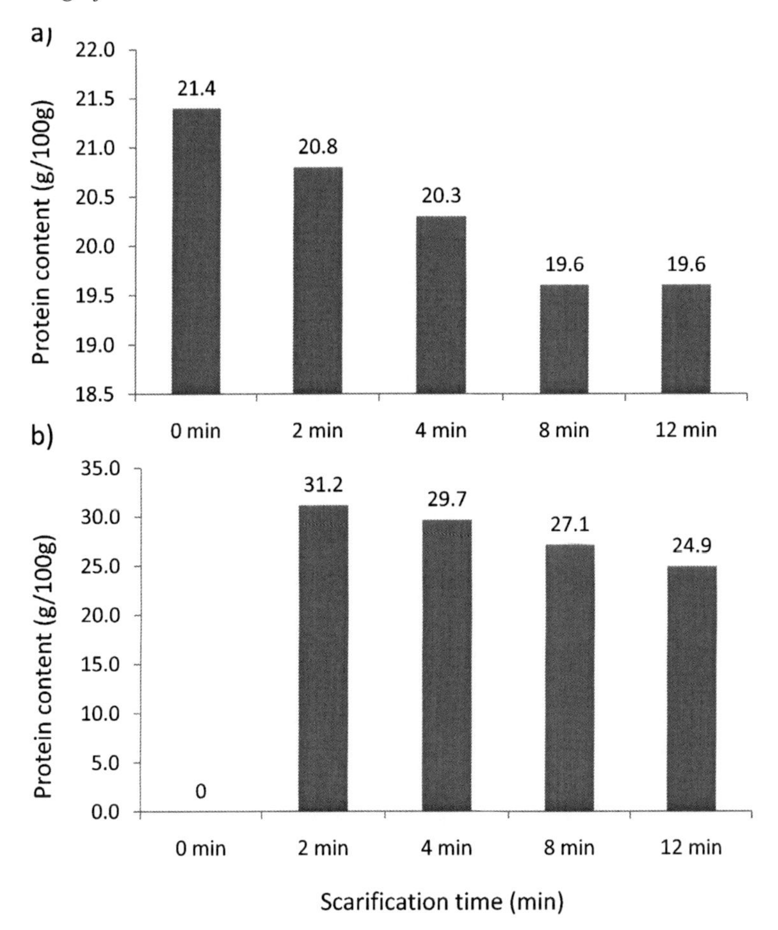

Figure 12.13 Effect of scarification on the chemical constituents of dehulled cotyledons and powder fractions of pigeon peas. Scarification time of 0 min indicates manual dehulling

12.6.2. Cooking Quality

Dehulling and splitting invariably improves the cooking quality of pulses, with some negligible loss of solids. Variation in cooking quality in whole grain from different genotypes can be reduced following splitting. Decortication and splitting reduces the cooking time. This is because dehulling removes the impermeable seed coat of pulses, which hinders water uptake during cooking. Further, similar benefits of reduction in cooking time can also be obtained by pre-soaking pulses (at least dry peas, chickpeas and lentils) in salt solutions (10 or 20 g kg^{-1}), particularly with sodium tripolyphosphate and sodium bicarbonate.[20] However, this may not be always true; sodium bicarbonate has been found to increase the cooking time in faba beans.[21] Enzyme pre-treatments have been found to be promising for dehulling, considering that they do not alter the cooking properties,[8] but the cost of the treatment remains a major question. The comment made on the enzymatic pre-treatment stems from the work of Sreerama *et al.*[8] who found xylanase-mediated degradation of cell wall polysaccharides of horse gram to result in expansion of the grain with improved nutritional and functional properties upon thermal treatment.

REFERENCES

1. P. P. Kurien, *Res. Ind.*, 1984, **29**, 207–214.
2. P. P. Kurien and H. A. B. Parpia, *J. Food Sci. Technol.*, 1968, **5**, 203–207.
3. W. Erskine, P. C. Williams and H. Nakkou, *J. Sci. Food. Agric.* 1991, 57, 77–84.
4. P. S. Phirke and N. G. Bhole, *Int. J. Food Sci. Technol.*, 1999, **34**, 107–113.
5. K. Krishnamurthy, G. K. Girish, T. Ramasivan, S. K. Bose, K. Singh and T. Singh, *Bull. Grain Technol.*, 1972, 10, 181.
6. G. Arora, V. Sehgal and M. Arora, *J. Food Eng.*, 2007, **82**, 153–159.
7. P. Verma, R. Saxena, B. Sarkar and P. Omre, *J. Food Sci. Technol.*, 1993, 30.
8. Y. N. Sreerama, V. B. Sashikala and V. M. Pratape, *J. Food Eng.*, 2009, **92**, 389–395.
9. A. Ehiwe and R. Reichert, *Cereal Chem.*, 1987, **64**, 86–90.
10. H. V. Narasimha, N. Ramakrishnaiah and V. Pratape, in *Handbook of Postharvest Technology: Cereals, Fruits, Vegetables, Tea, and Spices*, ed. A. Chakraverty, A. S. Mujumdar and H. S. Ramaswamy, Marcel Dekker, New York, 2003, pp. 427–452.
11. P. Burridge, A. Hensing, D. Petterson, *Australian Pulse Quality Laboratory Manual.* SARDI Grain Laboratory for GRDC, Urrabrae, SA, 2001.
12. J. A. Wood and L. J. Malcolmson, in *Pulse Foods*, ed. B. K. Tiwari, A. Gowen, and B. McKenna, Academic Press, San Diego, CA, 2011, pp. 193–221.
13. J. Ikebudu, S. Sokhansanj, R. Tyler, B. Milne and N. Thakor, *Can. Agric. Eng.*, 2000, **42**, 27–32.
14. E. E. Towo, U. Svanberg and G. D. Ndossi, *J. Sci. Food Agric.*, 2003, **83**, 980–986.
15. N. Wang, *J. Sci. Food Agric.*, 2008, **88**, 885–890.
16. A. A. Anton, K. A. Ross, T. Beta, R. Gary Fulcher and S. D. Arntfield, *LWT – Food Sci. Technol.*, 2008, **41**, 771–778.
17. U. Singh, *Plant Foods Hum. Nutr.*, 1993, **43**, 171–179.
18. U. Singh, P. V. Rao, R. Seetha and R. Jambunathan, *J. Food Sci.*, 1989, **54**, 974–976.
19. N. Wang, D. W. Hatcher and E. J. Gawalko, *Food Chem.*, 2008, **111**, 132–138.
20. R. G. Black, U. Singh and C. Meares, *J. Sci. Food Agric.*, 1998, **77**, 251–258.
21. K. Singh, W. Erskine, L. Robertson and P. Williams, *J. Sci. Food Agric.*, 1988, **44**, 135–142.

22. P. Kurien, in *Advances in Milling Technology of Pigeonpea*, Proc. Int. Workshop Pigeonpeas, 1980, ICRISAT Center, Patancheru, AP, December 1980, pp. 321–328.
23. S. Sokhansanj and R. T. Patil, *Handbook of Post-Harvest Technology: Cereals, Fruits, Vegetables, Tea, and Spices*, Marcel Dekker, New York, 2003, 397–420.
24. G. Milne, in *Pulse Cleaning and Processing Workshop*. Extension Division, University of Saskatchewan, Saskatoon, 1995.
25. B. K. Tiwari, R. J. Mohan and B. Vasan, *J. Food Process. Preserv.*, 2008, **32**, 610–620.
26. Saskatchewan Pulse Growers Pulse Production Manual Pulse production (2000) available online http://www.saskpulse.com/media/pdfs/ppm-general-production.pdf (accessed 12/01/2012).

CHAPTER 13

Pulse Products and Utilisation

13.1. INTRODUCTION

Processing of pulses, as mentioned earlier (Chapter 12), can be best studied under the classification of primary, secondary and tertiary processing. Primary processing of pulses and production of pulse flour has already been discussed, and in conjunction with pulse chemistry lays the foundation for the present chapter on utilisation of pulses and product development, commonly designated as tertiary processing.

It is a well-known fact and commonly observed scenario that the majority of pulse grains are consumed as whole or as split dhals (supporting data available in Chapter 11). However, the utilisation of pulses may not be just restricted to these channels. In fact, most of the fundamental principles of food processing and product development hold good for processing of most pulses. For example, drying, cooking, canning, extrusion, fermentation and puffing have all been applied with great success over periods for product development from pulses (Figure 13.1). Broadly speaking, the utilisation pattern of pulses can be grouped as follows.

- Green pulses (fresh, boiled and roasted)
- Sprouted and germinated (boiled and fried)
- Puffed and roasted (spiced/salted)
- Milled and cooked (steamed, boiled and fried)
- Fermented products (idli, dosa, dhokla, *etc.*)
- Canned pulses (baked beans).

The major underpinning reasons for processing of pulses may be summarised as follows:

- It renders palatability and better acceptance to the pulses.
- The antinutritional factors inherent to most pulses are considerably reduced and hence, improved nutrition can be achieved.

Pulse Chemistry and Technology
Brijesh K. Tiwari and Narpinder Singh

Published by the Royal Society of Chemistry, www.rsc.org

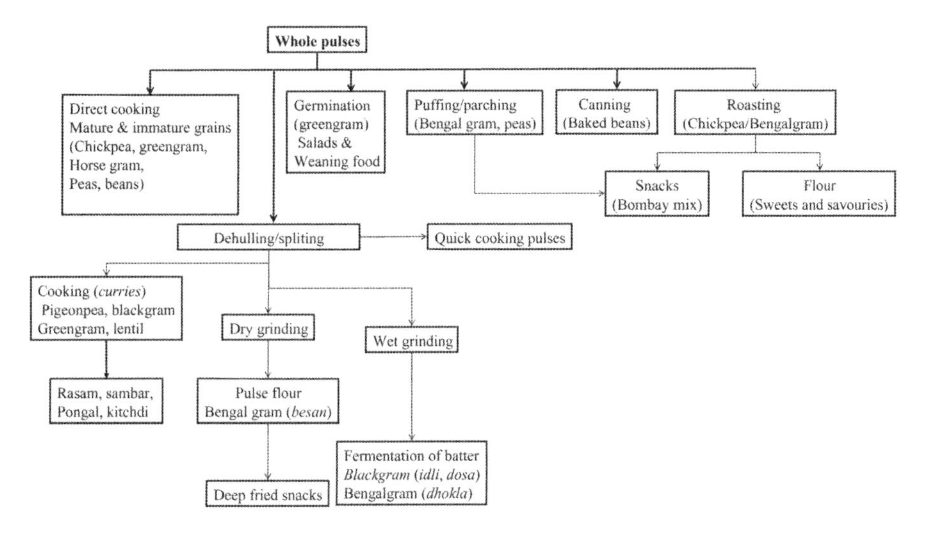

Figure 13.1 Pulse-based products

- The functional properties of the pulses can be utilised when tertiary processing is undertaken.
- The availability of pulses can be expanded to greater global horizons, for processed pulses (such as canned pulses) are easy to export and market. Such activities aid in value addition, although to a moderate extents.

A myriad of pulse-based processed products can be recorded in most pulse-growing nations, which includes mainly developing countries. However, the processing and utilisation of pulses has recently drawn the attention of the food industries of developed nations. The driving force for the increased utilisation of pulses in novel product development and as a food ingredient is primarily due to their nutraceutical and functional properties and the gluten-free nature of pulse flours. The significant change that could accrue from processing of pulses is that it could promote the consumption of pulses. It should be recalled that we have emphasised the declining trend of pulse consumption in the developing countries, earlier in our discussions. Table 13.1 provides an overview of the products developed or method of processing used for consumption of some common pulses. With this introduction, the later sections of this chapter focus on the home-scale and large-scale secondary and tertiary processing of pulses and pulse products.

13.2. COOKING OF PULSES

Cooking of most pulses is often preceded by a soaking step, wherein water gets distributed within the starch and protein matrix of each grain. Thus, water plays an essential role in the starch gelatinisation and protein denaturation during cooking. Soaking time usually varies among pulse species with an

Table 13.1 Method of utilisation of some common pulses as food

Pulse grain	*Utilisation*
Chickpeas	Dehulled pulses are consumed after boiling or frying or as crushed and cooked snack Used in curries Mature grains are used as whole seed or splits (dhal) or flour. Chickpea flour is used for making fermented dish dhokla and in a number of fried products/snacks
Black gram	Used as whole seed or dehulled splits (cotyledons) for making curry Extensively used in fermented products like idli and dosa Black gram flour is also used for making pancakes and papad
Green gram	Used in curry Sprouted beans are often consumed as snacks Roasting and frying of whole grains or splits (cotyledons) is popular in India to prepare snack products which are usually salted and spiced
Pigeon peas	Primarily used for making curries, *e.g.* cooked splits and sambar (cooked and spiced splits). Sambar is soup-like in consistency and appearance and is prepared from dehulled cotyledons that are cooked until tender and seasoned with spices. Sambar contains added vegetables (optional) such as eggplant, drum stick, carrots, tomato, potatoes, green peas, *etc.*
Peas, lentils, kidney beans, *etc.*	Generally used as whole seed for making curries Whole lentils and pea are used for making snacks after frying. Generally salted and spiced Dehulled beans are used for making flour

average time of order up to 10–20 hours. The hydration of pulses is a diffusion-driven mass transfer process, which facilitates use of any of the general principles enhancing diffusion to be applied for decreasing the soaking time. After soaking, grains become suitable for cooking. Cooking as a thermal process tenderises the pulse grains by facilitating deformation of starch and coagulation of proteins. Further, it also decreases the antinutritional factors (emphasised in earlier chapters with supporting numeric data) and improves nutritional value. Pulses are commonly cooked in boiling water for about 1–3 hours depending on the pulses. Soaked peas can be cooked in shorter times of about 50 min in boiling water. The size, composition and structure of pulse grain, water absorption and rate of heat transfer, have a large influence on the cooking time and quality of the cooked grain. Differences in gelatinisation pattern of the starch and the susceptibility of the cell constituents, notably the protein matrix, to softening may contribute to the overall textural characteristics of the cooked pulse grains.

13.3. QUICK-COOK DEHYDRATED PULSES

One major drawback associated with pulses which is responsible for their low acceptability among today's fast-moving consumers is their longer cooking time. In fact, pulses need long soaking times prior to cooking, though methods such as pressure cooking are also used alternatively. In this regard, the application of combined microwave and convective hot air drying of pre-cooked chickpeas has been shown to be promising.[1,2] The combination of high air temperatures with high levels of microwave power is an efficient method of dehydration, leading to faster drying and lower dried product moisture content, due to internal fluxes of water heated by microwaves and fast removal of surface water by high temperature convective currents. Drying chickpeas and soybeans to a water activity of 0.35 has been found to produce shelf-stable dry products. These quick-cook pulses rehydrate to moisture contents comparable with those of freshly cooked grains, with insignificant differences in terms of quality parameters.

13.4. CANNED PULSES (BAKED BEANS)

Canning is a heat sterilisation process during which all living organisms in food are killed to assure that no residual organisms can grow in the can. Properly sealed and heated canned foods should remain stable and unspoiled indefinitely in the absence of refrigeration. The hermetic sealing is critical to canning and application of heat is done under pressure for a specific temperature–time combination. During canning of most pulses, soaked or blanched beans are filled into cans into which tomato sauce (for baked beans) or salted water (for other bean varieties, *e.g.* chickpeas) is added. During soaking, dry beans should increase 80% in mass and reach 53–57% moisture content. The main purpose of blanching is the inactivation of enzymes, which might produce off-flavours, but also to soften the product and remove gases to reduce strain on can seams during retorting.[74] After filling, cans are heat-seamed and then thermally processed under pressure, during which the canned beans are cooked. Sterilisation is done in static retorts, agitating retorts or hydrostatic sterilisers. Rotation increases the rate of heat transfer, thereby reducing processing time and the gelation tendency of the sauce. The industrial canning process involves soaking of grains for 12–14 hours followed by blanching in water containing calcium chloride (10–15 ppm) at 85–90 °C for 5–8 minutes. The blanched grains are then filled in cans along with tomato sauce in equal amount and sealed followed by heat sterilisation at 121 °C for about 50 minutes. Texture is often used as an indication of the degree of consumer acceptance of canned beans,[3,4] as it affects the perceived stimulus of chewing.[3]

Baked beans are the most popular commercial canned product made from haricot beans, also known as navy bean – a variety of *Phaseolus vulgaris*. Baked beans in cans are a convenience food or ready-to-eat food consumed directly without heating or requiring a short heating time before consumption. Beans are processed in different cooking mediums such as brine or sauce of different types. A tomato and sugar based sauce is most commonly used for baked beans in the UK and Ireland. There are multiple styles of baked beans which are produced in different countries, for example Boston baked beans, a type popular in the USA, is prepared in a sauce made from molasses and salt pork. Beans cooked in tomato and brown sugar, sugar or corn syrup sauces are some other types. Maine and Quebec-style beans are made with maple syrup.

13.5. GERMINATED PULSES

Germinated pulses, also known as sprouted pulses, are marketed in most Asian countries in unit packs. Germinated pulses are obtained by germinating soaking grains under controlled temperature and relative humidity conditions. The germination process is particularly advantageous because it improves the nutritional quality of pulses. Germination increases vitamin concentration, bioavailability of trace compounds and minerals while reducing the antinutritional factors such as stachyose, verbacose and raffinose[5] as shown in Figure 13.2. Figure 13.2 shows the kinetics of reduction in antinutritional oligosaccharides over a germination period of 72 hours. It has been observed

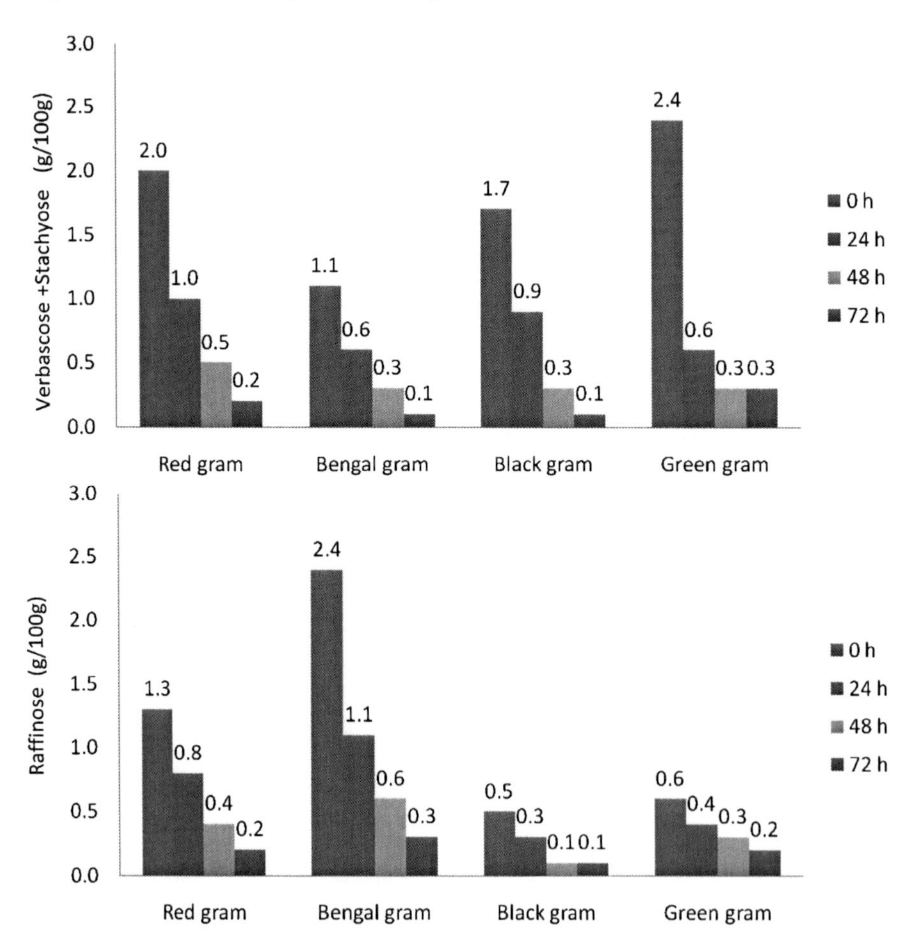

Figure 13.2 Effect of germination time on oligosaccharide content of some common pulses (Rao and Belavady)[75]

that germinated pulses contain significantly higher amounts of carotene, tocopherol, ascorbic acid, thiamine, riboflavin, pantothenic acid, biotin, niacin and choline than ungerminated pulses.[6] Following several outbreaks of foodborne illness associated with their consumption, sprouts have been classified as a source of foodborne microorganisms such as *Salmonella* and *Escherichia coli* O157. This can be linked to the optimal conditions for growth of microbes, which prevail during germination conditions. Germination under unhygienic and high temperature is responsible for foodborne illness.

13.6. PULSE PASTES

Pulse pastes are used for stuffing various products in the bakery industry (*e.g.* adzuki bean paste). Dried pastes are often used as extenders in the processing of meat products. The processing of adzuki paste includes soaking the beans, cooking to make them soft, grinding, and filtering through a pulp finisher to separate pulp from starch, washing the starch with water at least twice, pressing in a bag to separate water, and then cooking with sugar and oil. The processing of mung bean paste, mung bean starch vermicelli, pea paste and kidney bean paste are associated with similar problems to those in making adzuki bean paste.

13.7. ROASTED AND PUFFED PULSES

Roasting is a cooking method that utilises dry heat, whether an open flame, oven, or other heat source. Roasted chickpea (leblebi) is traditional snack food and is popular in the Middle East and Asian countries.[7] Roasted chickpeas may be dehulled or non-dehulled. The steps for all different kinds of roasted chickpeas (leblebi) production can be summarised as (1) cleaning and grading, (2) soaking, (3) tempering (preheating and resting), (4) boiling, (5) resting, (6) roasting and (7) dehulling. The roasting temperature can vary from 80 °C to 130 °C. Tempering in production of roasted pulses refers to holding the seed (resting) after heating to allow moisture penetration and stabilisation. These steps are summarised in Figure 13.3.

Puffed chickpeas are commonly used as snacks. For puffing, chickpea seeds are uniformly mixed with a calculated amount of water to bring the moisture content of seed to 12.5 $\pm$ 0.3% and rested for 10 min. Thereafter, it is roasted in hot sand for 20 s at 190 °C. Then the sample is taken out and rested for a period of 5 min and again roasted for 20 s in hot sand (190 °C). The roasted seeds are separated from the hot sand using a sieve (around 35 mesh size). The moisture content of roasted samples reaches around 3%. Seed volume, puffing capacity and puffing index are common quality attributes of puffed grains.[8] The expansion index or puffing index of roasted seeds can be determined using the following expression:

$$Puffing\ Index = \frac{Volume\ after\ roasting}{Volume\ before\ roasting}$$

Puffing capacity is associated with the textural properties of roasted pulses and therefore can be measured by employing a texture analyser. The percentage of

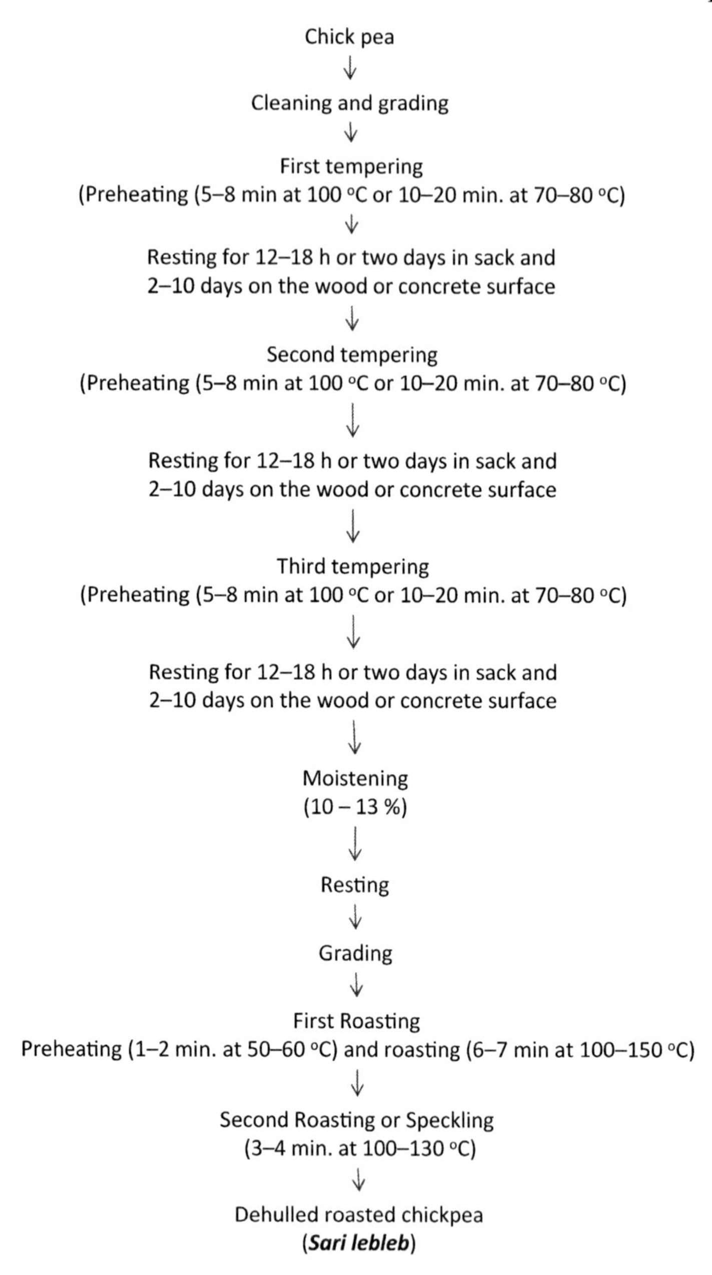

Figure 13.3 Flowchart for the production of dehulled roasted chickpeas (sari leblebi) Coskuner and Karababa[76]

unpuffed (hard shelled) grains can be determined manually by counting the number of unpuffed grains from a known sample. Roasting characteristics of different chickpea cultivars are shown in Table 13.2.

Table 13.2 Roasting characteristics of different chickpea cultivars

Cultivars	*Puffing capacity/100 seeds*	*Puffing index*	*Expansion index*	*Breaking strength (kg)*	*Hardshelled grains (%)*
Chickpea (desi type)					
PBG-1	5.2 ± 0.45	53.1 ± 1.10	1.53 ± 0.058	3.46 ± 0.32	46.0 ± 2.50
PDG-4	4.0 ± 0.43	31.7 ± 1.08	1.32 ± 0.062	4.17 ± 0.48	59.6 ± 2.30
PDG-3	6.0 ± 0.36	54.5 ± 1.24	1.55 ± 0.059	2.82 ± 0.36	42.1 ± 2.09
GL-769	4.0 ± 0.41	39.2 ± 0.98	1.39 ± 0.071	2.83 ± 0.39	54.0 ± 2.45
GPF-2	5.1 ± 0.39	51.0 ± 1.09	1.51 ± 0.067	2.54 ± 0.46	50.9 ± 2.39
Chickpea (kabuli type)					
L-550	2.5 ± 0.29	14.7 ± 1.23	1.15 ± 0.055	9.68 ± 0.42	69.2 ± 1.98

Source: Kaur *et al.*[8]

13.8. SNACK FOODS

Various snack products can be obtained from pulses. It has been shown that chickpeas can be texturised to obtain highly acceptable snack products.[9] Although the nutritional quality is generally high, the acceptability of pulse-based extruded products is usually far from that of traditional commercial ones. Extruded chickpea and bovine lung mixtures have also been used to produce snack foods, and these possess comparable attributes in terms of sensory quality to commercial brands, but with a superior nutritional quality.[10] There is high potential acceptability for flavoured snacks produced from blends of amaranth, chickpea and bovine lung mixtures. Thus, the high protein and iron levels present in these snacks make these novel products attractive to nutritional programmes aimed at alleviating anaemia and malnutrition. Incorporation of bean flours with cereal flours is widely used in the production of extruded ready-to-eat snacks. Extrusion cooking significantly decreases the content of antinutrients in the product.[11] Chickpea flour is also utilised for the preparation of chips (usually an extruded deep-fried product). Chips made from fermented chickpea flour contained significantly higher amounts of limiting amino acids than products prepared from non-fermented flours. The use of chickpea flour and starch as a thickening agent in the preparation of soup mixes has not been extensively explored. Such flour could be combined with vegetable concentrates or dehydrated vegetable shreds to manufacture soup mixes. Snack bar formulations with desirable texture properties can also be obtained with extruded chickpea flour and micronised flaked lentils.[12,13]

13.9. NOODLES AND PASTA

Starch noodles constitute the major category of Asian noodles. Starch noodles, also referred to as glass noodles or clear noodles, and are served both as the main course and in side dishes. These noodles are produced from starch of various botanical sources. Starches from the mung bean, yellow peas and potato are widely used in the production of these noodles.[14,15] Traditionally, mung bean starch is used as raw material for starch noodle production because it results in a product with the desired colour, glossiness, transparency and texture that are considered important by consumers for a high-quality starch noodles. Due to the relatively high price of mung bean starch, efforts have been made to substitute it with starches of other origins. The use of mung bean starch in the production of starch noodles leads to a very high-quality glass noodle, which appears white on drying. However, it turns transparent after boiling or cooking. These noodles have low water absorption capacity and high tensile strength and elasticity. The substitution of mung bean starch by different starches reduces the elasticity and increases the water absorption capacityof the noodle (softer consistency).[16] The mung bean starch is isolated by a wet-milling process. Mung bean seeds are first soaked in warm water (30–40 °C) for 6–9 h, followed by soaking in cold water (5–10 °C) for 8–16 h. After draining, grains are wet-milled to free starch granules from the cells (Figure 13.4). Starch is separated from proteins and fibres by mechanical separation and often further purified with enzyme treatment. The details of starch separation methods from pulses are given in Chapter 6. A small portion of starch (3–4%) is fully gelatinised and mixed with the raw starch and water to form dough. The dough is extruded through a die to make noodles of the appropriate size. The shaped noodles are cooked in boiling water for a few seconds, immediately cooled, and drained off. The noodles are refrigerated (5 °C) or frozen (-10 °C) for 12–24 h to allow the retrogradation of starch. They are then defrosted and air- or sun-dried before packaging. Mung bean noodles have short cooking time, less loss of solids on prolonged cooking and a distinctive chewy and elastic texture. The textural characteristics of noodles are contributed by the unique properties of the starch, which vary with the botanical source. Mung bean starch has greater hot-paste stability and show less breakdown viscosity during heating. It also has high amylose content and its properties are very similar to chemically cross-linked starches. It exhibits restricted swelling and solubilisation and higher retrogradation, which contribute to the desired quality traits. Fortified pasta (spaghetti) has also been produced by adding high amount of pea and faba bean flour (35%) to wheat semolina.[17] However, decline in some quality attributes (cooking loss) were observed and strategies for improving the products are being developed.

13.10. EXTRUDED SNACK PRODUCTS

Extrusion cooking is a high-temperature short-time (HTST) process that transforms raw ingredients to modified intermediate and finished products with the advantage of a continuous high-throughput processing with relatively dry viscous materials of foods with improved textural and flavour characteristics. Extrusion cooking is a very important food processing technology extensively employed for the production of breakfast cereals, ready-to-eat snack foods, baby foods, texturised vegetable protein, pet foods,

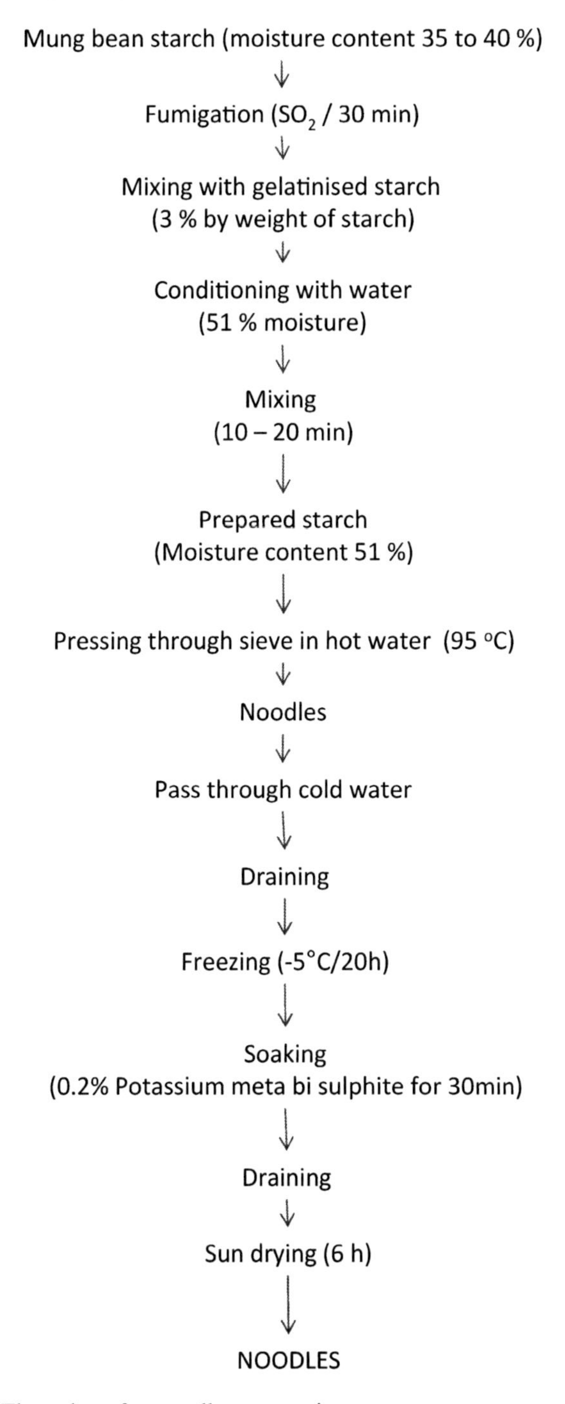

Figure 13.4 Flow chart for noodle preparation

dried soups and dry beverage mixes.[18] In addition to improving digestibility, extrusion cooking of pulses also improves bioavailability of nutrients in comparison to conventional cooking. Over the years, extrusion cooking has become the major processing method for food and feed industries, and it is rapidly evolving from an art into a science.[19] In addition to these properties, extrusion cooking is preferred to conventional cooking/processing techniques because of its ability to develop a range of products with distinct textural advantages including expansion, crispiness and general mouthfeel; being versatile, having high productivity, low operating costs, energy efficiency and shorter cooking times. Figure 13.5 shows the schematic diagram of a single screw extruder. Several extrusion process variables can influence the characteristics of finished products. These include raw material characteristics, mixing and conditioning of raw material, barrel temperature, pressure, screw speed, moisture content, flow rate, energy input, residence type, screw configuration, *etc.* Critical extrusion process variables such as temperature, screw speed, and moisture content may induce desirable modifications, thus improving palatability and technological properties of extruded products.[20,21]

These conditions have the ability to produce high-quality puffed extruded snacks. Pulse-based extruded snack products are generally obtained from pulse and cereal flour blends. Pulse flours or whole meals of lentils, peas, faba beans, chickpeas or kidney beans are often blended with cereal flours such as corn, rice and other grains. Extrusion cooking of cereal–pulse blends leads to the development of low-fat, high-fibre and protein-rich extruded products. Extruded products based on lentil and chickpea flour have been developed using a twin screw extruder.[22] Extrusion of common beans with cereal starch also enhances bioactive compounds such as polyphenol (Figure 13.6) and other antioxidants because coloured dry beans possess strong *in vitro* antioxidant activity.[23–25] Denaturation of pulse proteins during extrusion leads to open, loose structures which promote tannin–protein interaction, causing the

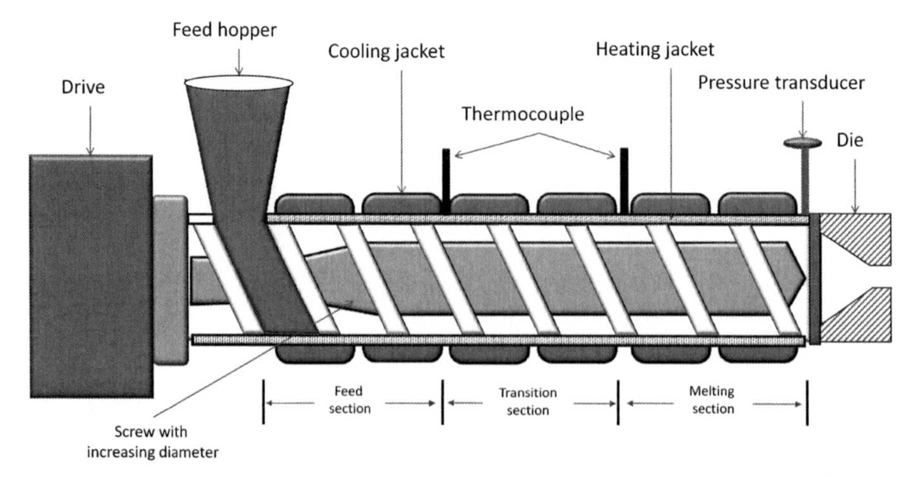

Figure 13.5 Schematic of extruder configuration indicating various zones

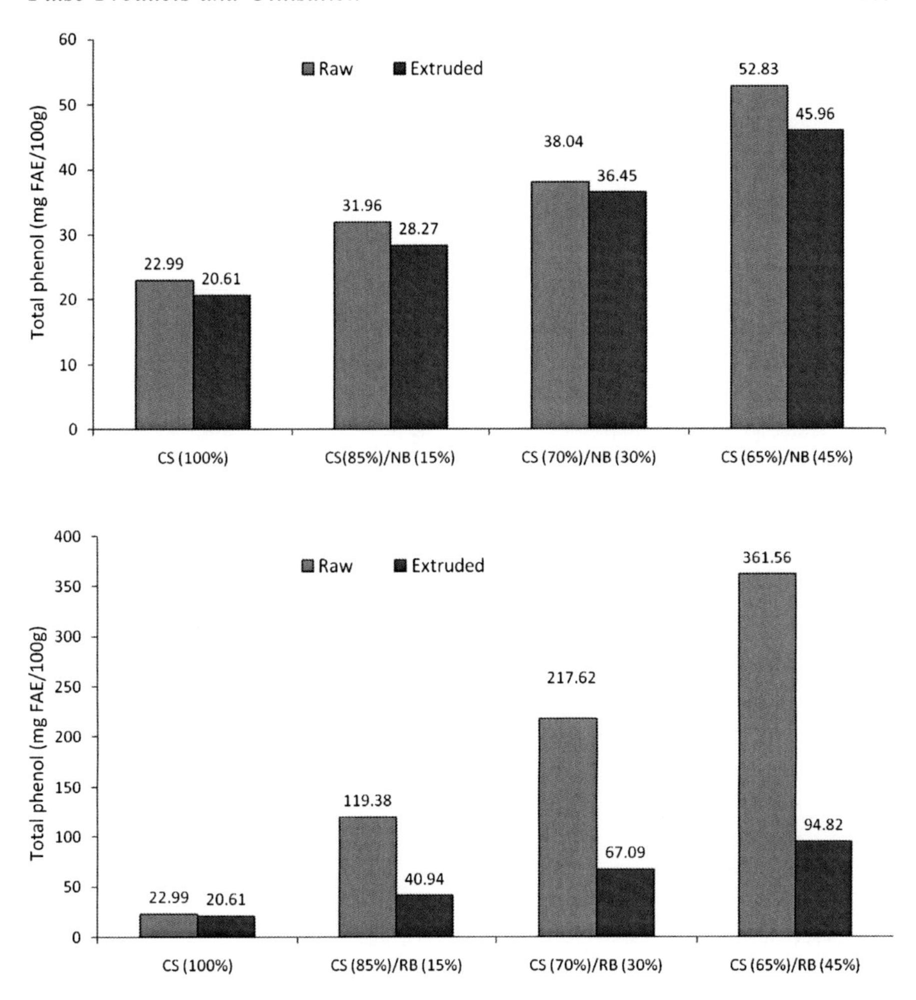

Figure 13.6 Effect of extrusion cooking on phenolic content (mg FAE/100 g) of corn starch added of bean flours (CS, corn starch; NB, navy bean; RB, red bean) Anton *et al.*[13]

formation of tannin–protein complexes that retain antioxidant activity. High temperature during extrusion cooking also helps in the destruction of non-nutrients largely present in commonly consumed pulses. Figure 13.7 shows some of the extruded snacks obtained from extrusion of 100% whole amaranth (a); oats (b); chickpea (c); lentil (d) and 50:50 blends of amaranth–chickpea (e); amaranth–lentil (f); oats–chickpea (g) and oats–lentil (h).

13.10.1. Extrusion-Cooking Parameters

Several extrusion-cooking parameters as shown in Figure 13.8 have to be adjusted depending on the raw material and desired product characteristics.

Figure 13.7 Extruded snacks obtained from extrusion of 100% whole amaranth (a), oats (b); chickpea (c); lentil (d) and 50:50 blends of amaranth–chickpea (e); amaranth–lentil (f); oats–chickpea (g); and oats–lentil (h)

Compared to other puffed products (*e.g.* roasted chickpea), extruded snack products experience a low moisture content, high shear, and high temperature during extrusion, resulting in a melt-in-the-mouth texture.[26] Screw speed is one of the main factors for the degree of barrel fill and therefore for the residence time, the shear stress and the volumetric output of the extrudates. With an increase in screw speed the measured torque and die pressure increase and the viscosity is reduced due to an increase in shear. If the screw speed and die area are increased it causes a decrease in both the feed rate and the barrel fill length. There has to be a balance between these parameters.[27–29] The configuration depends on the product and is usually a series of repeated conveying and mixing elements. Conveying screws drives the material flow and the mixing

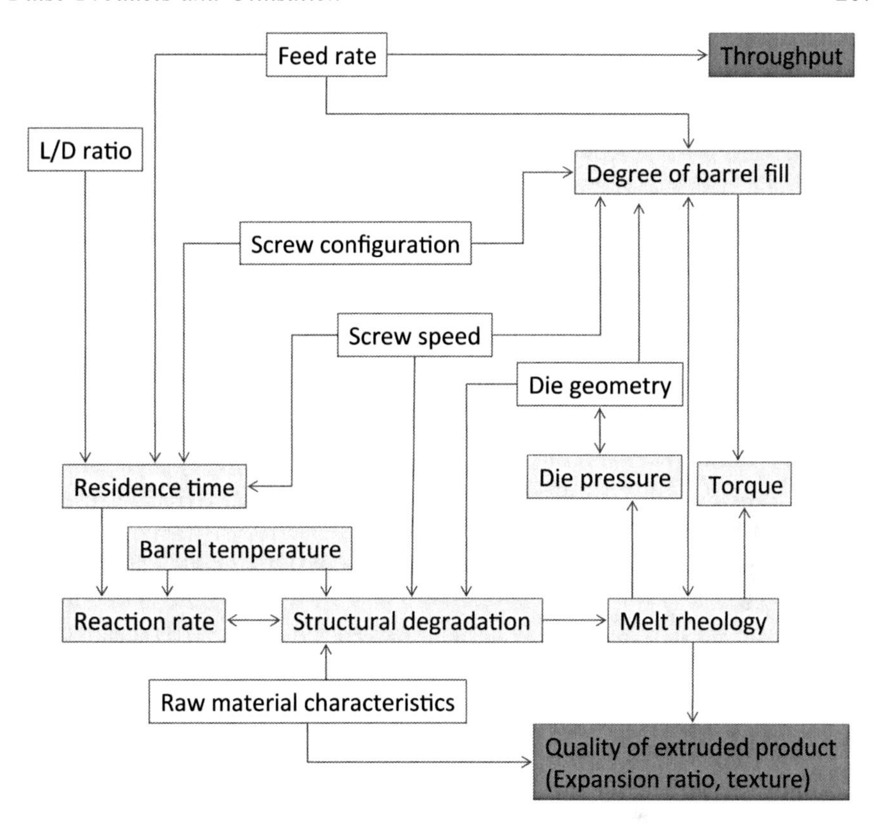

Figure 13.8 Flow diagram linking key process variables to output and product quality. (Modified from Frame[27])

restrictions create the biochemical conversions; both have an influence on the viscoelastic properties of the raw material.[27] Higher temperatures in the range of 100–180 °C allow a full transformation of raw material to its functional form in reduced processing time of 30–120 s with a minimum of 100 °C so that the product can expand.[30,31] There are two main thermodynamic processes during extrusion: autogenous reactions by mechanical conversion and isothermal operations such as cooling or heating. This is the reason that most extruders operate with temperature control.[27] For snack products the higher temperature may induce charring or burnt flavour. Therefore, the temperature has an influence on both the quantity and quality of extruded products, such as viscosity, expansion index and other physiochemical properties such as starch modification and protein denaturation.[27] Feed moisture content is one of the most important control points during extrusion and affects the density, texture, starch gelatinisation, and sometimes the flavour. During extrusion cooking, water lubricates the barrel, hydrates and solubilises the starch and protein polymers so that these can move freely in the mass of solids. Over 10%

moisture is sufficient for this movement, which is very much possible, and the physical nature of the extrudates changes from a glassy state to a viscous elastic fluid. Therefore, the screw conveying and shearing the fluid become larger and the mass heats up rapidly. If the moisture level becomes high the viscosity and friction decrease and therefore additional barrel heating will be required.[31] The feed rate has an influence on the residence time, barrel fill, pressure and the energy generated as well as on the barrel temperature. The primary effect of increased feed rate is a decrease in specific energy input and therefore the product changes. Feed rate and screw speed have to be adjusted relative each other so that the barrel is neither empty nor flooded.[29,33] In addition to the melt temperature, moisture content and feed rate, other variables such as pressure, specific mechanical energy (SME) and flow rate through the die are directly or indirectly controllable.[29] Pressure in the extruder is due to temperature build-up in the barrel of the extruder, the compression caused by the screw and the die restriction, and therefore the fluid viscosity and the feed rate as well.[22,30] The steam formation caused by temperature and moisture, and therefore the viscosity has a direct effect on die pressure which can range from 30–110 bars (3–11 MPa).[22,33] SME is related to the screw torque which can be adjusted by water injection. Water injection can also adjust the moisture content and is very often correlated to bulk density and the texture of the product.[27] The residence time depends on several parameters including feed rate, screw speed and configuration.[34] The shear rate is also a variable which increases the temperature and depends on several parameters like screw speed and configuration, rheological properties of the ingredients and other extrinsic and intrinsic parameters.[33,35]

Extrusion parameters such as barrel and die temperature, screw speed and geometry, and feed composition such as amylose/amylopectin ratio and moisture content have a particular effect on mechanical disruption and starch transformation.[36] The temperature, moisture and screw speed are the important factor during extrusion.[37,38] Moisture content alone does not significantly enhance starch transformation at low temperature (65–85 °C), but affects transformation significantly at high temperature (95–110 °C).[37] Dextrinisation is the major mechanism of starch fragmentation during low-moisture and high-shear extrusion.[39] Amylopectin loses overall molecular size without measurable changes, altering the percentage of 1→6 bonds,[40] and amylose is split at random locations in its chain.[41] Fragmentation of starch in an extruder is dependent upon specific mechanical energy input.[42] The thermal and mechanical energy affect amylopectin more than amylose during extrusion cooking and the effects are dependent upon screw speed, temperature and moisture content.[40,41] Under higher extrusion temperature and screw speed and lower moisture content, the dextrinisation is higher.[40] The amount of insoluble fibres decreases in most extruded products, including lentil and oat flour,[43–45] whereas soluble fibres, total sugars and free sugars are increased during the extrusion of black bean and oat flours.[44,45] Monosaccharides (glucose) are reduced due to Maillard and caramelisation reactions and oligosaccharides like raffinose, verbascose and stachyose, which are the main causes of flatulence in pulse-based foods are reduced due to dextrinisation.[43,46] Extrusion cooking at high temperatures causes structural changes in proteins mainly due to denaturation, association, and coagulation involving oxidation or reduction.[47] This is an important phenomenon in pulses which are relatively high in protein content.[48] During extrusion water-soluble albumins are denatured by heat and coagulated into a soft hydrated mass due to the shearing action. Larger globular proteins are also macerated to a smaller size, hydrates and form viscoelastic dough by agglomeration with water. Prolamins and glutenins, which are rich in cereals, also form a viscoelastic dough.[49] Proteins undergo the structural changes such as: hydrolysis of peptide bonds, formation of new covalent isopeptide cross-links and modification of amino side chains.[50] The major influence of heat during extrusion is to disassemble proteins and reassemble them into a fibrous, oriented structure leading to a characteristic texture. Extrusion or texturisation of material which has a higher protein than starch content, results in several stabilising forces that include non-disulfide covalent bonds, hydrophobic, hydrogen and disulfidc linkages. Although pulses have a high amount of sulphur-containing amino acids it cannot be concluded that intermolecular disulfide bonds contribute significantly to texture during extrusion.[51] This is in contrast to wheat flour which has a high amount of cystine and where disulfide bonds play a major role during extrusion.[52] It is also possible that proteins interact with oxidising lipids leading to polymerisation.[53] Protein nutritional quality during extrusion decreases mainly due to the Maillard browning which includes polymerisation of essential amino acids like lysine with reducing sugars due to a low moisture content. Generally, it is known that amino acid contents, as well as their availability can decrease during extrusion.[51,54,55] Therefore the temperature should be lower than 180 °C.[56]

13.11. PULSES AS INGREDIENTS IN OTHER FOODS

13.11.1. Weaning Foods

Soya bean, green gram and Bengal gram are commonly used in the preparation of weaning foods along with a main cereal component. The addition of cowpea results in reduced viscosity of the instant developed porridge, which could be very beneficial for infant feeding.[57] However, this benefit becomes a disadvantage when such products are aimed at adults, because reduced viscosity impacts negatively on the acceptability. The addition of cowpeas not only improves the protein content and quality of the extruded porridge, but but also makes it possible to feed this product to infants and thus enhance their food energy intake.

13.11.2. Breads and Cookies

The fortification of bakery goods with flours of peas, beans and chickpeas has been already employed and appears to be promising in particular for the functional food market.[58] Pea flour (5%) has been successfully incorporated into bread and cookies.[59,60] Both field pea and soy flours have been used in chemically leavened quick bread.[61] The primary benefits and role of pulses (or more specifically pulse flours) are due to their low glycaemic index (GI) and gluten-free nature. The low GI characteristic provides pulses with nutraceutical benefits, as exemplified by the role of low GI foods in the diet of diabetic patients.

Fortification with chickpea and broad bean flour as well as isolated soya protein (ISP) could be used in production of high-protein biscuits. Fortification with pulse flour decreases the spread factor may increase or decrease the lightness of biscuits. For example, fortification with chickpea increases the lightness, while fortification with broad bean or ISP increases the darkness.[62]

13.11.3. Meat Products

Non-meat ingredients derived from various plant and animal sources are used as fillers, binders, emulsifiers or extenders in meat systems to reduce cost and serve as functional ingredients. Chickpea flour is a potential source of high-protein flour for use as an extender in emulsified meat products due to its superior technological functionality and minimal effects on flavour.[63] Inclusion of chickpea flour improves the texture properties (whether measured instrumentally or sensorially) of low-fat pork bologna sausage. Bologna sausage with added kabuli and desi chickpea flour performs similarly to the usual formulations (with no added binder) for most flavour properties.[63] Similarly, soybean basic proteins have been added to egg albumen to improve foaming properties.[64] Pork and beef sausages could be significantly less acceptable at substitution levels above 30% with chickpea flour. However, the acceptability of mutton sausages containing chickpea flour may not be affected even at levels of substitution up to 40%.[65] The use of pea fibre concentrates in beef burger formulation improves their cooking properties, *i.e.* increases the cooking yield and decreases the shrinkage, and minimises production cost without degradation of sensory properties.[66]

13.12. TRADITIONAL PULSE-BASED FOODS

13.12.1. Papad

Papadum, also known as papad in northern India, is a thin, crisp Indian preparation sometimes described as a cracker. It is typically served as an accompaniment to a meal in India. It is also eaten as an appetiser or a snack and can be eaten with various toppings such as chopped onions, chutney or other dips and condiments. In some parts of India, it is served as the final item in a meal.

13.12.2. Wadi

Wadi is a pulse-based ready-to-fry product, popularly consumed in India, Pakistan and Bangladesh. Wadi is prepared from dehulled splits of black gram

(*Phaseolus mungo* L.), which is washed and soaked for about 10–12 h at room temperature. The water is drained off, and the splits are ground to a smooth thick batter. The batter is kept for natural fermentation at room temperature for 18–24 h, and then hand beaten for proper aeration. Finally, the batter is shaped into balls or cones and sun-dried for 4–8 days over bamboo or palm mats, smeared with oil. The product is marketed as hollow, brittle cones, which are fried briefly in a small amount of oil and used as an adjunct to cooked vegetables.

13.12.3. Pakora

Chickpea (desi) flour batter is prepared and blended with chopped vegetables. Onion, potato, spinach, cauliflower, green chilli and ginger are the most commonly used ingredients. The batter is then deep-fried in mustard oil. There are different varieties of pakora depending upon the ingredients blended. The most popular varieties of pakora are those made from spinach, acid-coagulated cheese (traditionally known as paneer in India), onion and potato. Pakoras are usually served as snacks or appetisers. Pakoras are also popular in Pakistan, where they generally resemble those found in India. Pakoras are sometimes served in a yoghurt (dahi)-chickpea flour-based curry known as kari-pakora rather than as separate snacks; this is consumed at lunch or dinners along with rice. In southern India, such preparations are known as bajji rather than pakora. Usually the name of the vegetable that is deep-fried is suffixed with bajji. For instance, potato bajji is sliced potato wrapped in batter and deep-fried. In northern part of India, pakora is taken to mean a mix of finely cut onions, green chillies and spices mixed in gram flour. This is rolled into small balls or sprinkled straight into hot oil and fried. These pakoras are very crisp on the outside and medium soft to crisp inside. There is also a variety that is softer overall, usually termed medhu pakora in restaurants, that is relatively softer and made of any other ingredients, such as potatoes.

13.12.4. Boondi

Boondi is a fried Indian sweet snack made from chickpea flour. Chickpea flour batter is poured through a ladle with holes into hot vegetable oil and fried to produce boondi. Boondi mixed with curd is popularly referred to as raita in North India. Boondi raita typically contains curd, boondi, chopped onions, green chillies, salt and black pepper. Boondi is also used to produce sweet snacks. After frying, the crisp boondi is dipped in sugar syrup and used to produce ladoos (laddu), a popular sweet Indian snack.

13.12.5. Bhujia

Chickpea flour is made into dough, moulded into vermicelli and fried in vegetable oil. Fried mung bean and chickpea dhal is one of the favourite snacks in India. Dhal is soaked in water in the presence of a small amount of sodium bicarbonate. The superfluous water is removed and then it is fried in vegetable oil.

13.13. TRADITIONAL INDIAN FERMENTED PULSE FOODS

13.13.1. Idli and Vada

Idli or Idly, a very popular fermented breakfast food staple consumed in the Indian subcontinent, consists mainly of rice and black gram (Figure 13.9). More specifically, idli is a traditional South Indian savoury and spongy pancake, which is usually 5–7 cm in diameter and is prepared typically using four parts uncooked rice to one part split black gram. The importance of black gram as the major contributing component in the fermentation of the batter is well known.[67] The black gram and rice are ground to a paste and fermented overnight, followed by steaming in special greased idli pans (see Figure 13.10). The fermentation batter prominently consists of *Saccharomyces cerevisiae*, in addition to *Bacillus amyloliquefaciens*, *Debaryomyces hanseni*, and *Trichosporon beigelii* (Table 13.3). CFTRI, India has developed an automatic idli-making unit that produces 1200 idlis per hour. The unit consists of an automatic idli batter depositor, a special idli pan conveyor, steam chamber and idli scooping system. Several attempts have been made to replace the black gram (*Phaseolus mungo*) component in idli with other legumes including common beans (*Phaseolus* spp.), soybeans (*Glycine max*) and great northern beans (*Phaseolus vulgaris* L.) for preparing certain novel idli-like products to add to the variety of cereal/legume fermented foods, but with little success as far as acceptability is concerned.[68]

Unfermented pulse batter is also used in the preparation of fried snacks product commonly known as vada, wada or vade) (Figure 13.11). Vada can also be prepared by using gram and/or dhal lentils. Vada prepared from black gram is shaped like a doughnut, with a hole in the middle as shown in Figure 13.11. Other variants of vada are masala vada or paruppu vadai, where

Figure 13.9 Steamed rice–black gram fermented product (idli)

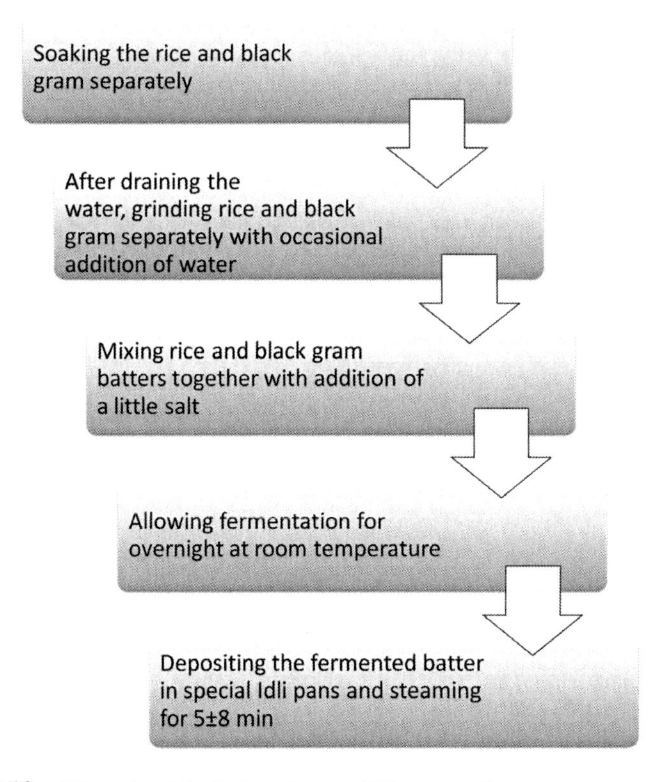

Figure 13.10 Flow chart depicting steps in idli preparation

the main ingredient is pigeon pea splits. There are several variants of vada, depending on the ingredients, shape and size.

13.13.2. Dosa (Indian Rice Pancake)

Dosa is another fermented crepe made from rice batter and black gram dehuleld splits popularly consumed in India, Malaysia, Myanmar and Singapore, and staple to southern parts of India. Dosa preparation is similar

Table 13.3 Prevalence of various microorganisms in fermented rice–black gram idli batter

Predominant bacteria (Total count: 10^6–10^9/g dry matter)	*Predominant yeast (Total count: 0–10^6/g dry matter)*
Leuconostoc mesentroides	*Saccharomyces cerevisiae*
Streptococus faecalis	*Debaryomyces hansenii*
Lactobacillus fermentum	*Hansenula anomala*
Pediococcus cervisae	*Trichosporon beigelii*
Lactobacillus delbrueckii	*Torulopsis candida*
Bacillus amyloliquefaciens	*Trichosporon pullulans*

Figure 13.11 Blackgram vada

to that of idli making. It involves the overnight pre-soaking (where fermentation occurs) of the two ingredients, rice and black gram, mixing them together after grinding, and incubating at room temperature for up to 15–17 h at a temperature close to 313–318 K (40–46 °C). The progress of the fermentation can be measured by studying the increase in acidity and leavening which are the deciding factors for the completion of the fermentation and for imparting the characteristic texture to the product. A thin layer of the batter is ladled on to a hot tava (griddle) greased with oil or clarified butter (ghee), and spread out evenly with the base of a ladle or bowl to form a pancake. For domestic baking the tava is heated at the bottom by any of the usual heat sources such as firewood, charcoal, coal, kerosene or liquefied petroleum gas (LPG).[69] Conduction from hot plate and radiation from the hood/top cover are considered most important for the desired product quality. It is a general feeling that heat transfer by radiation significantly contributes to the baking of dosa.

The pattern of craters formed due to the evaporation of moisture during the baking process has been analysed by Venkateshmurthy and Raghavarao[69] using scanning electron microscopy, as shown in Figure 13.12. It can be seen that materials such as stainless steel and Teflon-coated aluminium have large numbers of smaller and more uniform craters compared to the cast iron and alloy steel material. This can be attributed to the fact that the material with lower thermal conductivity has uniform distribution of heat and also lower rate of heat transfer compared to alloy steel and cast iron. The difference in density of the craters can be clearly seen in the better surface finish of the product prepared using stainless steel or Teflon-coated aluminium hot plates.

13.13.3. Dhokla

Dhokla is a popular fermented, steamed and fried dish native to the state of Gujarat in western India (Figure 13.13). For dhokla, rice and chickpea (or chickpea flour), in the ratio of 4:1 is soaked overnight. The soaked mixture is ground and fermented for 4–5 h, followed by addition of spices such as chilli

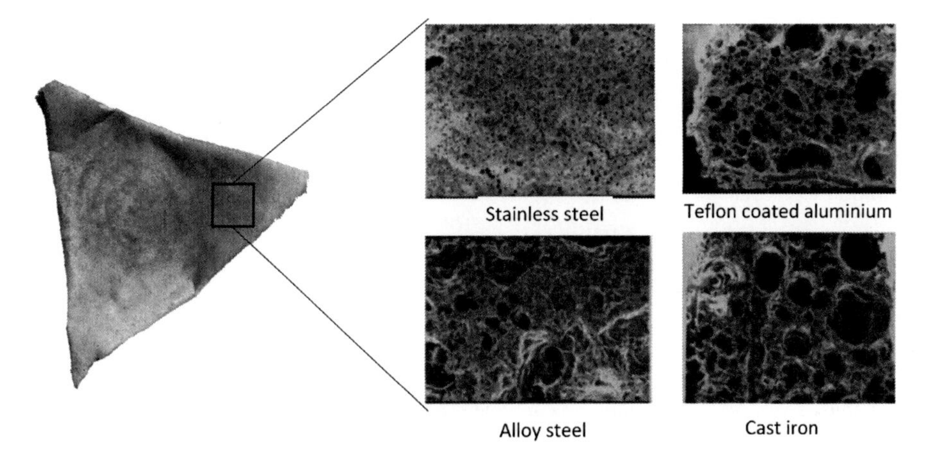

Figure 13.12 Rice pancake (dosa) and its microstructure made on different hot plate material. (Adapted from Venkateshmurthy and Raghavarao[69])

pepper, ginger, and baking soda. Pulses such as black grams can also be substituted for chickpeas. The fermented batter is steamed for about 15 min on a flat dish and cut into pieces, which are conventionally fried in mustard oil. Dhokla can also be prepared by complete replacement of the Bengal gram flour (chickpea) with soybean, to obtain improved nutritional, microbial and organoleptic qualities.[70]

13.14. UTILISATION OF BY-PRODUCTS

The several aspects of milling of pulses have already been discussed in this book; in brief, the milling of pulses involves removal of the outer husk and splitting the grain in two. In spite of the fact that soaking of pulses aids in loosening of husk, large abrasive force is usually applied for complete separation of husk. This leads to the generation of higher amounts of broken grains and powders, in addition to husk *per se*. To summarise, the by-products of pulse milling include broken grains (6–13%), germ, powder (7–12%) and

Figure 13.13 Dhokla

husk (4–14%). The by-products from pulse milling industry have both food and non-food uses, including use as animal feed. In the non-food sector, the starch fraction of peas has been used in the manufacture of adhesives and carbonless paper. An emerging non-food application of pulse by-products includes their use as a potential feedstock for bioethanol production, which can be accomplished post-hydrolysis.

A major application of minor fractions from pulse milling could be their incorporation into bakery formulations. Supplementation of bakery products with pulse fractions enhances their nutritive value, while at the same time allowing appropriate use of by-products. For example, incorporation of pigeon pea (*Cajanus cajan* L.) by-product flour into biscuit formulations considerably increases their protein content.[71] However, the success of such applications could be limited by the significant influence of the by-products on physical and sensory qualities of the product. In yet another example, pea hull fibre could be incorporated into dough of high-fibre wheat breads. The water-retention capacity of such composite flours increases, without markedly affecting the mixing properties, and crust colour of the breads prepared using such flour becomes lighter.[72] The water-retention capacity of the fibre is important because it helps to retain the firmness of crumb during storage. In general, high-fibre bread could be obtained by incorporation of pea hull at levels up to 15%. However, it should be noted that a decrease in loaf and specific volumes could be a major challenge associated with such composite flours.

Pulse milling by-products could also be incorporated into locally prevalent products, although on a small scale. To cite an example, Tiwari *et al.*[73] incorporated by-products of pulses and cereal processing into formulations of deep-fried snacks. Their study demonstrated that the incorporation of pulse flour (red gram, green gram and black gram) with cereal flour (rice) improves the nutritional value of the products. Further, sensory results and shelf life studies showed complementary results. In general it can be commented that the uses of pulse milling by-products remains unexplored and there is a great scope for new and potential food applications.

REFERENCES

1. A. Gowen, N. Abu-Ghannam, J. Frias and J. Oliveira, *Trends Food Sci. Technol.*, 2006, **17**, 177–183.
2. A. A. Gowen, N. Abu-Ghannam, J. Frias and J. Oliveira, *Innovative Food Sci. Emerging Technol.*, 2008, **9**, 129–137.
3. A. Ghaderi, G. Hosfield, M. Adams and M. Uebersax, *J. Am Soc. Hort. Sci.*, 1984, **109**, 85–90.
4. G. L. Hosfield, *Food Technol. (Chicago)*, 1991, 45.
5. H. A. Oboh, M. Muzquiz, C. Burbano, C. Cuadrado, M. M. Pedrosa, G. Ayet and A. U. Osagie, *Plant Foods Hum. Nutr.*, 2000, **55**, 97–110.

6. S. Banerjee, K. Rohatgi, M. Banerjee, D. Chattopadhyay And H. Chattopadhyay, *J. Food Sci.*, 1955, **20**, 545–547.
7. D. Y. Coşkuner and E. Karababa, *Food Rev. Int.*, 2004, **20**, 257–274.
8. M. Kaur, N. Singh and N. S. Sodhi, *J. Food Eng.*, 2005, **69**, 511–517.
9. J. P. Batistuti, R. M. C. Barros and J. A. G. Areas, *J. Food Sci.*, 1991, **56**, 1695–1698.
10. R. A. Cardoso-Santiago and J. A. G. Areas, *Food Chem.*, 2001, **74**, 35–40.
11. R. N. Chavez-Jauregui, R. A. Cardoso-Santiago, M. E. M. P. E. Silva and J. A. G. Areas, *Int. J. Food Sci. Technol.*, 2003, **38**, 795–798.
12. X. Meng, D. Threinen, M. Hansen and D. Driedger, *Food Res. Int.*, 2010, **43**, 650–658.
13. D. Ryland, M. Vaisey-Genser, S. D. Arntfield and L. J. Malcolmson, *Food Res. Int.*, 2010, **43**, 642–649.
14. S. Prabhavat, in *Proc. Mungbean Meeting 90*, Chiang Mai. Tropical Agriculture Research Center, Japan, Thailand Office, Bangkok, 1988, pp. 9–16.
15. R. Yang and S. Tsou, in *Proc. International Consultation Workshop on Mungbean. Asian Vegetable Research and Development Center, Shanhua, Tainan, Taiwan*, 1998, pp. 152–158.
16. R. W. Klingler, F. Meuser and E. A. Niediek, *Starch/Stärke*, 1986, **38**, 40–44.
17. M. Petitot, L. Boyer, C. Minier and V. Micard, *Food Res. Int.*, 2010, **43**, 634–641.
18. C. Brennan, M. Brennan, E. Derbyshire and B. K. Tiwari, *Trends Food Sci. Technol.*, 2011, **22**, 570–575.
19. M. N. Riaz, M. Asif and R. Ali, *Crit. Rev. Food Sci. Nutr.*, 2009, **49**, 361–368.
20. M. A. Brennan, J. A. Monro and C. S. Brennan, *Int. J. Food Sci. Technol.*, 2008, **43**, 2278–2288.
21. M. A. Brennan, I. Merts, J. Monro, J. Woolnough and C. S. Brennan, *Starch/Stärke*, 2008, **60**, 248–256.
22. J. Berrios, J. De, J. Tang and B. G. Swanson, U.S. Patent Application No. 2008/0145483 A1, 2008.
23. A. A. Anton, R. G. Fulcher and S. D. Arntfield, *Food Chem.*, 2009, **113**, 989–996.
24. T. Madhujith and F. Shahidi, *J. Food Sci.*, 2005, **70**, S85–S90.
25. C. W. Beninger and G. L. Hosfield, *J. Agric. Food Chem.*, 2003, **51**, 7879–7883.
26. J. M. Harper, in *Extrusion Cooking*, ed. C. Mercier, P. Linko and J. M. Harper, American Association of Cereal Chemists, St. Paul, MN, 1989.
27. N. D. Frame, in *The Technology of Extrusion Cooking*, Blackie, London, 1994.
28. E. A. Ozer, E. N. Herken, S. Guzel, P. Ainsworth and S. Ibanoglu, *Int. J. Food Sci. Technol.*, 2006, **41**, 289–293.

29. C. J. Chessari and J. N. Sellahewa, in *Extrusion Cooking – Technologies and Applications*, ed. R. Guy, Woodhead Publishing, Cambridge, 2000.
30. R. C. E. Guy and A. W. Horne, in *Food Structure – Its Creation and Evaluation*, ed. J. M. V. Blanshard and J. V. Mitchell, Butterworths, London, 1988.
31. R. Guy, in *The Technology of Extrusion Cooking*, ed. N. D. Frame, Blackie, London, 1994.
32. G. Moore, in *The Technology of Extrusion Cooking*, ed. N. D. Frame, Blackie, London, 1994.
33. P. Colonna, J. Tayeb and C. Mercier, in *Extrusion Cooking*, ed. C. Mercier, P. Linko and J. M. Harper, American Association of Cereal Chemists, St. Paul, MN, 1989.
34. F. Meuser and W. Wiedmann, in *Extrusion Cooking*, ed. C. Mercier, P. Linko and J. M. Harper, American Association of Cereal Chemists, St. Paul, MN, 1989.
35. J. M. Harper, *Extrusion of foods Vol. 1.*, CRC Press, Boca Raton, FL, 1981.
36. L. S. Lai and J. L. Kokini, *Biotechnol. Progr.*, 1991, **7**, 251–266.
37. B. Y. Chiang and J. A. Johnson, *Cereal Chem.*, 1977, **54**, 436–443.
38. J. Owusuansah, F. R. Vandevoort and D. W. Stanley, *Cereal Chem.*, 1983, **60**, 319–324.
39. M. H. Gomez and J. M. Aguilera, *J. Food Sci.*, 1983, **48**, 378–381.
40. V. J. Davidson, D. Paton, L. L. Diosady and G. Larocque, *J. Food Sci.*, 1984, **49**, 453–458.
41. P. Colonna, J. L. Doublier, J. P. Melcion, F. Demonredon and C. Mercier, *Cereal Chem.*, 1984, **61**, 538–543.
42. F. Meuser, R. W. Klingler and E. A. Niediek, *Starke*, 1978, **30**, 376–384.
43. J. D. J. Berrios, P. Morales, M. Cámara and M. C. Sánchez-Mata, *Food Res. Int.*, 2010, **43**, 531–536.
44. J. D. Berrios, M. Camara, M. E. Torija and M. Alonso, *J. Food Process. Preserv.*, 2002, **26**, 113–128.
45. D. G. Gualberto, C. J. Bergman, M. Kazemzadeh and C. W. Weber, *Plant Foods Hum. Nutr.*, 1997, **51**, 187–198.
46. S. E. Fleming, *J. Food Sci.*, 1981, **46**, 794–798, 803.
47. T. P. Shukla, *Cereal Food World*, 1996, **41**, 35–36.
48. N. R. Reddy, D. K. Salunkhe and S. K. Sathe, *Crit. Rev. Food Sci. Nutr.*, 1982, **16**, 49–114.
49. R. Guy, in *Extrusion Cooking – Technologies and Applications*, Woodhead Publishing, Cambridge, 2000.
50. J.-C. Cheftel, J.-L. Cuq and D. Lorient, in *Food Chemistry*, ed., O. R. Fennema, 2nd ed., Marcel Dekker, New York, 1985.
51. D. W. Stanley, in *Extrusion Cooking*, ed. C. Mercier, P. Linko and J. M. Harper, American Association of Cereal Chemists, St. Paul, MN, 1989.
52. M. Li and T. C. Lee, *J. Agric. Food Chem.*, 1997, **45**, 2711–2717.

53. J. W. Finlay, in *Chemical Changes in Food During Processing*, ed. T. Richardson and J. W. Finlay, AVI Publishing, Westport, CT, 1985.
54. R. D. Phillips, in *Protein Quality and the Effects of Processing*, ed. R. D. Phillips and J. W. Finley, Marcel Dekker, New York, 1988.
55. N.-G. Asp and I. Björck, in *Extrusion Cooking*, ed. C. Mercier, P. Linko and J. M. Harper, American Association of Cereal Chemists, St. Paul, MN, 1989.
56. J. C. Cheftel, *Food Chem.*, 1986, **20**, 263–283.
57. L. Pelembe, C. Erasmus and J. Taylor, *Lebensm.-Wiss. Technol.*, 2002, **35**, 120–127.
58. D. D. Dalgetty and B. K. Baik, *Cereal Chem.*, 2006, **83**, 269–274.
59. J. Sadowska, W. B½aszczak, J. Fornal, C. Vidal-Valverde and J. Frias, *Eur. Food Res. Technol.*, 2003, **216**, 46–50.
60. K. Kamaljit, S. Baljeet and K. Amarjeet, *Am. J. Food Technol*, 2010, **5**, 130–135.
61. M. Raidl and B. Klein, *Cereal Chem.*, 1983, 60.
62. T. M. Rababah, M. A. Al-Mahasneh and K. I. Ereifcj, *J. Food Sci.*, 2006, **71**, S438–S442.
63. W. Thushan Sanjeewa, J. P. D. Wanasundara, Z. Pietrasik and P. J. Shand, *Food Res. Int.*, 2010, **43**, 617–626.
64. G. Wang and T. Wang, *J. Food Sci.*, 2009, **74**, C581–C587.
65. M. Verma, D. Ledward and R. Lawrie, *Meat Sci.*, 1984, **11**, 109–121.
66. S. Besbes, H. Attia, C. Deroanne, S. Makni and C. Blecker, *J. Food Quality*, 2008, **31**, 480–489.
67. R. Radhakrishnamurty, H. Desikachar, M. Srinivasan and V. Subrahmanyan, *J. Sci. Ind. Res. C, Biol. Sci.*, 1961, **20**, 342.
68. S. Soni, D. Sandhu, K. Vilkhu and N. Kamra, *Food Microbiol.*, 1986, **3**, 45–53.
69. K. Venkateshmurthy and K. S. M. S. Raghavarao, *J. Food Sci. Technol.*, 2011, 1–7.
70. K. Gadhe, B. Jadhav and R. Kshirsagar, *Bioinfolet*, 2010, **7**, 309–312.
71. B. Tiwari, C. Brennan, R. Jaganmohan, A. Surabi and K. Alagusundaram, *LWT – Food Sci. Technol.*, 2011, **44**, 1533–1537.
72. F. Sosulski and K. Wu, *Cereal Chem.*, 1988, **65**, 186–191.
73. U. Tiwari, M. Gunasekaran, R. Jaganmohan, K. Alagusundaram and B. Tiwari, *Food Bioproc. Technol.*, 2011, **4**, 1172–1178.
74. H. F. Jones and S. T. Beckett, Physico-Chemical Aspects of Food Processing, Chapman and Hall, London, 1995.
75. P. U. Rao, and B. Belavady, Oliosaccharides in pulses: varietal difference and effects of cooking and germination. *J. Agr. Food Chem.*, 1978, **26**, 316–319.
76. Y. Coşkuner and E. Karababa, *Food Rev. International*, 2004, **20**(3), 257–274.

CHAPTER 14

Pulse Grain Quality Criteria

14.1. INTRODUCTION

Pulses add variety to the human diet, which is evident from their use in a number of food products. They serve as an economical source of supplementary proteins for the large human populations in developing countries. This contribution to nutrition is especially important for a country such as India where the majority of the population is vegetarian. Pulses are consumed both as whole grain and as decorticated splits, known as "dhal" in India and Pakistan. Pulses are milled into flours, which are used in a number of food products. Whole grains and dhal are generally consumed as curry, in which they are cooked along with pre-fried onion, garlic, ginger, green chillies, *etc.* Pigeon pea splits are cooked along with various vegetables, known as "sambar", which is a staple food of people living in south India. In northern regions of India, curries made from dhal or whole grains of different pulses are consumed with flat breads (chapatti) or cooked rice. The cooking of pulses is essential to make them edible and tender and to ensure the maximum consumer acceptability. Cooking time as well as cooked grain characteristics such as appearance, colour, taste, texture and flavour are considered to be important quality attributes. Cooking time is one of the main criteria used in evaluating cooking quality of pulses. Longer cooking time is a major constraint in the utilisation of whole pulse grains. Whole pulse grains require longer cooking time than dehulled splits. The dehulling of pulses removes the impermeable seed coat that prevents water uptake during cooking. The prolonged cooking of pulses causes reduction in the nutritive value of the proteins. Cooking of pulses cause starch gelatinisation, protein denaturation, enzyme inactivation, solubilisation of some of the polysaccharides, and softening and breakdown of the middle lamella, a cementing material found in the cotyledon.[1,2] Cooking also inactivates or reduces the levels of antinutrients such as trypsin inhibitors and flatulence-causing oligosaccharides, resulting in

Pulse Chemistry and Technology
Brijesh K. Tiwari and Narpinder Singh

Published by the Royal Society of Chemistry, www.rsc.org

improved nutritional quality. The cooking quality of pulses varies with cultivar, grain characteristics, chemical composition, location and environment.[1,3–5] Physical properties, such as weight, size and density, as well as seed coat and cotyledon characteristics, influence the cooking quality of pulses.[6] For example, the textural properties of cooked cowpeas (*Vigna unguiculata*) have a negative correlation with the amount of water absorbed.[7] Sefa-Dedeh and Stanley[6] reported that seed thickness and hilum size of cowpeas influenced water absorption during the initial stage of soaking while protein content became an important factor affecting water absorption in the later stages of soaking. These aspects have been discussed in detail in earlier chapters of this book.

14.2. PHYSICAL CHARACTERISTICS THAT DETERMINE PULSE QUALITY

- ***Broken/damaged grains:*** Pulse grains are marketed both as whole grains or splits. Pulse grains are broken down during post-harvest handling and dehulling operations. Pulses containing brokens are considered to be of inferior quality. The presence of brokens in pulses reduces the quality by reducing consumer acceptability and by increasing susceptibility to infestation during storage.
- ***Impurities:*** The presence of foreign impurities in pulses reduces their value and affects handling and processing. Foreign matter of the following classes may be present:

animal origin – insects, insects residues, rodent excreta, *etc.*

vegetable origin – straw, weeds, seeds of other crops, dust, microorganisms/toxins.

mineral origin – stones, mud, dust, glass, metals, oil products, pesticide residues.

- ***Infested, infected grain:*** Grain yield is reduced by insect infestation. The contamination of pulses besides causing food hygiene problems, may lead to acute or chronic illness.
- ***Discoloured grains:*** Discolouration of grains is caused by heat, fermentation, moulds, weathering or disease.
- ***Purity:*** Pulses of different cultivars should not be mixed because of difference in their physical, composition and cooking properties.
- ***Grain weight:*** Grain weight is generally expressed in grams per 100 grains. It is a function of grain size and density. It is determined by randomly counting the grains and weighing.
- ***Grain volume:*** Grain volume is determined by randomly selecting 50 seeds, transferring them to a 50 mL measuring cylinder and adding 25 mL distilled water. The difference in the volume is noted. The volume per seed is calculated and expressed.

- ***Bulk density:*** This is defined as the mass per unit volume. Bulk density is affected by insect infestation, foreign matter and moisture content. Grains are gently added to a 100 mL graduated cylinder. The bottom of the cylinder is gently tapped several times, until there is no further diminution of the sample level after filling to the 100 mL mark. Bulk density is calculated as weight per unit volume of sample (g mL^{-1}).
- ***Colour:*** Colour is an important parameter by which a consumer judges pulse quality before purchase. The colour of pulse grains is a genetic characteristic and is affected by storage and environmental conditions. Colour is measured by various instruments, such as Ultra Scan VIS (Hunter Lab). This is a visible-range high-performance colour measurement spectrophotometer that measures the full range of human colour perception. This instrument measures both reflected and transmitted colour as well as transmission haze in the wavelength range 360–780 nm. L^*, a^* and b^* values of the grains are measured. L^* indicates the lightness whereas a^* and b^* values indicate the redness-greenness and blueness-yellowness, respectively (Figure 14.1). Hunter colour parameters of different pulses are shown in Table 14.1.Field pea lines with the L^* and

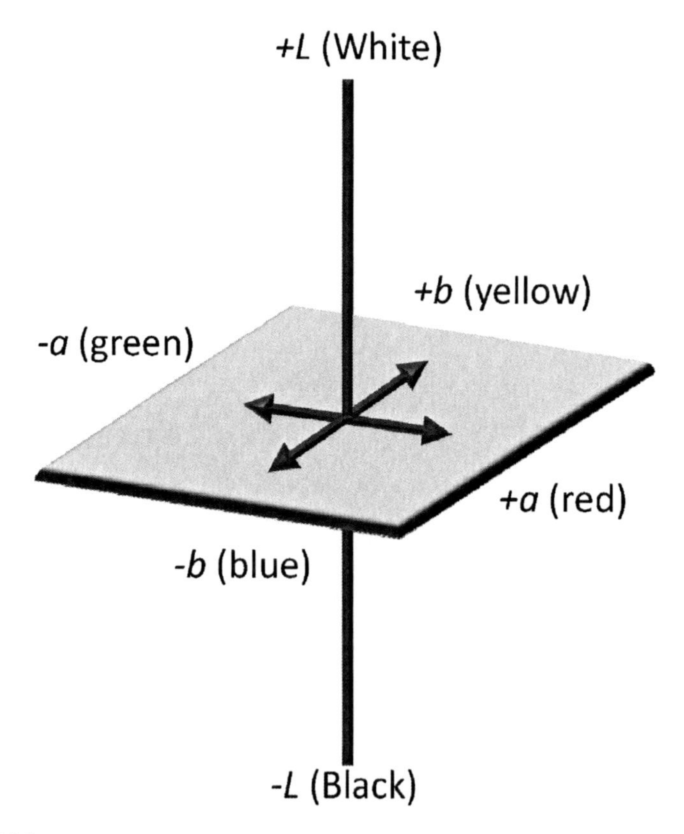

Figure 14.1 Colour coordinates

Table 14.1 Colour values of some pulses

Pulses	*L**	*a**	*b**
Kidney beans (red)	37.49	5.92	0.83
Kidney beans (white)	55.60	8.68	11.93
Chickpeas (kabuli)	59.34	7.50	16.77
Chickpeas (desi)	50.44	9.71	14.85
Lentils	47.60	3.61	7.15
Horse gram	43.3–51.5	4.4–10.9	6.1–14.8
Peas	43.6–67.1	-2.3–6.2	5.8–17.4
Adzuki beans	38.72–41.53	0.18–11.22	0.68–4.07
Lima beans	46.38	4.42	4.58
Navy beans	70.28	1.69	10.68
Cow peas	36.44–45.77	-0.15–2.37	-0.8–3.16
Scarlet beans	40.72	4.57	1.79

*b** values of less than 50 and 10 respectively, are darker and have lower commercial significance.[8] Higher *L** values indicate that the pea lines are lighter in colour. Negative *a** (indicator of greenness) values ranging between -0.8 and -2.3 have been reported for some pea lines, indicating that these lines are slightly greenish and have higher consumer acceptability. Similarly, mung bean varieties with negative *a** value (more greenness) and kidney beans with greater redness (higher *a** value) are preferred. This technique is very good for screening a large number of samples.

14.3. HYDRATION CHARACTERISTICS THAT DETERMINE PULSE QUALITY

Pulses are generally soaked before cooking to reduce the cooking time. Soaking assists in uniform expansion of the seed coat and cotyledon and is also essential for uniform cooking and tenderness. The ability of grains to hydrate is related to cooking quality. Higher hydration capacity of pulse grains is associated with shorter cooking time. Hydration capacity of kidney bean and field pea has been observed to be positively correlated with both seed weight and volume.[8,9] In general pulses with higher seed weight have higher hydration and swelling capacity but they have the drawback of longer cooking time. Seed coat thickness and hilum size of cowpeas influence their water absorption during the initial stage of soaking, while protein content becomes an important factor affecting water absorption in the later stages of soaking.[6] Pulses that possess a thin, amorphous and porous seed coat have a higher rate of water absorption, while those with a relatively thick seed coat have a slower water absorption rate. Some of the hydration properties and associated physical properties of pulses are shown in Table 14.2. The grains are soaked in water overnight and the following properties of hydrated seeds are measured:[10]

Table 14.2 Some hydration properties and associated physical properties of pulses

Pulse	*100 grain weight (g)*	*100 grain volume (mL)*	*Bulk density (g/mL)*	*Hydration capacity*	*Hydration index*	*Seed volume*	*Swelling capacity*	*Swelling index*
Kidney beans (red)	23.01	28	0.82	0.23	0.98	0.2	1.16	4.14
Kidney beans (white)	46.72	62	0.75	0.45	0.96	0.42	1.22	1.97
Chickpeas (kabuli)	41.44	56	0.74	0.41	0.98	0.3	1.2	2.14
Chickpeas (desi)	17.32	24	0.72	0.19	1.09	0.12	1.1	4.58
Lentils (masur)	3.21	4	0.80	0.03	1.09	0.02	1.04	26.00
Mung beans	3.34	4	0.84	0.06	1.80	0.04	1.02	25.50
Horse gram	2.83–3.88		0.8–0.9	0.03–0.04	0.96–1.28		0.03–0.07	0.07–0.14
Pigeon peas								
Peas	4.26–29.3	–	0.7–0.83	0.05–0.31	0.85–1.37	–	0.02–0.76	0.62–2.60
Adzuki beans	7.48–14.8	5.0–10.0	0.75–1.0	0.05–0.12	0.42–0.94	–	0.04–0.15	0.5–1.87
Black beans								
Lima beans	23.8	38	0.73	0.25	0.89	0.26	0.08	0.20
Navy beans	20.87	28	0.75	0.13	0.64	0.18	0.10	0.64
Great northern beans								
Cowpeas	10.28–15.93	14–22	0.64–0.73	0.08–0.13	0.74–0.87	0.10–0.14	0.11–0.17	0.51–1.22

- *Hydration capacity:* this is measured by transferring 50 seeds to a 125 mL flask and adding the distilled water to 100 mL. The flask is lightly stoppered and left overnight at room temperature. The next day, the water is drained, superfluous water removed with an absorbent paper, and the seeds are reweighed. Hydration capacity per seed is recorded as:

$$\text{Hydration capacity} = (\text{weight after soaking} - \text{weight before soaking})/50$$

- *Hydration index:* is calculated as:

$$\text{Hydration index} = \text{Hydration capacity per seed}/\text{weight of one seed (g)}$$

- *Hydration coefficient* = mass of soaked beans/mass of dry beans
- *Swelling capacity:* Swelling capacity is determined by calculating the difference in seed volume before and after soaking the seeds. Fifty seeds are transferred to a 100 mL measuring cylinder and 50 mL distilled water is added. Swelling capacity per seed is recorded as:

$$\text{Swelling capacity} = (\text{volume after soakingvolume before soaking})/50$$

- *Swelling index:* Swelling index is calculated as:

$$\text{Swelling index} = \text{Swelling capacity per seed}/\text{volume of one seed (mL)}$$

14.4. CHEMICAL CHARACTERISTICS THAT DETERMINE PULSE QUALITY

The chemical composition of pulses varies due to their genetic make-up. Factors such as soil type, agronomic practices (such as plant density, weeds or soil fertility), climatic conditions (rainfall, light intensity, duration of growing season, day and night temperature) and storage also contribute to the variation in chemical constituents of grain. Technological treatments also have significant effect on chemical constituents of pulses. The chemical composition of different pulses has already been discussed in Chapter 3 and 4.

14.4.1. Moisture Content

The moisture content is the amount of water present in the grain and is usually expressed as in percentage. It is expressed either on wet weight basis (where it is expressed as a percentage of the fresh weight of the seed) or on a dry weight basis (where it is expressed as a percentage of the dry weight of the seed). The grain moisture content influences the storage life of the grain. Grain moisture content is closely associated with several aspects of physiological seed quality. For example, it is related to grain maturity, optimum harvest time, mechanical damage, drying cost, grain longevity, and insect and pathogen infestation. Storage of pulses at high moisture leads to development of discolouration. The moisture content of grains can be measured by direct and indirect methods.

14.4.1.1. Direct Methods

In the direct method, the grain moisture content is measured directly by loss or gain in seed weight. Direct methods are the desiccation method, phosphorus pentoxide method, oven-drying method, vacuum drying method, distillation method, Karl Fisher's method, direct weighing balance, microwave oven method. In the oven method grains are ground and heated to dryness by evaporating the water into the atmosphere, a direct form of total volatile extraction, and the loss in mass is measured. The temperature of evaporation has to be carefully selected so that thermal decomposition of labile substances is minimised to avoid adding to the volatile loss, assumed to be water. The sample is kept in oven at 110 °C for 2 h. The long time taken to reach a constant residual weight has stimulated the search for other methods.

14.4.1.2. Indirect Methods

Indirect methods do not provide the exact value, but are convenient and quick. These are frequently used at procurement centres and processing plants. These methods involve measurement of physical parameters like electrical conductivity or electrical resistance of the moisture present in the grain. Values are measured with the help of seed moisture meters, and these values are converted into grain moisture content with the help of calibration charts, for each species, against standard air-oven method or basic reference method. Moisture can also be measured by NMR, NIR and microwave spectroscopy techniques.

14.4.2. Protein Content

The Kjeldahl method has been almost universally used to determine nitrogen content.[10] The total protein content of a grain is estimated as total nitrogen after digestion, salt neutralisation and titration of the ammonia released against standard acid. Nitrogen content is then multiplied by a factor to determine the protein content. The method depends upon the fact that most organic nitrogen compounds are converted into $(NH_4)_2SO_4$ when heated with concentrated H_2SO_4. In this method the organic matter is digested with hot concentrated H_2SO_4. A "catalyst mixture" is added to the acid to raise its boiling point, usually a catalytic agent (mercury, copper or selenium) together with potassium sulphate. All organic nitrogen is converted to ammonia, which is usually measured by titration or, rarely, colorimetrically.

14.4.3. Lipid Content

Total lipid (fat) content may be calculated simply as the material extracted into diethyl ether. However, there are concerns over the availability of many chemically different forms of fat, and at least a digestion of the protein and carbohydrate would ensure the efficient release of fat from the tissue. The classical method is based on continuous extraction performed on dried samples of food in a Soxhlet extractor, sometimes preceded by acid hydrolysis. This technique is time consuming and subjects the extracted lipids to high temperature for longer periods. The extractant/solvent used is often petroleum spirit (which is less flammable that diethyl ether and less likely to form peroxides), which requires completely dry analytical portions and the removal of mono- and disaccharides.

14.4.4. Ash Content

Ash content represents the mineral matter and is recorded as the inorganic material remaining after the removal of all vaporisable material by high temperature combustion in a furnace (*e.g.* at 500 °C). The majority of methods used for determination of inorganic constituents require the elimination of organic matrix of the foods. During grain proximate analysis, the organic matrix is incinerated (usually in a muffle furnace at a controlled temperature) and the resultant inorganic residue is weighed to determine the ash content. The organic matrix can also be destroyed by heating in concentrated acids. This procedure reduces the losses caused by the oxidation and avoids any reaction between the inorganic constituents. After the organic matrix has been removed the inorganic constituents can be measured using a variety of techniques. These include classical gravimetric or volumetric methods, polarimetry, ion-selective electrodes, colorimetric procedures (which may or may not be highly specific) and instrumental methods (which offer an increase in speed of analysis, automation and good precision).

14.5. COOKING CHARACTERISTICS THAT DETERMINE PULSE QUALITY

14.5.1. Cooking Time

Several methods for measuring the cooking time of pulses are used. There is hardly any universally accepted standard method available for determining optimum cooking time of pulse grains. For commercial and research purposes, the cooking time is evaluated by measuring the softness of grains by a sensory panel. Another method is a tactile method that involves squeezing of cooked grains between the forefinger and thumb to know the cooking time;[2] pulse grains are considered to be cooked when they can be easily squeezed. Both the sensory panel and tactile methods are subjective and time consuming and are not suitable when large numbers of samples have to be evaluated. A texture analyser which involves measuring resistance to compression after cooking grains for a fixed time is a more objective method.[11] However, this method compares relative cooking times among the samples and does not give the optimum cooking time. This method is particularly not suitable when samples have large differences in cooking time or when hard to cook (HTC) grains are present in the sample. Another objective method for determining the cooking time involves the use of a "Mattson cooker" (Figure 14.2). This cooker was developed by Mattson[12] with 100 plungers and was later modified by reducing

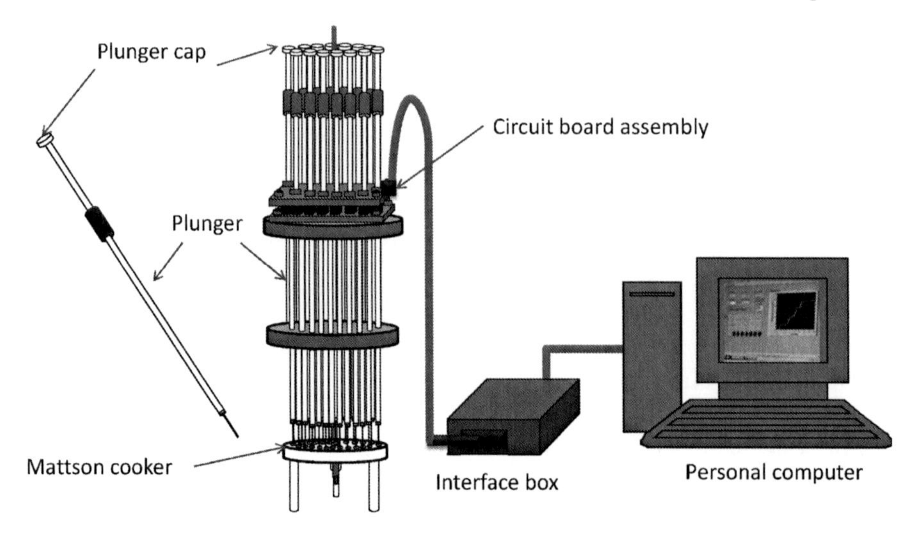

Figure 14.2 Schematic of Mattson bean cooker

the number of plungers to 25 by Jackson and Varriano-Marston.[13] Soaked grains are placed in the saddles of the rack so that the tip of each plunger just touches the grain. The lower portion of the cooker rack is kept immersed in the boiling water bath during the measurement. The plunger penetrates the grain when it is cooked and has become sufficiently soft. The plunger then drops through the hole in the saddle and the time taken by a plunger to drop is recorded manually.

Cooking times have been expressed in different ways in scientific literature. The number of grains the plungers penetrate at the end of each minute is counted. In one approach, the cooking time is expressed as the time taken by the plungers to penetrate 50% and 92%, respectively, of the beans as identified by Burr *et al.*[14] and Proctor and Watts.[15] Hsieh *et al.*[16] reported cooking time as the time required for 100% of the grains to be penetrated. The cooking time determined using this method also corresponds to the results of sensory evaluation scores. The Mattson cooker provides objective data and is simple to use, but it has some limitations. It requires continuous attention by the operator in order to record the penetration of the plungers into grains during cooking. The test is tedious, particularly when samples have longer cooking time due to genetic differences or due to the presence of HTC grains, or develop HTC phenomenon due to exposure to unfavourable storage conditions or some other factors. Also, the measurements become difficult when several plungers drop together during cooking. Both size and composition are responsible for the differences in cooking time. The cooking time of pulses is dependent upon grain weight and volume. The longer cooking time in pulse grains of larger size may be due to the greater distance that the water has to penetrate in order to reach the innermost portion of grains.[6] The difference in cooking time among the various pulses has been related to the rate at which cell separation occurs due to loosening of the intercellular matrix of the middle lamella upon cooking.[17] Pulse cotyledon with higher density has slower water uptake and consequently has longer cooking time. A positive relation of cooking time with hydration and swelling capacity for pea germplasm has been observed.[8] It was reported that that the pea lines that absorb more water as well as swelling more during soaking require less cooking time. Varietal differences in cooking quality have been reported to exist amongst pulses.[9,18]

14.5.2. Canning Quality

The canning process involves soaking, blanching and thermal processing of beans in brine or tomato sauce. The canning of pulses is discussed in details in Chapter 13. Physical factors such as seed size, colour, hydration capacity, swelling capacity and cooking time are the quality parameters that indicate the canning quality of beans. Pulses with uniform size grains, less cooking time and higher hydration as well as swelling capacity are preferred for canning. Pulses with such characteristics lead to higher drained weight. Organoleptic properties of canned products are also major quality evaluation criteria. Pulse cultivars vary in organoleptic quality attributes such as appearance, taste, flavour and texture. Discoloured grains, greater hardness and presence of grains with broken seed coat are the main factors which adversely affect the quality of canned beans.[19] Seed coat splitting is considered an important factor in canned beans because it indicates the thermal resistance of the grains and processing conditions that have been used. Canned products with large number of grains with split coats are considered undesirable by consumers. Generally, canned products having large number of seeds with split coats exhibit starchy and excessively viscous consistency. Beans that have a tendency to splitting during canning result into more exudation of starch into the canning medium, causing graininess of the sauce, and may result into lumping of grains.[20] Pulse cultivars with superior agronomical attributes may or may not have good processing and quality traits. The canning quality of pulses is related to many factors, such as cultivars, maturity, environmental condition, storage condition, processing variables and handling methods.[21–23] Non-hydrated bean seeds are negatively correlated to hydration capacity; presence of such grains in higher amounts result in poor quality canned products. Beans with higher hydration capacity exhibit light colour after canning whereas non-hydratable beans darken the canned product.[20] Beans with higher hydration capacity and swelling capacity also show a higher tendency to split. The hydration capacity and swelling capacity of beans are strongly correlated.[9] Various chemical constituents present in the seed coat of the beans are also related to canning quality. The tendency of the seed coat to split during canning varies depending upon soluble pectin content and calcium content in the seed coat as well as starch gelatinisation behaviour.[20] Calcium chloride is used in the canning industries to enhance the firmness of canned vegetables. The use of calcium chloride during canning of beans results in the formation of a metal–pectin complex, which may contribute to the toughening of seed coat and the turgidity of cell walls of the cotyledon tissue.[23–26] Pectin substances cross-link with divalent cations, such as calcium, and form intercellular polyelectrolyte gels. The presence of calcium cross-links retards water intake, and increases firmness due to the formation of calcium pectate.[25] This is a major factor contributing to the textural quality in food.[26] Calcium or other divalent ions have the ability to form salt bridges between adjacent polymer chains in the middle lamella in beans, leading to lower water uptake and harder beans.[58] Moscoso *et al.*[24] reported that in HTC red kidney beans, higher calcium and magnesium contents in the seed coat are associated with higher firmness in cooked beans. High mineral profiling is often related to high value in both total pectin substances and high water-soluble pectin substances. The calcium content in brine or soaking water has also been found to markedly affect the quality of canned kidney bean product. Canned beans with greater firmness yield lower split seeds during canning and are considered to be of good quality. The addition of calcium chloride at 66 °C during the soaking process has the greatest effect on reducing seed coat splits. However, insignificant differences in splitting can be found for a given pulse, when soaked in brine added with sodium chloride.[27] Percentage washed drain weight (PWDWT) is measured as:[23]

$$\text{PWDWT} = [\text{WDWT(g)} \times 100/\text{weight of contents of can (g)}] \times 100$$

where WDWT is washed drained weight, which refers to the mass of rinsed beans, rinsed for 2 min, on 8 mesh screen positioned at a 15° angle.[28] Regulations in many countries require a minimum of 60% PWDWT. Pulses with excessive gruel solids loss indicate lower WDWT, whereas those with higher swelling capacities indicate higher WDWT. WDWT varies with the cultivars, blanching and processing conditions. Blanching of beans in the presence of calcium decreases the WDWT.[29] The canning medium also has significant effect on PWDWT; beans canned in tomato sauce have lower PWDWT than those canned in water.[30]

14.5.3. Textural Properties

Textural properties of pulse grains measured using a texture analyser are listed in Table 14.3. Textural properties of cooked pulses are related to granular structure. Pulse grains with higher grain weight require longer cooking time and have higher hardness than those with lower weight, and this might be due to difference in compactness in granular structure. Pulses with large size grains have higher hardness than those with smaller size grains. Texture profile analysis (TPA) is performed on a single soaked or cooked grain using a texture analyser to determine various textural parameters. The grains are subjected to 50–75% compression with a probe at a constant speed. The textural profile curves of soaked seed of different pulses are illustrated in Figure 14.3. The textural parameters such as hardness, springiness, cohesiveness, gumminess and chewiness are usually reported. A brief description of these is as follows:

Table 14.3 Textural properties of soaked pulses

Pulse	*Hardness (N)*	*Cohesiveness*	*Gumminess*	*Springiness*	*Chewiness*
Kidney beans (red)	184.09	0.29	53.27	0.49	25.85
Kidney beans (white)	246.78	0.25	61.92	0.41	25.49
Chickpeas (kabuli)	117.03	0.15	17.24	0.32	5.60
Chickpeas (desi)	98.39	0.16	15.97	0.38	6.08
Lentils (masur)	30.87	0.16	5.00	0.34	1.69
Mung beans	24.98	0.20	4.95	0.43	2.11
Horse gram	24.1–46.3	0.23–0.35	6.68–11.71	0.51–0.66	3.86–7.63
Pigeon peas					
Peas	39.1–138	0.14–0.42	6.44–45.6	–	3.43–17.8
Adzuki beans	68.7–120.3	0.21–0.26	14.30–28.56	–	5.21–9.85
Black beans					
Lima beans	115.6	0.21	24.1	0.39	9.5
Navy beans	110.8	0.19	20.5	0.37	7.58
Cow peas	74.6–96.5	0.23–0.29	17.17–27.64	0.40–0.38	6.9–10.37
Scarlet beans	547	0.42	230	0.56	130

Source: Singh *et al.*[8] Yadav *et al.*[33]

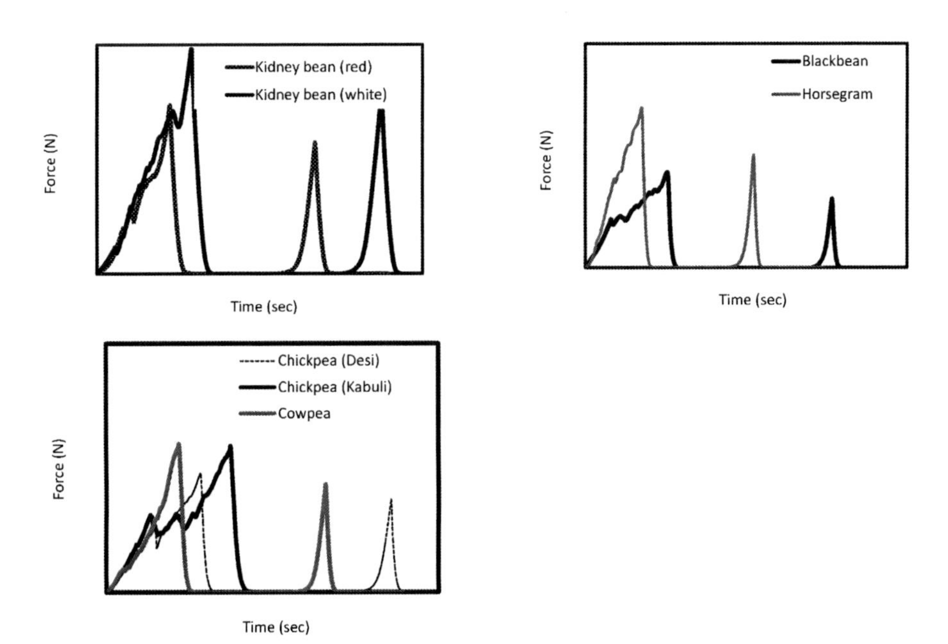

Figure 14.3 Texture profile curves of soaked pulse grains

- Hardness: maximum height of the force peak on the first compression cycle.
- Springiness: ratio of the time elapsing between the end of first and the start of second.
- Cohesiveness: ratio of the positive force areas under the first and second compressions.
- Gumminess: product of hardness and cohesiveness.
- Chewiness: product of gumminess and springiness.

14.5.4. Gruel Solids Loss

The gruel solids loss during processing indicate amount of the solids that leaches in soaking or cooking water. Beans with smaller size that imbibe more water at a faster rate generally show more loss of solids during soaking. The thickness of seed coat and cotyledon cell wall influences the leaching of solids. Smaller size grains that are hydrated and loosen the seed coat quickly are usually susceptible to higher leaching of solids.

14.5.5. Hard to Cook Defect

Hard to Cook (HTC) grains do not absorb sufficient water and require longer time to soften under normal cooking conditions. HTC grains can be identified by soaking a known number of grains in water for 12–16 hr and then counting the harder grains which do not swell and imbibe water. Grains with the HTC defect are observed both in stored and freshly harvested pulses. This defect is usually associated with hard shelled grains. Sometimes, HTC grains absorb sufficient water but do not soften upon cooking. Storage of pulses under adverse environmental conditions (high temperature and relative humidity, RH) makes them susceptible to the hardening phenomena characterized by longer cooking time. The losses due to the presence of HTC grains are high in developing countries due to inadequate storage facilities and prevalence of high temperature and RH conditions. This phenomenon is of great concern in developing countries because the pulse grains are consumed in large quantities and are cooked on a domestic scale. Hard shelled beans also have lower protein efficiency ratio as well as protein digestibility. The extended cooking time required by HTC pulses also increases fuel consumption. In developed countries, the pulse grains are marketed as pre-processed or canned products. These pre-processed or canned grains do not require prolonged soaking and cooking by the consumer at home before consumption. HTC grains have lower nutritional value than normal grains. The HTC phenomenon affects the antioxidant activity of grains which is a potential nutritional characteristic of pulses. The storage of dry pulse grains under high temperature (37 °C) and RH (76%) conditions has a severe effect on the rate of hardening and also lowers the protein quality and the availability of essential amino acids.[31]

Several theories have been suggested to explain the HTC defect in pulse grains, such as (1) lipid oxidation and/or polymerisation, (2) formation of insoluble pectates, and (3) lignification of the middle lamella.[32] Changes in the phenolic acid content and profile in HTC beans have also been shown.[33–35] The processes that cause HTC are not fully understood because a decrease in proportion of HTC grains in some pulses during storage has been observed. Hardening of the HTC grain has also been attributed to the presence of compounds such as tannin and phytate. Tannins, a group of polyphenols present in plants that are composed of high-molecular-weight compounds with hydroxyl clusters, form stable linkages with proteins.[36] The tannins are concentrated in the seed coat portion of pulses and range from 0 to 2% depending on the species and colour of the seed coat.[36] Tannins may migrate from the seed coat to the cotyledons, and cross-link with macromolecules from the cell wall or middle lamella during the storage period at high temperature (40 °C), thus resulted in a hardening effect.[32] The phytate has the ability to form complex with minerals such as Ca, Zn and Mg, as well as with proteins, and is degraded neither during cooking nor during digestion. Phytate content decreases during storage due to the gradual action of the phytase enzyme, which hydrolyses a portion of the phosphate ester groups. Therefore, the phosphate esters are no longer available for complex formation with Mg and Ca ions. As a result, these ions diffuse into the middle lamella and cause insolubilisation of pectinic acid and affect the softening process of grains during cooking since the middle lamellae of the cotyledon cells do not separate. The different genotypes vary in susceptibility to the development of HTC and this change is dependent upon environmental factors. The genetic variability amongst common beans in tannins[19] and phytate content is well reported.[37,38] The phytate content of pulse grains is also influenced by the application of phosphorus fertilisers during plant growth; a higher supply of phosphorus can lead to an increase in phytate content in the grain.

14.6. FUNCTIONAL CHARACTERISTICS OF FLOURS THAT DETERMINE PULSE QUALITY

Functional properties are defined as the physical and chemical properties which affect the behaviour of proteins in food systems during processing, storage, preparation and consumption.[39]

14.6.1. Pasting Properties

Pasting properties of the pulse flours is studied using the Rapid-Visco Analyzer and Brabender Viscoamylograph. These instruments continuously measure and record the viscosity of a flour suspension under controlled conditions of heating, holding and cooling. The pasting curves thus obtained represent the change in the behaviour of flour viscosity with change in temperature. During heating, the viscosity of a flour suspension increases due to the swelling of the granules to several times their original size, as a result of loss of crystalline order and absorption of water.[40] The temperature at which the viscosity of flour suspension begins to rise is taken as the pasting temperature and represents the minimum temperature required for cooking. The variation in chemical composition is mainly responsible for differences in the pasting behaviour among different flours. The peak viscosity represents the point of maximum swelling of starch granules. Further heating of gels at higher temperatures results in a drop in the viscosity, commonly attributed to the rupturing of starch granules. The lower breakdown is attributed to lower disintegration of the swollen starch granules in the presence of higher amylose content.[41] The amylose and lipids have been reported to assist in maintaining granule integrity during heating.[42] During cooling, the viscosity of cooked gels increases and this is attributed to the retrogradation or tendency of gel components to reassociate. Pasting properties of some common pulses are given in Table 14.4.

14.6.2. Protein Solubility

Protein solubility is one of the most critical characteristics that influence emulsification, foaming and gelation of flours. The solubility of protein in the system depends upon pH, heating conditions and ingredients used. Protein solubility varies with pH and hence flour of one type shows variable performance in different food systems when pH is changed. The extent of protein denaturation in flours varies with heat or chemical treatments. Pulse proteins show higher solubility at low and high pH values and have minimum solubility at pH 4–5, which is the isoelectric pH of most plant proteins. At this pH range there is no net charge on the proteins. This decreases the repulsive interactions as well as the protein–protein interactions that cause a reduction in solubility. Whereas above or below isoelectric pH, highly charged proteins tend to remiansoluble and cotribute to higher solubility. Pre-treatments (soaking, dry heating, wet heating, *etc.*) of pulse grains have variable effects on the solubility of their flours. The soaking of pulses before milling has less influence on protein solubility of flour as compared to soaking followed by germination or boiling. During germination of pulses the activity of proteolytic enzymes increases many-fold, which leads to increase in protein solubility. The soaking followed by boiling decreases the protein solubility. Extensive heat treatments such as boiling of pulses cause greater changes in protein solubility that is attributed to the exposure of hydrophobic groups and aggregation of the unfolded protein molecules. The decrease in protein solubility after heating is ascribed to biochemical changes in protein structure. Both intramolecular hydrogen bonds and non-polar bonds are disrupted during heating and reform on subsequent cooling. This occurs because heat increases the kinetic energy that causes the molecules to vibrate so rapidly and violently that the bonds are disrupted. The resultant changes in protein conformation reduce the solubility. The disulphide–sulfhydryl interchange reactions also contribute to the decrease in solubility. The cross-linking between protein and starch molecules during heating, and particularly during boiling, is attributed to the formation of aggregates that renders the protein insoluble.

Table 14.4 Pasting properties of flours from different pulses

Pulse	*% suspension*	*Pasting temperature (°C)*	*Peak viscosity (cP)*	*Trough viscosity (cP)*	*Breakdown viscosity (cP)*	*Final viscosity (cP)*	*Setback (cP)*
Kidney beans	12.50%	79–95.0	402–3235	–	31–414	862–5310	362–2488
Chickpeas	10%	75.0– 87.1	564–853	516–790	32–123	573–969	84–185
Lentils	11.9 %	70.0–71.1	1185–1359	–	140–239	1651–1781	605–662
Mung beans	11.9%	76.5	591	–	–	1060	–
Pigeon peas	10%	83.45	1994	1771	224	2347	577
Peas	12.50%	73.5– 81.5	533–3000	507–1390	8–187	121–2276	183–998
Adzuki beans	10%	75.0–78.3	1918–2373	1880–2308	38–282	2754–4004	931–1666

Data compiled from Kaur *et al.*;[9] Singh *et al.*;[60] Singh *et al.*;[8] Yadav *et al.*;[55] Kaur *et al.*[56]

14.6.3. Fat Absorption Capacity

The fat absorption capacity (FAC) is an important functional property that contributes to the improvement in sensory attributes such as mouthfeel and the retention of flavour of the product.[43] Pulse flours are used for coating different food materials such as fish, chicken, eggs, *etc.* before deep-frying. Pulse flour batters alone or in combination with different ingredients such as potato, onion and green chillies are deep-fried to produce a number of products. A similar product made from chickpea flour is known as pakora in northern India. A large number of fried snacks are produced from pulses, such as fried dehulled mung and bhujia. Bhujia is a famous crisp fried traditional Indian snack product prepared from different pulse grains and flours (see Chapter 13 for details). The FAC of pulse flour is important in such products and flour with less FAC is preferred. FAC indicates emulsifying capacity, and in some products such as mayonnaise this is a highly desirable. FAC can be explained as physical entrapment of lipids by the non-polar side chains of proteins. Both the protein content and the type contribute to the FAC of food materials.[47] Besides these factors, the FAC of flours is dependent upon the pre-treatments given to pulse grains before milling into flour. For example, the boiling of soaked cowpeas leads to a slight increase in the FAC.[54] The improvement in fat retention as a result of heat treatment is attributed to enhancement in hydrophobic properties of proteins as well as to the superior fat-binding performance of non-polar amino acid side chains.[43] The change in fat absorption caused by thermal processing depends upon the extent to which the non-polar residues from the interior of the protein molecules are exposed during thermal processing. The physical structural changes such as porosity caused by processing methods also contribute to FAC. The processing treatments that cause greater porosity in flours generally result in greater entrapment of fat.

14.6.4. Water-Holding Capacity

The water-holding capacity (WHC) is defined as the ability of a system to physically hold water against gravity.[44] It is defined as the amount of water that can be absorbed by a known weight of flour. The WHC of flour varies with the composition and pre-treatment given to pulse grains. Boiling of pulses before milling into flour increases the WHC. Heating causes denaturation of proteins, which improves the WHC. Winged bean flour,[45] chickpea flour[46] and lentil flour[47] have been reported to exhibit significantly increased WHC after autoclaving, cooking in water, $NaHCO_3$, and citric acid, and dry heating under pressure. Different processing methods bring about variable effects in WHC due to different changes caused in physical structure. The extent to which the amino acid residues are exposed by heat denaturation also affects the WHC properties of flours. Heating changes the conformation of proteins from globular to random coil and exposes buried amino acid side chains. These buried amino acids are made available to interact with water, result in increased WHC.[48] Starch gelatinisation and the amount of crude fibre as well as swelling of crude fibre during heating also contribute to increased WHC.[49] The polar amino acid residues of proteins have an affinity for water molecules. The differences in the amount of these amino acids contribute to the variation in WHC amongst different pulses. Flours with high WHC are considered suitable ingredients for some bakery applications (*e.g.* bread formulation), since flours with higher WHC result in a dough which gives a larger number of breads with larger volume and better shelf life.

14.6.5. Gelling Ability

The least gelation concentration (LGC) is the lowest concentration required to form a self-supporting gel. This is considered as an index of the gelation capacity of flour. A lower LGC of flours indicates the presence of proteins with better gelling ability. Higher gelling ability is considered desirable in food products such as puddings and sauces where thickening and gelling is required. Gelation is described as a process in which denatured molecules cross-link to form aggregates stabilised by different bonds including electrostatic interactions, hydrogen, hydrophobic and/or disulfide bonds. LGC for pulse flours is affected by protein content, type of protein and the presence of non-protein components. The variations in the gelling capacity among pulse flours are due to the variation in proportion of different chemical constituents (proteins, carbohydrates, lipids, enzymes) as well as the interactions between these constituents. The physical competition for water between protein gelation and starch gelatinisation also influence the LGC of flours.

14.6.6. Emulsifying Properties

The emulsifying properties of flours are contributed by proteins and other amphoteric molecules. Emulsifying properties are referred to as the emulsifying activity index (EAI) and emulsifying stability index (ESI). Proteins act as emulsifiers by forming a film surrounding the oil droplets dispersed in an aqueous medium, thereby preventing structural changes such as coalescence, creaming, flocculation or sedimentation. The EAI indicates the ability and capacity of a protein to assist in the formation of an emulsion. This is related to the protein's ability to absorb at the interfacial area of oil and water in an emulsion. The ESI indicates the ability of the proteins to impart strength to an emulsion against stress and changes and is therefore related to the consistency of the interfacial area over a defined time period.[50] Pulse flours show variable emulsifying properties due to difference in composition. The emulsifying properties are affected by protein content, fat content, protein structures and starch composition (amylose/amylopectin ratio). Starch and fibre may also enhance emulsion stability by acting as bulky barriers between the oil droplets, preventing or slowing down the rate of oil droplet coalescence.[51] Pulse flours with superior emulsifying properties are preferred in food products such as salad dressing, beverages, and meat analogues. The EAI is significantly affected by thermal processing; an improvement in some pulse grains has been attributed to the dissociation and partial unfolding of globular proteins, leading to exposure of hydrophobic amino acid residues which consequently increase the surface activity and adsorption at the oil and water interface.[52] To cite an example, the emulsifying capacity of yam beans is reduced from 50.7% to 20% by roasting.[53] Decreased EAI after boiling and autoclaving of cowpeas, winged beans, chickpeas, and lentils has been reported.[45,49,54]

REFERENCES

1. D. W. Stanley and J. M. Aguilera, *J. Food Biochem.* 1985, **9**, 277–323.
2. O. L. Vindiola, P. A. Seib and R. C. Hoseney, *Cereal Foods World*, 1986, **31**, 538–552.
3. S. Bishnoi and N. Khetarpaul, *Food Chem.*, 1993, **47**, 371–373.
4. G. H. Gubbels and S. T. Ali-Khan, *Can. J. Plant Sci.*, 1991, **71**, 857–859.
5. G. H. Gubbels, B. B. Chubey, S. T. Ali-Khan and M. Stauvers, *Can. J. Plant Sci.*, 1985, **65**, 55–61.
6. S. Sefa-Dedeh and D. W. Stanley, *Food Technol.*, 1979, **33**, 77–83.
7. S. Sefa-Dedeh, D. Stanley and P. Voisey, *J. Food Sci.*, 1978, **43**, 1832–1838.
8. N. Singh, N. Kaur, J. C. Rana and S. K. Sharma, *Food Chem.* 2010, **122**, 518–525.
9. S. Kaur, N. Singh, N. S. Sodhi and J. C. Rana, *Food Chem.*, 2009, **117**, 282–289.
10. Official Methods of Analysis (15th ed.) Washington DC: Association of Official Analytical Chemists, 2000.
11. R. G. Black, U. Singh and C. Meares, *J. Sci. Food Agric.*, 1998, **77**, 251–258.
12. S. Mattson, *Acta Agric. Suec.*, 1946, **2**, 185–231.
13. G. M. Jackson and E. Varriano-Marston, *J. Food Sci.*, 1981, **46**, 799–803.
14. H. K. Burr, S. Kon and H. J. Morris, *Food Technol.*, 1968, **22**, 336–338.
15. J. P. Proctor and B. Watts, *Can. Inst.Food Sci. Technol. J.*, 1981, **20**, 9–14.
16. H. M. Hsieh, Y. Pomeranz and B. G. Swanson, *Cereal Chem.*, 1993, **69**, 244–248.
17. L. B. Rockland and F. T. Jones, *J. Food Sci.*, 1974, **39**, 342–346.
18. H. V. Narasimha and H. S. R. Desikachar, *J. Food Sci. Technol.* 1978, **15**, 47–50.
19. N. N. Wassimi, G. L. Hosfield and M. A. Uebersax, *Crop Sci.*, 1988, **28**, 452–458.
20. W. Lu and K. C. Chang, *Cereal Chem.* 1996, **73**, 785–787.
21. C. R. Wang, K. C. Chang and K. Grafton, *J. Food Sci.*, 1988, **53**, 772–776.
22. L. G. Occena, M. A. Uebersax, and A. Shirazi, Canning quality characteristics of Anasazi beans. Annual report ofthe Bean Improvement Cooperative, 1992, **35**, 198–199.
23. P. Balasubramanian, A. Slinkard, R. Tyler and A. Vandenberg, *J. Sci. Food Agric.* 2000, **80**, 732–738.

24. W. Moscoso, M. C. Bourne and L. F. Hood, *J. Food Sci.*, 1984, **49**, 1577–1583.
25. M. A. Uebersax and S. Ruengsakulrach, *ACS Symp. Ser.*, 1989, **405**, 111–124.
26. A. D. Lange, M. Labuschagne, *J. Sci. Food Agric.*, 2000.**81**, 30–35.
27. J. van buren, M. Bourne, D. Downing, D. Quele, E. Chase and S. Comstock, *J. Food Sci.* 1986, **51**, 1228–1230.
28. G. L. Hosfield and M. A. Uebersax, *Michigan Dry Bean Digest*, 1979, **3**, 4–9.
29. D. M. Larsen, M. A. Uebersax, J. G. Wilson, S. Ruengsakulrach and G. L. Hosfield, *Michigan Dry Bean Digest*, 1988, **13**, 6–7.
30. C. L. Nordstrom and W. A. Sistrunk, *J. Food Sci.*, 1977, **42**, 795–798.
31. P. L. Antunes, V. C. Sgarbieri, *J. Food Sci.*, 1979, **44**, 1703–1706.
32. C. Reyes-Moreno and O. Paredes-Lopez, *Crit. Rev. Food Sci. Nutr.* 1993, **33**, 227–286.
33. N. Srisuma, R. Hammerschmidt, S. Uebersax, S. Ruengsakulrach, M. Bennink and G. Holsfield, *J. Food Sci.*, 1989, **54**, 311–314.
34. E. Garcia, T. Filisetti, J. Udaeta, F. Lajolo, *J. Agric. Food Chem.* 1998, **46**, 2110–2116.
35. P. Rao, Y. Deosthale, *J. Food Sci. Agric.*, 1982, **33**, 1013–1016.
36. N. R. Reddy, M. D. Pierson, S. K. Sathe, D. K. Salunkhe, *J. Am. Oil Chem. Soc.* 1985, **62**, 541–549.
37. C. M. M. Coelho, J. C. P. Santos, S. M. Tsai, and V. A. Vitorello, *Brazilian J. Plant Physiol.*, 2002, **14**, 51–58.
38. N. Wang and J. K. Daun, *J. Sci. Food Agric.* 2004, **84**, 1021–1029.
39. J. E. Kinsella, in *Food Proteins*, ed. P. F. Fox and J. J. Condon, Applied Science Publishers, New York, 1982, p. 51.
40. J. Bao and C. J. Bergman, in *Starch in Food: Structure, Function and Applications*, ed. A.-C. Eliasson, Woodhead Publishing, Cambridge, 2004, pp. 258–294.
41. A. K. Singh, M. Bhattacharyya-Pakrasi and H. B. Pakrasi, *J. Biol. Chem.*,2008, **283**, 15762–15770.
42. W. R. Morrison, R. F. Tester, C. E. Snape, R. Law and M. J. Gidley, *Cereal Chem.* 1993, **70**, 385–391.
43. J. E. Kinsella and N. Melachouris, *Crit. Rev. Food Sci. Nutr.* 1976, **7**, 219–280.
44. J. E. Kinsella, *J. Am. Oil Chem.Soc.*, 1979, **56**, 242–258.
45. K. Narayana and M. S. Narasinga Rao, *J. Food Sci.*, 1982, **40**, 1534–1538.
46. M. C. Bencini, *J. Food Sci.*, 1986, **51**, 1518–1521.
47. B. Nagmani and J. Prakash, *Int. J. Food Sci. Nutr.* 1997, **48**, 205–214.
48. C. W. Hutton, A. M. Campbell, in *Protein Functionality in Foods*, ed. J. P. Cherry, ACS Symp. Ser., 1981, pp. 177–200.
49. Y. Aguilera, R. M. Esteban, V. Benitez, E. Molla and M. A. Martin-Cabrejas, *J. Agric. Food Chem.* 2009, **57**, 10, 682−688.
50. K. N. Pearce, J. E. Kinsella, *J. Agric. Food Chem.* 1978, **26**, 716–723.

51. R. E. Aluko, O. A. Mofolasayo and B. M. Watts, *J. Agric. Food Chem.* 2009, **57**, 9793–9800.
52. I. Nir, Y. Feildman, A. Aserin and N. Garti, *J. Food Sci.*, 1994, **59**, 606–610.
53. V. A. Obatolu, VS. B. Fasoyiro, L. Ogunsunmi, *J. Food Process. Preserv.*, 2007, **31**, 240–249.
54. W. Prinyawiwatkul, L. R. Beuchat, K. H. McWatters and R. D. Phillips, *J. Agric. Food Chem.* 1997, **45**, 480–486.
55. U. Yadav, N. Singh, A. Kaur, S. Singh and J. C. Rana, Presented at *ICFOST-XXI*, 20–21 January 2012, Pune, India.
56. M. Kaur, K. S. Sandhu and N. Singh, *Food Chem.*, 2007, **104**, 259–267.
57. N. Singh, S. Kaur, N. Isono and T. Noda, *Food Chem.*, 2010, **122**, 65–73.
58. L. R. Nelson and K. H. Hsu, Effects of Leachate Accumulation During Hydration in a Thermalscrew Blancher on the Water Absorption and Cooked Texture of Navy Beans. *Journal of Food Science*, 1985, **50**, 782–788.

Subject Index